Vigor

Vigor

Neuroeconomics of Movement Control

Reza Shadmehr and Alaa A. Ahmed

The MIT Press
Cambridge, Massachusetts
London, England

This book was set in Times New Roman by Westchester Publishing Services. Printed and bound in the United States of America.

Library of Congress Cataloging-in-Publication Data

Names: Shadmehr, Reza, author.
Title: Vigor : neuroeconomics of movement control / Reza Shadmehr and Alaa A. Ahmed.
Description: Cambridge, Massachusetts : The MIT Press, [2020] | Includes bibliographical references and index.
Identifiers: LCCN 2019041223 | ISBN 9780262044059 (hardcover)
Subjects: LCSH: Reaction time--Physiological aspects. | Reward (Psychology)--Physiological aspects. | Decision making--Physiological aspects.
Classification: LCC BF317 .S48 2020 | DDC 152.8/3--dc23
LC record available at https://lccn.loc.gov/20190412

10 9 8 7 6 5 4 3 2 1

Contents

Preface vii
Acknowledgments xv

1 **The Effort of Movement 1**
1.1 Choosing, Moving, Then Harvesting 1
1.2 A Measure of Effort 3
1.3 Utility of a Movement 13
1.4 Additive versus Multiplicative Interaction of Reward and Effort 20
1.5 Temporal Discounting of Effort 21
1.6 Alternative Representations of Effort 24
1.7 Cognitive Effort versus Movement Effort 27

2 **Movements and Decisions in a Normative Framework 33**
2.1 Optimal Foraging Theory 35
2.2 Marginal Value Theorem 43
2.3 Optimizing Movement Vigor 51
2.4 Example: Walking Speed 56
2.5 Walking Speed at Various Cities 60
2.6 Foraging with Our Eyes 62
2.7 Salience versus Utility 69

3 **Reaction Time and Deliberation 77**
3.1 Latency to Start a Movement: The LATER Model 80
3.2 The Variable Threshold Model 83
3.3 Linking Latency and Utility 84
3.4 Reward Modulates Vigor 88
3.5 Effort Modulates Vigor 95
3.6 Rate of Reward Modulates Vigor 98
3.7 Deliberation and Decision-Making 101
3.8 Modulation of Saccade Vigor during Decision-Making 107
3.9 Decision Uncertainty and Vigor of Movements 112
3.10 Between-Subject Differences in Vigor 114
3.11 Decision-Making and Movement Traits of Individuals 119

4 Neural Prelude to a Movement 129

4.1 Superior Colliculus 130
4.2 Bursting to Move 134
4.3 Bursting to Hold Still 138
4.4 Movement Latency and the Activity in the Fovea-Related Cells 144
4.5 Movement Latency and the Rate of Rise in the Buildup Cells 148
4.6 Reaction Time as an Interaction between Utility of Holding versus Utility of Moving 152
4.7 Salience versus Utility 156
4.8 Effect of Reward on the Colliculus 157
4.9 Reacting Much Earlier 162
4.10 Influencing Decisions with Manipulation of the Colliculus 169

5 Cortical Computation of Utility 175

5.1 Moving without Computation of Utility 177
5.2 Attention and the Effort Cost of Movement 179
5.3 Inhibiting Reflexive Behavior 184
5.4 Activity in the FEF during the Reaction Time Period 188
5.5 FEF and Deciding Where to Direct Gaze 193
5.6 Setting a Threshold to Move: Omnipause Neurons 202
5.7 The Premotor Cortex and the Decision to Reach 206
5.8 Encoding Utility and Salience in the Parietal Cortex 212
5.9 Good-Based versus Action-Based Models of Decision-Making 220
5.10 Deciding When to Leave 224

6 Basal Ganglia and the Motivation to Move 231

6.1 Anatomy of the Basal Ganglia 233
6.2 Substantia Nigra Reticulata 234
6.3 Caudate 244
6.4 Dopamine and the Pleasure of Moving 256
6.5 Dopamine and the Willingness to Exert Effort 278

7 Serotonin and the Promotion of Sloth 289

7.1 Anatomy of the Serotonergic System 290
7.2 Serotonin and the Average Reward Rate 292
7.3 Serotonin and the Willingness to Wait 295
7.4 Serotonin and Movement Vigor 306
7.5 Serotonin, Dopamine, and the Willingness to Work 312
7.6 Serotonin and Motoneuron Activity 316

Conclusions 323

Appendix A: Effective Mass of the Human Arm 329
Appendix B: Alternate Forms of Utility 331
Appendix C: Algebra of Random Variables 333
References 337
Index 349

Preface

> Because of our wisdom, we will travel far for love,
> As all movement is a sign of thirst,
> And speaking really says
> "I am hungry to know you."
> —Hafez, 14th-century Persian poet

At the airport, waiting outside the security area, you might notice a young family, mom and child, spotting the person they are waiting for: dad. There is a smile on the dad's face as he exits the security area and finds them. They run toward each other.

At around the same time there is another man, about the same age, also coming out, but not being greeted by his family. Rather, he notices that a limo driver is holding a tablet with his name on it. Not surprisingly, he does not run toward the driver.

Why do we run toward people we love, but merely walk toward others? What brain mechanism compels us to move with more vigor towards things we value more?

To illustrate the importance of these questions, imagine that you are at a nice restaurant and the waiter rolls the dessert cart to your table. You look over the options, focusing on a couple of plates, one a chocolate cake with raspberries, the other cheesecake with sliced peaches. As you deliberate, you shift your gaze by making saccadic eye movements that go from one plate to another. After a few seconds, you make up your mind and choose the chocolate cake.

Now suppose that as you were deliberating, the restaurant had a camera that measured your eye movements. It recorded the velocity with which your eyes traveled toward each plate and noted that your saccades toward the chocolate cake had velocities that were higher than your saccades toward the peaches. From this measurement, the machine estimated the value that your brain assigned to each option, then predicted that you preferred the chocolate cake. The waiter did not need to ask you which dessert you wanted. You preferred the dessert for which your saccades had the higher velocity.

If the vigor of our movements is driven by how we value the thing we are moving toward, then our movements give away one of our big secrets: the subjective value that we assign to things and people. Notably, because velocity is a real-valued variable, one can compare velocities between two options and measure precisely how much faster we move toward one than the other.

This means that someday soon, we might be able to estimate not just which dessert you prefer by the velocity of your eyes, but how much more you prefer one dessert to the other. The restaurant would certainly be interested in using this information to set their prices.

As we like to say to our students, if you want to get a sense of how someone feels about you, you should probably pay attention to how they move toward you.

Preferences Gleaned from Vigor

If we want to know whether you prefer A or B, we proceed by presenting you with A and B and then ask you to pick the one you prefer. But suppose we wanted to measure how much more you preferred A to B. Here, the typical approach is to give you choice lotteries and ask you to make your selection from a menu of probabilities. It would go like this: you can have A with 100% probability, or choose B, for which you can get B1 or B2 with 50% probability. The result is an abstract quantity called utility. Utility is an estimate of your subjective valuation of an outcome.

Over the past decade, it has become clear that there is another way with which we might measure preference. This research has revealed that factors that affect preference, such as reward and effort, also affect movements. For example, if you prefer A to B, then you are likely to move with a shorter reaction time and greater velocity towards A. Vigor provides a real-valued scale to quantify the difference between movements toward A and B. Therefore, vigor adds a continuous dimension to the ordinal scale of choices, with the intriguing possibility that this dimension may overcome some of the limitations inherent in inferring utility from choice behavior.

The discovery that we move more vigorously toward things we prefer was serendipitous. It came about as an unexpected outcome of experiments that focused on patterns of decision-making, and used movements as a read out of choice. For instance, in studies employing eye movements as the choice-reporting method, if the subject preferred A to B, it expressed its choice by making a saccade toward A. However, it soon became clear that movements were more than just a proxy for choice. Rather, humans and other animals moved faster and with shorter reaction times toward items that they preferred. This raised the possibility that vigor could serve as a real-valued scale with which subjective value may be inferred.

The Link between Vigor and Utility

Why should the way we move be affected by how we value the item at our destination? After all, we could easily imagine a scenario in which the brain assigns values to the various stimuli, picks the one that has the greatest subjective value, and then passes on the chosen action to the motor system, which simply executes a stereotypical movement to acquire that stimulus. Indeed, many text books still separate decision-making and motor control into separate brain regions, and imagine a hierarchy in which the decision-making circuits make choices, and the motor circuits produce the action needed to acquire the chosen option.

In this book, we closely review the experimental results and arrive at a different conclusion: brain regions that were thought to be principally involved in decision-making also affect the vigor of movements, and brain regions that were thought to be principally responsible for moving a part of the body also bias patterns of decision-making.

As the promised reward increases, animals react earlier and start their movements with a shorter latency, and then move with greater velocity toward reward. In contrast, as the effort required for acquiring reward increases, subjects tend to become slower in starting their movements and take longer to complete their actions. When they arrive at their reward site, their past effort expenditure encourages them to stay longer and acquire more of the reward before moving on to the next reward opportunity.

That is, reward and effort, factors that were thought to be the domain of decision-making, have a clear effect on vigor. Furthermore, these same factors also affect what we do when we get to our destination: harvest duration depends on the effort we have spent to get there. We linger longer at the restaurant if we had to spend effort securing a reservation.

Now why should the way we move be influenced by the same factors that influence our decisions? That is, why should you not only prefer to spend time with your best friend rather than your school teacher, but also walk faster to meet your friend? To answer this question, we imagine the problem from the point of view of an ecologist: what is the currency that the brain is trying to optimize via its choices and movements?

In their natural environment, animals make choices on the basis of a desire to maximize a specific currency: the global capture rate, which is defined as the sum of all rewards acquired minus all efforts expended, divided by total time. This currency plays a fundamental role in the longevity and fecundity of animals, suggesting that living one's life in a way that increases the capture rate has evolutionary advantages.

Because movements require expenditure of effort, and utility-based decisions affect reward accumulation, if we wish to maximize the capture rate, then we must find policies that are informed by both the effort of making movements and the joy of acquiring reward.

Thus, the link between movements and decisions arises because both are elements of behavior that the brain must control to maximize a single currency: the capture rate. To optimize this currency, you cannot simply make good choices; you must also move with vigor that is consistent with those choices. When rewards are low, it is not cost effective to pay for time with expenditure of effort. However, when rewards are larger, it is worthwhile to move faster because the time that is saved in acquiring that reward increases the capture rate. Effectively, by moving faster, you are buying time through expenditure of effort, thereby raising your capture rate when you get the reward.

Neural Basis of Vigor

A good example of the neural link between systems that assign value to things and systems that control our movements is illustrated in Parkinson's disease.

Patient C1 was a 62-year-old retired florist who had suffered from Parkinson's disease for about 14 years. He exhibited severe tremor and rigidity, and he was totally dependent on his wife. She started her day by dressing him, laying out his breakfast, and making his lunch. She went to work, then came back in the afternoon to make his dinner and finally got him undressed and ready for bed. One evening, she had severe abdominal pain and had to be taken to the hospital by ambulance for emergency surgery. The next day she woke at the hospital worried about her husband. However, the nurse informed her that he had come to visit her. He had dressed himself, made his own breakfast, and then walked to the hospital. At the hospital his neurologist noticed him and upon examination found that he walked considerably faster than in past examinations. In his case report (Schwab and Zieper, 1965), the neurologist wrote the following: "All his motor tests were improved in spite of the presence of the same amount of rigidity and tremor that had been present before."

It was clear that this patient could, under increased urgency, dress himself, make his own breakfast, and move fairly normally. But why wasn't this potential available to him until he was put in the extraordinary circumstance of having his wife taken to the hospital? Pietro Mazzoni, Anna Hristova, and John Krakauer (2007) were among the first neuroscientists to examine this question. They named the phenomenon motor motivation and suggested that movement slowness in Parkinson's disease was not due to a loss of ability to make normal movements, but rather an altered economic evaluation of movements that resulted in reduced vigor.

Indeed, research in the past decade has demonstrated that control of vigor is influenced by release of dopamine. Dopamine is like the two faced god Janus. On the one hand, when the brain experiences more reward than expected, there is a burst of dopamine, and the dopamine release teaches us to increase the value that we assigned this stimulus: it was better than expected. On the other hand, just before a movement starts, dopamine levels rise and stay higher if we are spending effort. For example, if an animal has to press a lever to get a pellet of food, the levels of dopamine in the basal ganglia are greater than if the same animal has to simply stay still and be given that pellet of food. Thus, on the one hand, dopamine teaches us about the value of things, and on the other hand it supports expenditure of effort. As we will see, this bridge that dopamine builds between reward and effort has much to do with the fact that we place greater value on things that we have worked for.

Arrival of dopamine affects the excitability of neurons, particularly those in the striatum, which constitutes the input stage of the basal ganglia. The striatal neurons react to dopamine by altering how they respond to cortical inputs, eventually affecting how much inhibition the output structures of the basal ganglia produce upon downstream motor structures. When an act is expected to be more rewarding, the inhibition produced by the basal ganglia output structures upon the motor system is reduced. This modulation of inhibition is one of the principal ways with which our perception of reward affects the vigor of our movements.

A second principal factor is the excitation that the motor structures receive from cortical neurons that are involved in computing utility. Utility is a complicated quantity that reflects how we subjectively evaluate reward and effort. Many cortical regions participate in evaluating utility of each option, and then compete to produce a choice, which is followed by a movement that expresses that choice. The most valuable option produces neural activity that rises quickly, and reaches threshold earlier, thereby triggering a movement. The rate at which this activity builds influences reaction time of the movement that follows.

Thus, movement vigor is a result of the combined excitation and inhibition from disparate neural structures that converge upon the motor system, compelling it to not only move toward a particular reward, but do so with a vigor that depends on the subjective value of reward. When the reward system is affected because of disease, or because of a history of low rewards, as in depression, our movements become slowed, and vigor drops.

Organization of the Book

In this book, our goal is to consider a simple question: why do we move faster toward things we value more? The first part of the book considers (chapters 1–3) this question from a behavioral and mathematical perspective, while the second part (chapters 4–7) considers it by looking at the neural mechanisms that control our choices and movements.

To formulate a rationale for why subjective value of an option should affect vigor, we consider the mathematical framework of optimal foraging theory. This theory was designed to provide an understanding of the various factors that animals consider as they choose between their options, and was motivated by the observation that animals sometimes behave unexpectedly: as they search for food, they sometimes pass on an inferior piece, even though they have already spent effort trying to find it.

An example of this is behavior of crows as they search for clams on the beaches of the Pacific Northwest of the United States. They spend effort digging up a clam, but if it is small, they abandon it and go try to find another clam. Optimal foraging accounts for this decision by suggesting that the option that the animal prefers often accords with a desire to maximize their capture rate. By passing up the small reward, thereby not spending additional time in trying to open the clam, the crows make a series of decisions that in the long run increases their global capture rate.

When we move fast, we are expending greater effort. It turns out that moving vigorously, and thereby spending more energy, makes sense in the framework of optimal foraging. If we use the energy that it takes to perform a movement as a proxy for effort expenditure, then there is a natural link between movements and decisions: moving faster is a good policy because it improves the capture rate, even though it also requires greater effort expenditure. Effectively, by moving faster toward more valuable options we are spending energy to buy time (and thus get the reward sooner). This is a worthwhile purchase because it improves our capture rate currency.

Animals that achieve increased capture rates tend to have an evolutionary advantage through greater longevity and fecundity. Therefore, this evolutionary benefit links control of movements with control of decisions.

The resulting framework provides a way to consider many fascinating observations, including the data that show that people in certain cities tend to walk faster than those that live elsewhere, and data that show that after a period of low reward rates, we not only tend to move more slowly but also linger longer at the next reward site.

In the second part of the book we consider the neural link between control of movements and subjective valuation of options. There, we focus on the simplest of voluntary movements: saccades. There are two reasons for focusing on saccades. First, we make about 100,000 saccades during each day, directing our fovea so that we acquire information from various items in our visual space. Each saccade takes around 50 ms, during which time we are effectively blind. As a result, we are blind for a total of 1.5 hours during each waking day, making it particularly important for the brain to optimize duration of each saccade.

With each movement of our eyes, we express what we value in our visual scene, and how much we value it. If the image that we are making a saccade toward is valuable, the brain generates that saccade with greater velocity, reducing saccade duration by a few milliseconds. The sensitivity of saccades makes them a particularly useful behavior with which to test the links between variables that affect choice and variables that affect vigor.

The second reason that this book delves deeply into control of saccades is because the entire neural system of saccades is within the cranium, affording neurophysiologists the ability to study the neurons that control eye movements. In comparison, arm movements and walking partly rely on neural structures within the spinal cord, making it more difficult to study vigor of these behaviors by using standard techniques.

Our analysis of the neural basis of vigor will begin with the superior colliculus. We will incorporate contributions of the frontal eye field and other cortical regions, then shift to the basal ganglia, where we will focus on dopamine and the neural basis of reward evaluation and vigor modulation. We will conclude our journey with serotonin, a neurotransmitter that in many ways functions to oppose dopamine: whereas dopamine signals the rewarding value of the stimulus and encourages vigor, serotonin signals the punishing value of the stimulus and encourages sloth.

Overall, the results suggest that the neural circuits that determine our choices are deeply influenced by the circuits that control our movements. From a scientific perspective, this implies that by studying vigor, we may discover a new way with which to measure individual preferences and thus provide economists a behavioral tool that can objectively estimate subjective utility. From a clinical perspective, vigor may act as a proxy for our current affective state. And from a technological perspective, with the increasing power of smart phones and presence of surveillance cameras, we would not be surprised if someday soon, the results of this research encourage invention of machines that measure our movements and gather vigor-based estimates of our personal preferences. These machines would gather these data even when we are not overtly making a choice, thereby unwittingly revealing one of our secrets: how much we value the thing we are looking at or moving toward.

Acknowledgments

We thank Michelle Harran and Juan Camilo Cortés who edited this book, pointing out things that needed to be explained, things that were missing, and things that were wrong. We are grateful to Timothy Carroll, Bilal Bari, and Thomas Reppert who read certain chapters and provided insightful comments.

Reza Shadmehr: I am grateful to Tehrim Yoon, Scott Albert, Simon Orozco, Ehsan Sedaghat-Nejad, Jay Pi, Kaveh Karbasi, Paul Hage, Tara Palin, and David Herzfeld. These graduate students are the pillars of the Laboratory for Computational Motor Control at Johns Hopkins School of Medicine. I thank Oleg Komogortsev for his devotion to the study of individual differences in eye movements. I thank John Krakauer for his lifelong friendship and deep reflections on science. I thank David Zee who taught me oculomotor control, and demonstrated to a generation of doctors how to be a good physician-scientist, as well as a good human being. I thank Jeremiah Cohen, Bilal Bari, and Marshall Shuler for wonderful discussions about serotonin, reward and vigor. I am grateful to Paul Glimcher for building the field of neuroeconomics and teaching me about the art of writing. I thank Okihide Hikosaka, whose lifetime of work on the basal ganglia sets the standard of how to investigate the shared neural mechanisms of movement and reward.

Alaa Ahmed: I am grateful to Rodger Kram, Art Kuo, Max Donelan, Manoj Srinivisan, and Andy Ruina for teaching me about locomotion energetics and for the many discussions that followed. I am thankful to my postdoctoral advisor Daniel Wolpert for initially leading me down this path of neuroeconomics and its links with movement. Most of all, I am grateful to the members of the Neuromechanics Lab, especially Megan O'Brien, Erik Summerside, Gary Bruening, Shruthi Sukumar, Colin Korbisch, Robbie Courter, Dan Apuan, and Hannah Doris, whose thoughts and insights helped develop many of these ideas.

1

The Effort of Movement

> Love goes toward love as schoolboys from their books
> But love from love, toward school with heavy looks
> —William Shakespeare

When one of your authors was a child living with his three siblings in a small apartment in Tehran, his parents would occasionally go out for walks in the evening. When they returned, sometimes they were a bit frazzled; his mom would complain about how fast his dad walked, and his dad would say that he couldn't help it. Indeed, once when they were visiting a city in the eastern part of the country, his mom lost track of his dad for a good portion of the afternoon.

Years later, when the same author was a student in Boston, he shared an office with a brilliant young man who already had an algorithm named after him. However, the pace with which the officemate spoke reminded the author of his dad's walking speed: rather vigorous.

What controls the vigor with which we move, and why are there differences among people in their vigor? In this chapter, we will begin building a mathematical framework with which we can consider these and similar questions.

1.1 Choosing, Moving, Then Harvesting

There are two interesting puzzles regarding the question of how the brain controls behavior. The first is regarding which action to perform: should one reach for the apple or the doughnut? The second is regarding how to perform the chosen action: should one reach slowly or quickly?

The first puzzle, which action to perform, is usually studied in the field of decision-making by using a framework in which a utility is assigned to each potential option. Utility is the value that the mind assigns to the proposed option, allowing one to measure its subjective worth. If you prefer the doughnut, that preference is reflected in the greater utility that your brain assigns to it. At its simplest form, utility depends on two factors: the reward at stake and the effort required for its acquisition. The basic assumption is that the brain chooses the option that has the highest subjective utility.

The second puzzle, vigor of the movement (tentatively defined as the inverse of the time it took to reach for the doughnut), is usually studied in the field of motor control by using a framework in which there is a goal for the movement, often specified as a desired state that needs to be reached (e.g., the position of the doughnut), and a cost that is assigned to the sequence of motor commands that can achieve that goal (Todorov and Jordan, 2002). The motor commands that are chosen, which affect the speed of the movement and its trajectory, are the ones that achieve the goal (acquire the doughnut) while minimizing the cost of executing the commands.

In a sense, the field of decision-making is concerned with understanding how the brain decides between various options, whereas the field of motor control is concerned with the understanding how the brain produces movements. In this way of thinking, in our apple versus doughnut example, the two questions are answered in sequence: the brain uses the decision-making machinery first to decide what it wants, and then once this decision is made, it relies on the motor-control machinery to perform the movement.

However, factors that affect the process of decision-making also affect the process of motor control. Suppose that you are at the airport waiting for the arrival of someone you know. As you scan the arriving passengers, you decide which one is your target (the decision-making component of the task), and then you walk to greet that person (the motor-control component of the task). However, you will walk faster if the arriving passenger is your child than if the passenger is a distant relative. Both the decision of choosing the target (where to walk to) and the vigor of the ensuing movement (the speed of walking) are influenced by the purpose of the action. An action that culminates with a high amount of reward (walking to greet your child) is not only preferred, that is, it has higher utility (as compared to walking to greet a distant relative), but is also performed with greater vigor. Hence, when the goal of the movement is more rewarding, your brain is willing to spend the extra energy required so that you arrive sooner.

The idea is that when we are deliberating our options, we compare utilities. Each utility specifies our subjective evaluation of the merit of that option, which in turn depends on the value of the reward we believe that option can provide and on the effort that it will cost. However, once we decide what to do (reach for the doughnut, not the apple), shouldn't the movement that we make reflect what is at stake in terms of the expected reward?

Thinking about decision-making and motor control in this way raises the possibility that behavioral and neurophysiological data in these two fields may be better understood if there is a common mathematical framework in which we can consider both the choice that the brain makes between options and the details of the movement that ensues. One of our aims here is to suggest a unified mathematical framework. There are key experimental results that can guide us: given a goal that promises high reward (i.e., the doughnut), subjects will start their movement sooner and move more rapidly toward that goal. That is, the subjective value that we assign our goal affects both the reaction time and the speed of the ensuing movement. If we define vigor as the inverse of the sum of reaction time and movement time, we find that we move more vigorously toward things that we value more.

To consider such results, we need to look at the relationship between the reward that the brain expects to acquire upon completion of the movement and the effort it takes to make that movement. The key idea is that because it takes time to acquire the rewarding state, time acts as a kind of penalty, discounting the reward, encouraging us to move more vigorously so we arrive at the rewarding state sooner. However, moving faster requires greater effort, and saving effort encourages sloth. As a result, there is competition between the desire to acquire the reward soon and the desire to avoid the expenditure of effort. This competition suggests that goodness of a movement may be measured via the rate of positive input (reward) that we will experience if we successfully reach our goal, minus the rate of loss (effort) that we will experience while trying to achieve that goal.

Ultimately, we will build a framework with which we can consider four kinds of behavioral data.

- which option was chosen (doughnut or apple)
- how long was the period of deliberation (short or long reaction time)
- how vigorous was the ensuing movement (fast or slow)
- how much time was spent lingering at the reward site harvesting the reward after the movement ended (short or long harvest periods)

1.2 A Measure of Effort

We do not know how the brain represents effort. A reasonable place to start is to use an objective measure of effort, the energetic cost associated with the movement.

For many kinds of movements, there is a speed that minimizes the total energy consumption. Animals usually move near this speed.

Henry Ralston (1958) was among the pioneers who estimated the rate of energy expenditure during walking. He had people walk with a respirometer in a backpack and measured the oxygen concentration of the expired air as they walked for about 10 minutes at various speeds. He computed the average rate of metabolic energy during walking $\dot{e}_w$ from of the average rate of oxygen consumption (figure 1.1A) and noted that when the rate was normalized by body weight m, it increased as a quadratic function of the average walking speed v:

$$\frac{\dot{e}_w}{m} = a + bv^2 \tag{1.1}$$

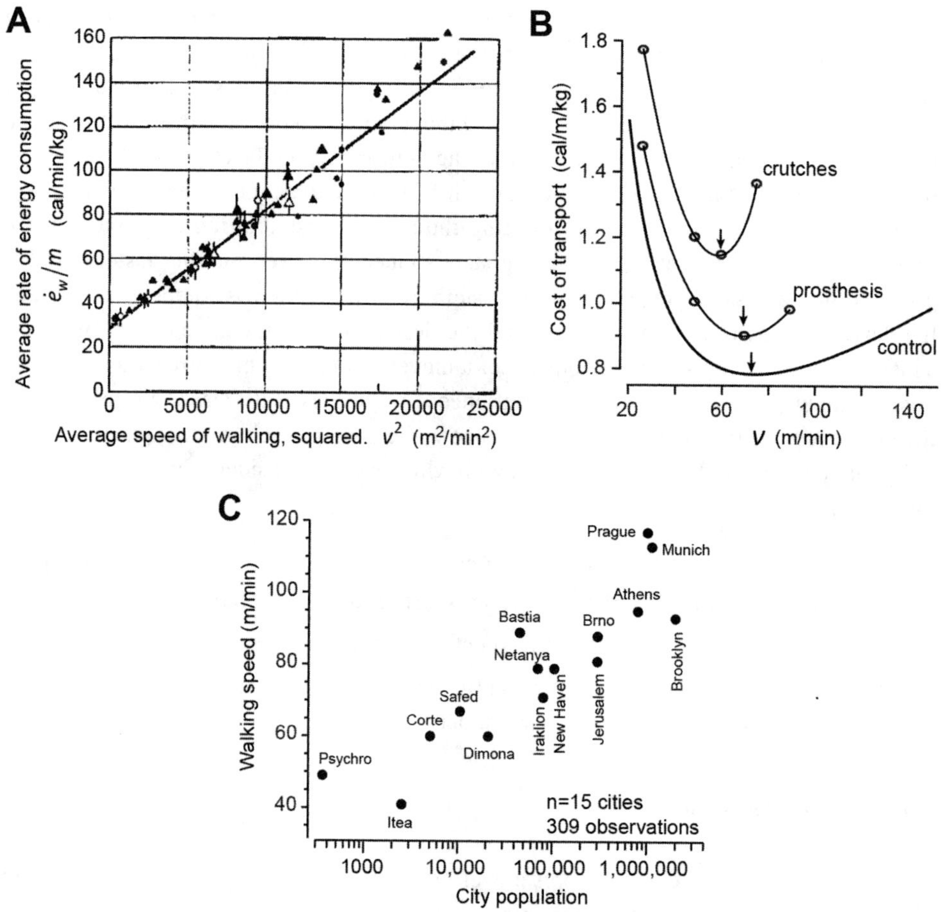

Figure 1.1
Energy consumption during walking and natural walking speeds. **A**. Rate of energy consumption (normalized to weight) as a function of average walking speed, as measured over a period of 10 minutes while the subjects (healthy volunteers, average age of 32 years) walked around a track. (Data from Ralston, 1958.) **B**. Cost of transport: rate of energy consumption divided by mass and divided by average velocity. This is the energy expended to transport 1 kg of mass a distance of 1 cm. The curve labeled "Control" is for the data shown in part A. Curve labeled "Prosthesis" is for an individual who walked with a well fitted prosthetic leg. Curve labeled "Crutches" is for the same individual who walked with crutches. The circles are the measured data points. The arrows indicate the walking speed that minimizes cost of transport. **C**. Preferred walking speed of pedestrians as measured in 15 cities around the world. Preferred walking speed of pedestrians in large cities is twice as fast as those in small towns. (Data are from Bornstein and Bornstein, 1976.)

The units of $\dot{e}_w/m$ are in cal/min/kg, and a and b are constants. Ralston noted that if we divide both sides of the equation by average speed, we arrive at units of cal/m/kg, which is the weight-normalized cost of transport during walking. This cost estimates the total energy that is consumed to transport (via walking) 1 kg of mass for a distance of 1 meter (m) as a function of average speed of the movement:

$$c_w = \frac{\dot{e}_w}{vm} = \frac{a}{v} + bv \tag{1.2}$$

For walking, the cost of transport is composed of a linear term that increases with speed and a hyperbolic term that decreases with speed. The sum of the two terms is a function that has a minimum, indicating the speed at which the cost of transport is smallest (i.e., the optimum walking speed). This speed is optimum in the sense that it minimizes the energetic cost of walking. We have plotted cost of transport as a function of average walking speed in figure 1.1B (labeled control). Ralston (1958) noted that the cost had a minimum at around 74 m/min for healthy people. He hypothesized that "if a subject is told to walk at a speed which is natural or comfortable for him, he adopts a speed which lies at or near the minimum" of this curve. Therefore, he proposed that the brain controls our movement speed in such a way to minimize gross energy consumption.

To test this idea, Ralston had an above-knee amputee walk with a well fitted prosthesis at various speeds, and then he examined that person again without the prosthesis and with forearm crutches. He found that with the crutches, the energy expenditure was higher than with the prosthesis (figure 1.1B). Importantly, the two curves had slightly different minima. Indeed, when the subject was asked to walk at a natural speed, he walked with a speed that roughly corresponded to the minimum of each curve, slower with the crutches than with the prosthesis.

Similar to that for walking, the cost of transport for jogging also has a global minimum at a particular speed (which varies between individuals). Karen Steudel-Numbers and Cara Wall-Scheffler (2009) measured oxygen consumption of nine individuals during 5 minute jogs of various speeds and found that the average rate of energy consumption (in units of kcal/s) for each subject was a quadratic function of the average speed of jogging:

$$\dot{e}_j = a_1 + a_2 v + a_3 v^2 \tag{1.3}$$

(Note that unlike equation 1.1, the rate of energy consumption in the above equation is not normalized by weight of the subject.) The terms a_1, a_2, and a_3 are constants. The cost of transport for jogging, that is, the total energy expended to jog for a distance of 1 kilometer is expressed by this equation:

$$c_j = \frac{a_1}{v} + a_2 + a_3 v \tag{1.4}$$

Like the cost of transport for walking, the cost of transport for jogging has a hyperbolic term that decreases with speed and a linear term that increases with speed. This implies that there exists a running speed for which the energetic expenditure is a minimum.

Subjects experienced various jogging speeds and were asked to pick two speeds that were considered slow, two speeds that were comfortable, and two speeds that were fast. Each subject's comfortable speed was around the minimum of the cost of transport.

The energy that is consumed during locomotion has also been measured in other animals. Donald Hoyt and Richard Taylor (1981) trained horses to walk, trot, or gallop on a treadmill at various speeds and recorded their rate of oxygen consumption. Importantly, the researchers were able to train the animals to produce different patterns of locomotion at a given speed (e.g., acquire data for both walk and trot at 2 m/s, acquire data for both trot and gallop at 5 m/s). For every speed and pattern of locomotion, they divided the rate of oxygen consumption by speed of motion to determine cost of transport, as shown for one horse in the line graph in the top part of figure 1.2. Because the rate of energy consumption increased faster than linearly with speed and had a non-zero bias at 0 m/s (as in equation 1.1), the cost of transport (line graph, figure 1.2) exhibited a minimum of around 1 m/s for walking and of around 3.5 m/s for trotting. For galloping, the authors felt that they could not train the animals to run fast enough to be able to see whether the cost exhibited a minimum. Interestingly, when the horses were allowed to select their own speed while running on ground, they chose a speed that roughly agreed with the energetically optimal speed for each gait (bar plot, figure 1.2). Together, these results suggested that humans and other animals "select speed within a gait in a manner that minimizes energy consumption" (Hoyt and Taylor, 1981).

Effort of walking depends on speed and stride length.

Your walking speed is determined by the length of your steps as well as the number of steps you take per minute (step frequency). Walking speed v is the product of step length x and step frequency q, that is, $v = xq$. This means that if you want to walk a little faster, you could do so either by taking longer steps or by increasing your step frequency. It turns out that the effort of walking depends not only on walking speed, but also on the step pattern that generates that speed. You tend to not only pick the speed that minimizes effort, but also pick the stride length that minimizes the effort at that speed.

While people are walking at a fixed speed, their energetic costs vary with step frequency (Atzler and Herbst, 1927; Elftman, 1966). Mohamed Zarrugh and C.W. Radcliffe (Zarrugh and Radcliffe, 1978) quantified the effect of step length and step frequency on oxygen consumption by testing seven people under two sets of walking conditions: free and forced. Under the free conditions, the subject walked on a level treadmill at five test speeds for duration of 5 minutes each. Under the forced conditions, the speed of the treadmill was maintained at 1.5 m/s, and the subjects were asked to match their step frequency to a metronome. The enforced step frequencies included a step frequency close to that

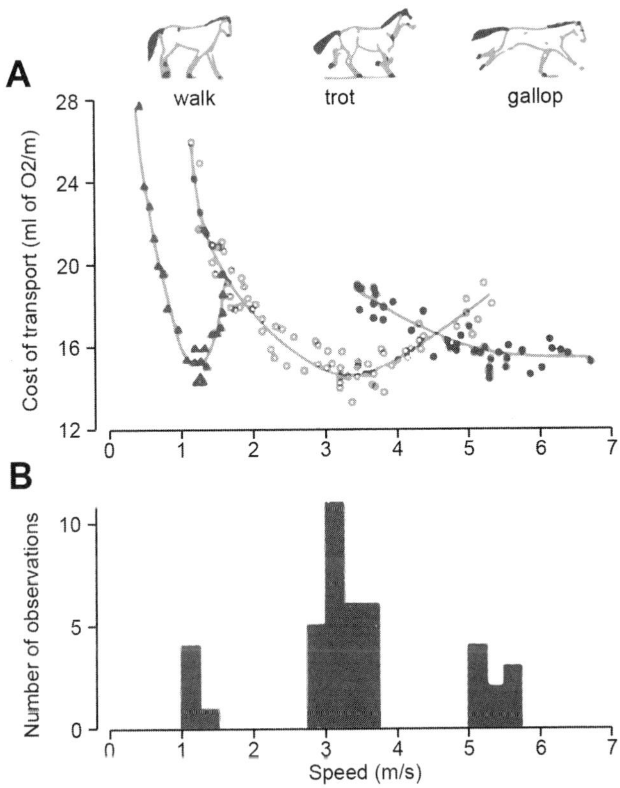

Figure 1.2
Speed of locomotion and its energetics in horses. **A**. Oxygen cost to move a unit distance (cost of transport) is illustrated for one horse. The minimum oxygen cost to move a unit distance was approximately the same in all three gaits. **B**. Gaits for which the horse was allowed to select its own speed while running on ground. For each gait, the animal chose to move near the speed that minimized cost of transport. (Data from Hoyt and Taylor, 1981.)

chosen by the subject under the free walking conditions at that speed (i.e., the subject's preferred walking pattern) as well as three or four more step rates.

As people performed the tests, the experimenters measured oxygen consumption and found that the rate of oxygen uptake increased as a quadratic function of walking speed (as in the studies by Ralston and others). However, as speed requirements increased under the forced conditions, the subjects changed their freely chosen step frequency and step length. Notably, at a fixed speed, the energetic rate increased both at step frequencies that were lower than those under freely chosen conditions as well as at step frequencies that were greater than those under freely chosen conditions (figure 1.3). Thus, even at a given speed, subjects preferred a step frequency and step length combination that minimized the energetic cost.

Running speed is also a product of step length and step frequency. If you like to run long-distances, you might wonder what the appropriate step length should be. Paul

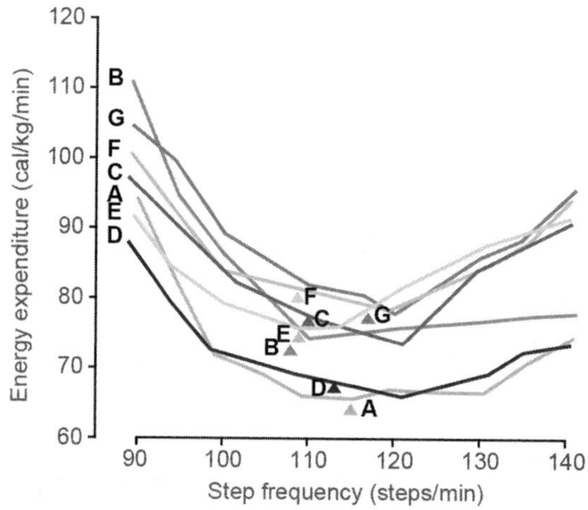

Figure 1.3
Energetics and step frequency. Energy expenditure was measured in seven male subjects as they walked on a treadmill at a fixed speed of 1.5 m/s. Subjects walked at prescribed step frequencies both greater and less than their freely chosen step frequency. Triangles indicate the energetic expenditure for each subject at their freely chosen step frequency. Subjects tended to freely select a step frequency that approximated the energetic minimum. (Data from Zarrugh and Radcliffe, 1978.)

Hogberg (1952) approached this question from the perspective of energetics. He measured oxygen consumption in a single elite runner (himself) running at a range of fixed speeds but at different step lengths. Similar to that when walking, his preferred step length when running corresponded to the energetic minimum. In 1982, Peter Cavanagh and Keith Williams tested this more rigorously in a group of 10 male runners (Cavanagh and Williams, 1982). Each subject ran at a 7 min/mile pace at their freely chosen step length and at six additional step length values (+/- 6.7%, 13.4%, and 20% of their freely chosen value). While there was large variability between subjects in the amount of oxygen consumed, all exhibited an energetically optimal step length that minimized oxygen intake. When compared to the freely chosen step length, seven subjects were over-striding (choosing step lengths longer than optimal), and the remaining three were under-striding. Despite these deviations, the freely selected step lengths were remarkably close to the energetically optimal step lengths, with an average difference of 4.2 cm or 4.5% of leg length. The authors concluded that this energetically favorable behavior may be "an adaptation to the chosen stride length through training or a successful process of energy optimization."

Effort of walking also depends on stride width.

When we take a step, we freely select how far ahead to place our foot (which determines step length), but we also unconsciously select how wide a step to take. Healthy adults generally choose a step width corresponding to 12% of their leg length. Is the energetic minimization principle also at play in dictating this choice?

Max Donelan, Rodger Kram, and Art Kuo approached this question using a combined theoretical and experimental approach (Donelan et al., 2001). They developed a mathematical model of walking to describe the relationship between step width and metabolic cost. Walking can be modeled as a series of controlled falls, in which at every step the body must redirect the downward velocity of the center of mass upwards into the next step. The model predicted that a significant component of the effort cost of walking was the step-to-step transition cost. The model also predicted that this cost would increase with increasing step width. However, walking with a narrow step width is not energetically favorable because that will incur a different type of cost: the cost of swinging the leg laterally to avoid the other leg. Thus, the authors hypothesized that there would be an energetically optimal step width. They tested 10 subjects while walking on a treadmill at a comfortable, fixed speed (1.25 m/s). Subjects walked at their freely chosen step width and under seven more conditions corresponding to various percentages of leg length. As the authors had hypothesized, energetic cost increased significantly for step widths wider than the freely selected width (13% of leg length). At the narrowest step width, the cost was also greater than that at the preferred step width. Taken together, the results demonstrate that each subject had an optimal step width and that the subjects tended to walk with a step width that minimized energetic cost.

Humans can continuously adapt their movements to reduce energetic cost.
Walking and running are actions that we do all of our lives. When we want to walk toward a friend, we have a lifetime of experience with which to estimate the effort it takes to get there. However, what happens when the energetic cost of our action changes transiently, that is, when we need to rely on learning to estimate the effort of moving? If the way we move is driven in part by energetic costs, then we should change how we move when our energy requirements change. For example, walking up an incline alters energetic costs and lowers the energetically optimal walking speed. Indeed, animals walk slower up inclines (Wickler et al., 2000). More generally, if the brain is concerned with energy costs, then we should be continuously adapting our movements to changing energetic requirements in order to track the energetic minima and reduce energetic expenditure.

Jessica Selinger, Shawn O'Connor, Jeremy Wong, and Max Donelan considered this question with a novel experimental design (Selinger et al., 2015). They used a custom exoskeleton that applied torques at the knee to create energetic landscapes for subjects as they walked on a treadmill at a fixed speed. Although the treadmill speed was fixed, the choice of step frequency was left to the subject. As mentioned earlier, at a given speed there is an energetically optimal step frequency. The step frequency that people prefer roughly corresponds with this energetic minimum. The authors programmed the exoskeleton so that it reshaped the relationship between energetic cost and step frequency. Subjects were tested under one of two conditions: penalize-high and penalize-low. Under the penalize-high conditions, the exoskeleton applied an increasing amount of resistive knee torque with increasing step frequency, thus energetically penalizing higher step frequencies. Under the penalize-low conditions, resistance increased with decreasing step frequency, penalizing lower step frequencies. Critically, under both sets of conditions, when subjects walked at their preferred step frequency, the exoskeleton would apply a high resistive knee torque, elevating the energetic cost above normal.

Under these novel energetic conditions, the energetic minimum was shifted to a step frequency that was lower (penalize-high) or higher (penalize-low) than their original preferred step frequency. After an initial baseline walking period to determine preferred step frequencies, the exoskeleton motors were turned on and the subject walked on the treadmill for 15 minutes. Surprisingly, despite the fact that the energetic cost of walking with the preferred step frequency was now elevated, subjects did not shift their step frequency. During the next phase, the authors used a metronome to enforce a range of step frequencies and systematically expose the subjects to the novel energy conditions. The subjects experienced the lower or higher energetic costs associated with each step frequency. After this period of forced exposure, the metronome was turned off and subjects were allowed to settle upon a step frequency of their choosing. Subjects did not settle back upon their original preferred step frequency. Rather, they walked at a higher step frequency under the penalize-low conditions and a lower step frequency under the penalize-high conditions.

Thus, the results illustrated that when the energetics of walking changed transiently, through exploration people converged onto a new walking pattern, one that tended to reduce energetic expenditures.

There exists a reaching speed that minimizes energetic cost.

This is a powerful hypothesis because in principle, it can predict the preferred speed of motion for any animal in any modality of movement. To explore this question, Helen Huang measured the rate of metabolic energy consumed during reaching by having young healthy volunteers reach at various distances (10 cm, 20 cm, and 30 cm) and durations (indicated in figure 1.4A) (Shadmehr et al., 2016). The subjects were seated and held the handle of a robotic manipulandum that moved in a horizontal plane. The weight of the forearm was supported by the robot. When the subjects were sitting quietly, their resting rate of energetic consumption was $a_0 m$ ($a_0 = 1.16$ J/s/kg and m is the subject's body weight). As subjects made reaching movements, she measured the rate of energy consumption and then subtracted the resting rate, resulting in the net rate of energy consumption with respect to rest. This is shown in figure 1.4A. As movement duration T increased, the net rate decreased via a function that depended on distance d: $a_1 + bd/T^{2.7}$, with values of $a_1 \approx 30$ J/s, and $b \approx 154$. If we combine the net rate with the resting rate, we arrive at a function that describes the total energy expended during the reach:

$$e_r = aT + b\frac{d}{T^{1.7}} \tag{1.5}$$

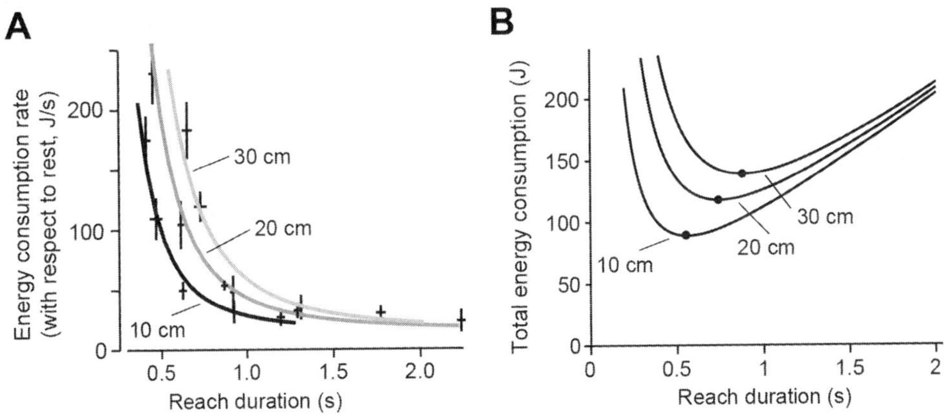

Figure 1.4
Energetic cost of reaching. **A**. Rate of energetic consumption during reaching movements to various distances and durations. Movements were made in a horizontal plane with the weight of the arm supported. **B**. Total energy consumed to reach a given distance as a function of reach duration. The optimum duration is specified by the filled circle. (Data from Shadmehr et al., 2016.)

In the above equation, $a \approx 100$ J/s. This function is plotted in figure 1.4B. Note that like the energetic costs of other kinds of movements such as walking or jogging, the energetic cost of reaching is a concave-upward function of movement duration. Thus, like walking, the act of reaching has an energetic cost that has a global minimum: given a reach distance, there exists a reaching speed that minimizes the energetic consumption.

For example, for a 20 cm reach, the energy consumption is minimized when the movement duration is around 0.75 seconds. People perform a 20 cm reach only somewhat faster, in approximately 0.6 seconds (de Grosbois et al., 2015).[1] Therefore, the duration of natural reaching movements of people is fairly close to the duration that minimizes the total energetic cost of the movement.

Let us consider a very different kind of movement: driving a car. Just like walking and reaching, the energetic cost of driving is also concave-upward. For a typical car with a gasoline engine (i.e., not hybrid), fuel efficiency (miles per gallon) increases with driving speed up to around 50 miles per hour (figure 1.5A). However, as one drives faster than 50 miles per hour, fuel efficiency decreases. As a result, in order to drive a given distance, the total energy that one must consume is a concave-upward function of drive duration (figure 1.5B). This implies that there is an optimum driving speed that minimizes energy consumption.

However, people do not always choose the energetically optimal solution. Jesse Dean and colleagues demonstrated this in walking (Hunter et al., 2010). Dean and his students

1. Reach speed depends on width of the target, which specifies an accuracy constraint (Fitts, 1954): as target size increases, requiring less endpoint accuracy, reach speed decreases. For a very large target width of 5 cm at a distance of 20 cm, reach duration is around 0.6 seconds.

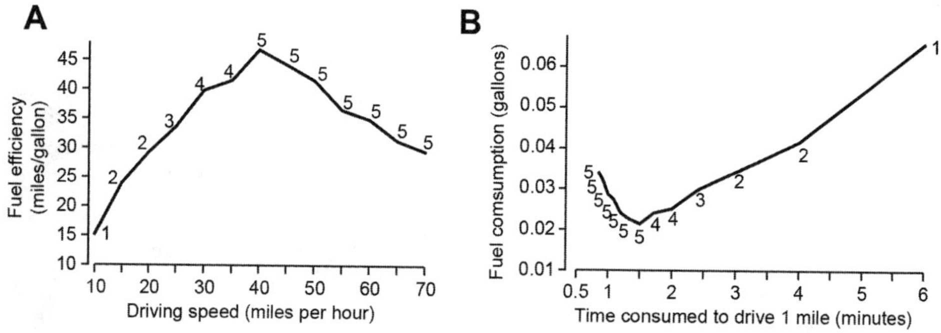

Figure 1.5
Energetic cost of driving a car (1986 Volkswagen Golf GTI). **A**. Fuel efficiency Miles per gallon as a function of driving speed. Numbers indicate the gear ratio. **B**. Total fuel consumption to drive 1 mile as a function of time. (Data from www.metrompg.com/posts/speed-vs-mpg.htm.)

asked 10 subjects to walk on a treadmill at a fixed speed, but downhill at different gradients. They tested the subjects under two sets of conditions: one in which the subjects were asked to walk naturally and another in which they were asked to walk in a relaxed manner. Surprisingly, for the same gradient, the natural gait incurred greater energetic cost than the relaxed gait. The critical question here is why did subjects not freely select the more energetically optimal pattern of walking?

Thus we arrive at a major problem with our story. Energetic expenditure, our proxy for effort, cannot be the only concern for the brain: when you are late to class, you will walk faster. To illustrate this idea more formally, consider the observation that the natural walking speed for people in large cities is faster than for those in small cities (Bornstein and Bornstein, 1976). In their study, the authors measured a set distance (15 m) on the sidewalk in downtown or commercial areas of various cities around the world and then timed individuals unobtrusively from a vantage point. They found that natural walking speed increased with the logarithm of the population of the city (figure 1.1C). People in large cities walked twice as fast as people in small cities. A similar concern is present in reaching movements; when people are asked to reach for a candy bar, they reach faster for their favorite candy bar than for another type of candy bar (Sackaloo et al., 2015). These curious observations demonstrate that while the brain may be concerned with energy consumption during movements, this is not the only concern that influences the decision of how fast to move.

So, we arrive at a puzzle. People vary the speed with which they perform an action. Their speed depends on the purpose of the movement and its context. For example, people walk faster in cities than they do in small towns. They reach faster for things that they prefer. We will suggest that in deciding how fast to move, the brain not only considers the energetic cost of the ensuing action, representing an objective measure of effort, but also the reward for that action, which is the purpose of the movement.

1.3 Utility of a Movement

It seems rational that we move for a purpose and that purpose must play a role in how our brain controls our movements. To express this idea mathematically, assume that the purpose of a voluntary movement is to change the state of the body to one that is more valuable than the current state. The value of the movement may be determined by two factors: the reward that we expect to acquire when we complete the movement and the effort that we will spend in generating that movement. The former serves as a gain; the latter serves as a loss. Together, these two elements form a measure of goodness. In principle, the utility should not only describe which movement to perform (reach for the apple or the doughnut; walk toward my aunt or my daughter), but also how to perform that movement (reach fast or slow; walk or run).

Utility of an action may be defined as the reward expected when the action is completed minus the effort required to complete the action, divided by time to acquire the reward. Let us return to the airport example where you are waiting arrival of someone you know. Meeting this person has some rewarding value $\alpha > 0$, implying that there exists a scale with which we can compare the value of meeting this person with that of meeting another person (e.g., your child may be associated with a large amount of reward while a distant relative may be associated with small reward). We can imagine that this value is not constant, but changes as a function of time: it is better to receive reward sooner rather than later. Time discounts the reward such that it is preferable to acquire the valuable state sooner rather than having to wait and receive it later. Psychologists and economists have quantified the shape of the reward temporal discount function. In a typical experiment, the subject decides whether to receive a small reward α_S immediately or a larger reward α_L after delay T. The delay is manipulated until it becomes as likely for the subject to choose α_S as it is for the subject to choose α_L. The choices that people and other animals make suggest that time discounts reward as a hyperbolic function (Jimura et al., 2009; Kobayashi and Schultz, 2008):

$$\alpha_S = \frac{\alpha_L}{1+\gamma T} \tag{1.6}$$

The temporal discount factor γ determines how rapidly reward is discounted by time. A larger value produces faster discounting, indicating a preference to take the immediate, less valuable reward α_S, and a reluctance to wait for the larger reward α_L. An individual whose decisions indicate fast temporal discounting is sometimes called impulsive.

Although the above hyperbolic temporal discount function has been used to describe tasks in which people and other animals make choices between small immediate rewards and large delayed rewards (like food and money), such a temporal discount function may also be relevant to the control of movements (Shadmehr et al., 2010). The idea is to view duration of a movement as an implicit delay in the acquisition of a reward, and the act of moving fast or slow as a decision between the acquisition of a large reward soon in exchange for payment of large effort and the acquisition of smaller, discounted reward later in exchange for payment of small effort. For now, let us assume that for control of movements, the brain assigns a value to the goal state and discounts it hyperbolically as a function of the duration of the movement.

Generating a movement involves expenditure of effort, which depends on the duration of the movement T. Let us represent effort with function $U(T)$. For example, $U(T)$ may represent the energetic cost of the action. We can now write the utility of the movement as the sum of the (temporally discounted) reward and effort:

$$J = \frac{\alpha}{1+\gamma T} + U(T) \tag{1.7}$$

An issue that remains unresolved is the way effort should be represented. As we have seen from our discussion above, a movement carries a cost via its metabolic energy expenditure. For walking, the energy expenditure is calculated as follows (integrating equation 1.1 with respect to duration):

$$e_w = amT + bm\frac{d^2}{T} \tag{1.8}$$

Returning to our airport example, suppose we see the person we are waiting for and walk a distance of 50 meters to greet them. Our utility depends on both the energetic cost of the action (our proxy for effort) and the reward that we expect to attain at the end of that action. If we assume that effort of the movement is proportional to the energy required to make the movement, and further assume that like reward, effort is discounted with time (an assumption that we will return to shortly), we arrive at the following representation of effort associated with walking:

$$U_w = -\frac{e_w}{1+\gamma T} \tag{1.9}$$

We now incorporate reward and effort into a single function describing the utility of walking:

$$J_w = \frac{\alpha - e_w}{1+\gamma T} \tag{1.10}$$

The unit of reward α is the calorie, the same as the unit of effort (energy that it takes to produce the movement). The unit of utility is calorie per unit of time (γ is unitless). Therefore, in this formulation of utility, the numerator represents the net energetic intake (energy gained by reward minus energy lost by effort), and the denominator represents a temporal discount of that intake. Utility of the action becomes approximately the average rate of net energetic intake.

Representation of utility in terms of rate of caloric intake, reward as food, and effort as calories spent to acquire food will be useful (in the next chapter) when we consider decision-making and movements of animals during foraging. However, reward is not calories when we are walking to meet someone we love. Rather, in that case reward is computed via its neural proxy, dopamine. Similarly, effort is not calories spent walking, but its neural proxy, which at this point is poorly understood. Thus, the objective is to perform actions in such a way as to maximize the neural proxy for the rate of reward intake, while minimizing the neural proxy for the rate of effort expenditure.

As reward increases, moving faster increases utility of the action.

The energetic cost of walking a distance of 50 meters is plotted in figure 1.6A. To minimize this cost, the optimum duration of travel should be around 40 seconds, resulting in an average walking speed of 75 m/min. In contrast, if we use the utility formulation and assume that the purpose of the walk is to acquire reward (one valued at 5 kcal), then the optimum duration of travel is shorter, around 32 seconds, resulting in an average walking speed of 94 m/min. The reward and effort of the utility as a function of walk duration are plotted in figure 1.6B. A fast walk lasts a short duration, resulting in small discounting of reward (because we arrive at our goal sooner). However, this rapid pace will require a large effort (a larger negative value). A slower pace results in a longer duration, requiring a smaller effort, but will produce large discounting of reward. To identify the optimum duration, we take derivative of the above expression with respect to time and after setting it to 0, finding the following:

$$T^* = \frac{bmd^2\gamma + d\sqrt{m^2ab + \alpha bm\gamma + b^2m^2d^2\gamma^2}}{ma + \alpha\gamma} \tag{1.11}$$

The movement duration T^* is optimum in the sense that at this duration, the movement has the highest utility, optimizing the interaction between reward and effort. An increase in reward value α makes the numerator increase at slower rate than the denominator (because $bm < 1$), allowing us to infer that as the value of the reward increases, movement duration should decrease and hence result in a faster walk. In contrast, if our objective is to minimize the energetic cost of walking, then from equation 1.8, we infer that the optimum duration is described as follows:

$$T^* = d\sqrt{b/a} \tag{1.12}$$

Suppose you are given the option of walking in order to receive a valuable reward α_L (doughnut) or walking in order to receive a less valuable reward α_S (apple). The utilities for these two options are shown in figure 1.7A. Clearly, the utility associated with the act

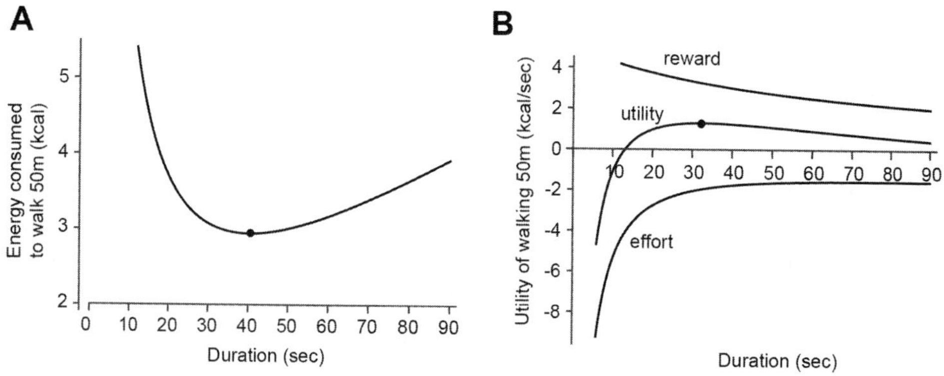

Figure 1.6
Energetics and utility of walking. **A**. Energy consumed to walk a distance of 50 m for a person of weight 75 kg, a=29 cal/min/kg, and b=0.0053 cal.min/kg/m^2. **B**. Utility of the movement. The curves are the temporally discounted reward and the metabolic cost of walking, plotted as a function of movement duration. The center curve is the utility, which is the sum of the temporally discounted reward and metabolic cost. The utility is mostly a positive function of duration, indicating that the movement will produce a greater amount of reward than effort and therefore is worth performing. It has a peak (black circle) corresponding to the duration of movement that maximizes the utility. This plot was generated by using the metabolic cost shown in part A, with γ= 1 and reward α= 5 kcal.

that brings the greater reward is higher (more calories for the doughnut), and therefore you should choose that option. However, notice that the movement duration that maximizes the utility in each case is different. As the reward value increases, the optimum duration of the movement decreases. This means that we should not only prefer to move toward the stimulus that promises greater reward, but also move with greater speed toward it. The reward that is expected at the conclusion of the movement makes it worthwhile to spend the extra energy needed to move faster.

Suppose you are given a choice between performing an act that requires a large metabolic cost and performing one that requires a small metabolic cost. Importantly, the rewards associated with the two acts are the same. Clearly, you should pick the option that requires a smaller metabolic cost. We can increase the metabolic cost by asking the subject to carry a mass while keeping all other variables (including reward) constant (figure 1.7B). Between a small mass and a large one, the utility of moving the smaller mass is greater, and therefore one should choose the option of moving the smaller mass. Additionally, the optimal movement duration for moving a small mass is shorter than that for moving a larger mass. An increase in m causes a larger increase in the numerator than in the denominator of equation 1.12, a result which implies that an increase in mass results in an increase in the optimal duration of the movement. This means that everything else being equal, one should not only prefer the action that requires moving a smaller mass, but also should perform that action with greater vigor.

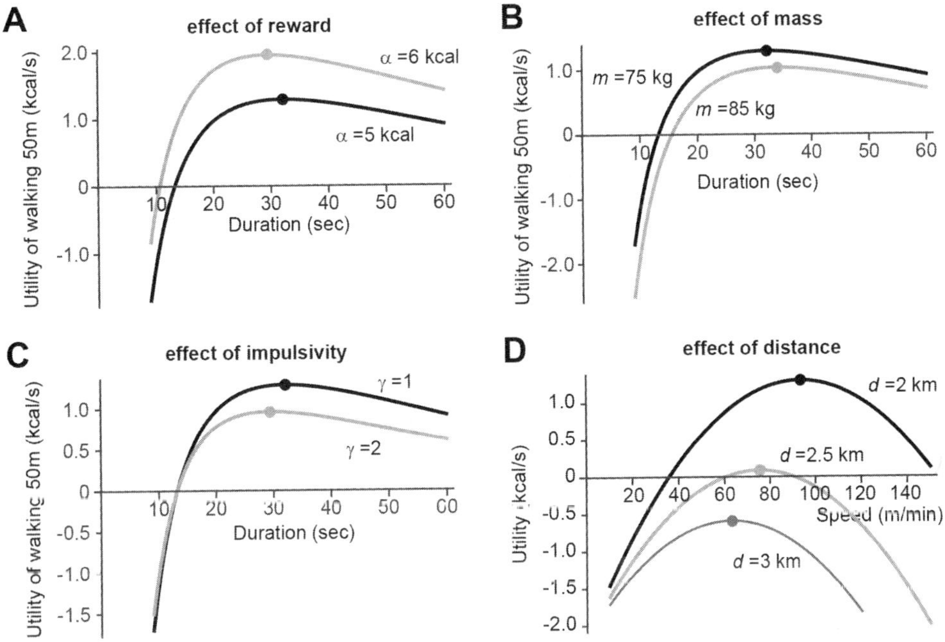

Figure 1.7
Utility of walking as temporally discounted sum of reward and effort. The baseline simulation (black curve) is the utility of a 75 kg person to walk distance of 50 m, for reward $\alpha=5$ kcal, with $\gamma=1$. **A**. With increased reward, the maximum utility increases. In addition, the optimal duration shifts to a smaller value. As a result, a stimulus that promises greater reward not only carries a greater utility, but produces a movement that has greater velocity (reduced duration). **B**. The effort of the movement is increased by increasing the mass of the moving object. An increased mass decreases the maximum utility of the movement, but also shifts the optimal duration to a larger value, thereby decreasing the speed of the resulting movement. **C**. The effect of increased rate of temporal discounting. Increasing the rate of temporal discounting decreases the maximum utility of the movement, but also shifts the optimal duration to a smaller value, thereby increasing movement velocity. **D**. Utility as a function of average speed of walking (all variables except distance are kept constant). As distance increases, the maximum utility decreases. The optimum speed of walking is slower for longer distances. At large distances, the entire utility function is negative. In this case, the best option is to reject the reward and choose not to walk.

We can compare decision making and movements of two individuals, one who has a steep temporal discount function (large γ), and one who has a shallow discount function (small γ). A steep temporal discount means that passage of time greatly affects reward and effort. Everything else being equal, the individual who has a steep discount function will be more impulsive, preferring the immediate reward. A reward obtained after a certain period of time has less value for an individual with a steep temporal discount function (impulsive) than for another individual with a less steep function (patient). As a result, the utility of acting to acquire a given reward is generally lower for the impulsive person (figure 1.7C), and the optimal movement duration is shorter for the more impulsive person.

This makes the interesting prediction that everything else being equal, the individual who has a steep temporal discount function (impulsive decision-maker) will generally walk faster than the individual who has a shallow temporal discount function (patient decision-maker). In chapter 3, we will explore data regarding this conjecture.

Finally, consider walking toward a goal, such as an ice cream cart. In one scenario, the cart is nearby, while in another case, it is farther away. Beyond a certain point, we judge the distance as too far to walk, and we opt not to walk. We can use the utility framework to consider this example. In figure 1.7D, we have plotted utility as a function of walking speed for a reward value of $\alpha = 150$ kcal (representing the ice cream that we want to acquire). If the distance to walk is 2 km, the utility is maximized at an average speed of about 90 m/min. When the distance to walk is 2.5 km, the utility is maximized at an average speed of about 75 m/min. So as the distance to goal increases, the optimal average walking speed decreases. When the ice cream vendor is close (≤ 2.5 km), the utilities are positive at their maximum. However, if we now set the distance at 3 km, the utility is no longer positive at its maximum. At this distance, the utility becomes less than 0 kcal/s, implying that performing this action will result in a net loss. We would choose to not walk to the ice cream vendor and perhaps find an option requiring less effort, such as driving.

Because the energetic cost of reaching is also concave-upward with respect to duration (figure 1.4B), the theory predicts similar effects of reward and effort on the speed of reaching. Let us represent effort as the temporally discounted energetic expenditure during reaching:

$$U_r = -\frac{e_r}{1+\gamma T} \tag{1.13}$$

The utility of reaching becomes as follows:

$$J_r = \frac{\alpha - e_r}{1+\gamma T} \tag{1.14}$$

The effort, reward, and utilities associated with reaching a distance of 20 cm are plotted in figure 1.8A. As the value of the reward increases, the utility of the movement increases, but the optimum duration of the reach decreases (figure 1.8B). This suggests that one should not only reach for the most valuable object, but also reach faster for that object. As distance to the reward site increases, the utility decreases, and the optimum duration increases (figure 1.8C). Given the option of reaching for two identical objects placed at different locations, one should reach for the closer object. However, whereas optimum speed decreases with walking distance (figure 1.7D), it increases with reaching distance (figure 1.8E). Indeed, experimental data support the prediction that reach speed increases with distance (dark heavy line in figure 1.8E). Furthermore, the preferred reach speed in humans (de Grosbois et al., 2015) is faster than the speed that minimizes the

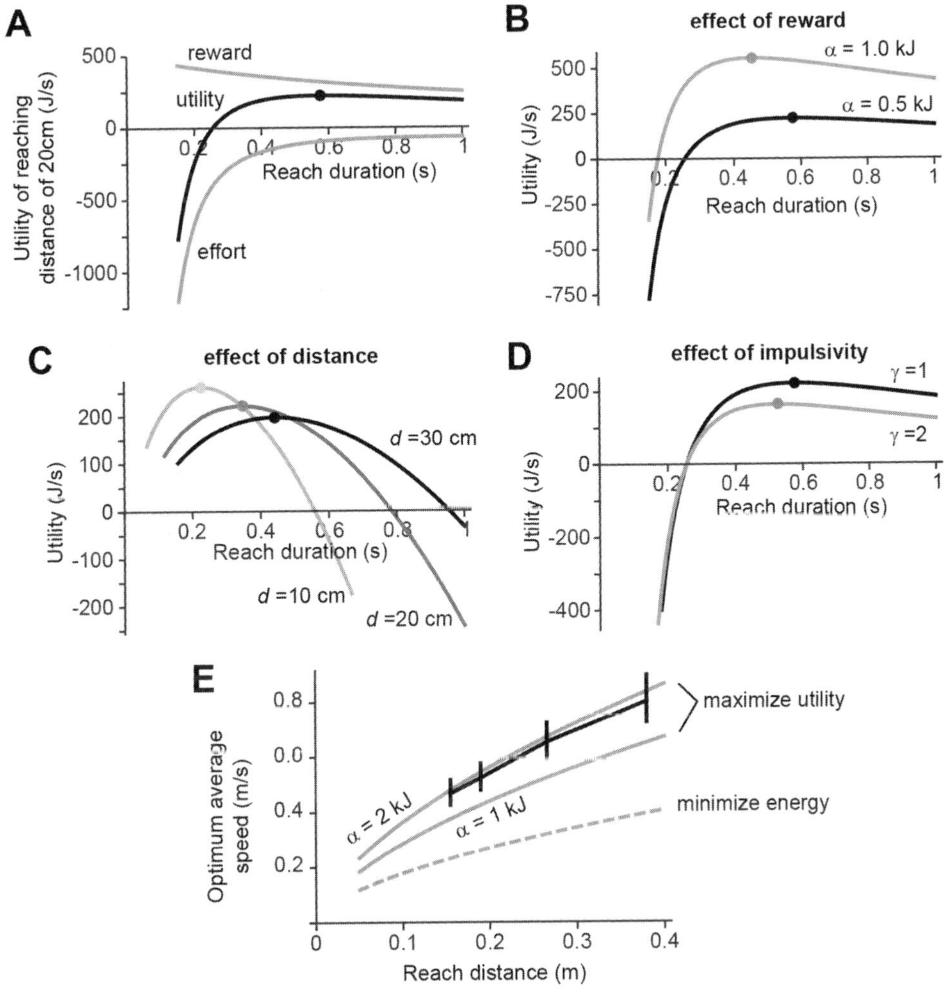

Figure 1.8
Utility of reaching as temporally discounted sum of reward and effort. The baseline simulation (black curve) is the utility of reaching a distance of 20 cm for reward $\alpha = 0.5$ KJ with $\gamma = 1$. **A**. Temporally discounted reward, effort, and utility. **B**. When reward value increases, the maximum utility increases, but there is also a reduction in the optimum duration. That is, an increase reward value causes a faster reaching movement. **C**. An increase in distance reduces the maximum utility and increases the optimum duration. **D**. Increased impulsivity reduces the maximum utility and decreases the optimum duration. **E**. The speed that minimizes the energetic cost of reaching (dashed line), and the speed that maximizes the utility of reaching (solid gray lines). The dark black lines display preferred reaching speeds from human subjects, from (de Grosbois et al., 2015). Reaching speeds are typically faster than the speed that minimizes energetic expenditure. That is, the value of the reward makes it worthwhile to spend energy.

energetic cost (dashed line, representing the average speed of reach that minimizes the energy expenditure e_r during the movement). Finally, let us consider the effect of temporal discounting. As temporal discounting increases (passage of time rapidly devalues reward), utility decreases. That is, the utility of reaching for a given reward is less for someone who has a larger temporal discount rate. However, the optimum reach duration for the person with the larger temporal discount rate is smaller (figure 1.8D). Hence, an impulsive person (large γ) should reach faster than a patient person.

In summary, if we use energetic expenditure as a proxy for effort, we find that in both reaching and walking, the total energy expended as a function of duration is concave-upward; this finding implies that there exists a speed of movement that minimizes the effort during the period of movement. In general, the speed with which people (and other animals) walk is close to the minimum of the energetic cost of walking. Similarly, the speed with which people reach is close to the speed at which the energetic cost of reaching is minimum. However, reward modulates the speed with which people and other nonhuman animals move: we tend to move faster when there is a greater reward at stake. One way to account for these observations is to imagine that the brain selects the speed of a movement by considering both the required effort and the expected reward. In this framework, the optimum speed of movement is one that maximizes a utility, which is defined as the expected reward minus the required effort, divided by duration of time required to acquire that reward. The utility of an action provides a currency with which we can select among the various actions (reach for the doughnut or for the apple) as well as describe how to perform the selected action (reach slowly or quickly).

1.4 Additive versus Multiplicative Interaction of Reward and Effort

In composing our utility, we chose to combine effort and reward additively. This is in contrast to other work in which it is assumed that effort discounts reward multiplicatively (Sugiwaka and Okouchi, 2004; Prevost et al., 2010; Klein-Flugge et al., 2015). Let us compare these two approaches. Consider an arbitrary function $U(T)$ that specifies how effort varies with duration of movement. In the case of multiplicative interaction between reward and effort, we have the following representation of utility:

$$J = \frac{\alpha}{1+\gamma T} U(T) \tag{1.15}$$

In the above formulation, an increase in reward produces an increase in the utility of the action. This matches the observation that as the value of the reward increases, a person is more willing to choose that option. However, in this formula, reward has no effect on the optimal movement duration. This is because the effect of reward is to scale the utility function, which has no effect on the value of T^* that maximizes the utility. Therefore, if reward and effort interact multiplicatively, reward has no effect on the speed of movement that maximizes the utility.

A utility in which reward is multiplied by a function of effort generally fails to predict dependence of movement speed on reward. In contrast, an additive interaction between reward and effort can predict that an increased reward increases movement speed. As we will see in the next chapter, there is a wealth of evidence that increasing reward increases the speed with which people and other primates move their eyes (Kawagoe et al., 1998; Xu-Wilson et al., 2009; Reppert et al., 2015) or their arms (Opris et al., 2011; Sackaloo et al., 2015). This reward-dependent modulation of movement vigor appears consistent with an additive but not a multiplicative interaction between reward and effort.

1.5 Temporal Discounting of Effort

We represented effort as the temporally discounted metabolic cost of performing an action. This allowed us to view utility as reward rate, a measure that described goodness of movements as a function of reward minus effort, divided by time. Let us examine the merits of the conjecture that like reward, effort is discounted by time.

Consider the task of producing an isometric force $f(t)$ for duration T. To estimate the metabolic energy consumed during isometric force production, we can consider the data collected by David Russ et al. (2002). These authors electrically stimulated the human gastrocnemius muscle at 20 Hz and 80 Hz and measured the resulting force and metabolic cost. They measured the metabolic energy consumed during this task by using nuclear magnetic resonance (NMR) spectroscopy. The 20 Hz stimulation produced a force that increased with a time constant of around 0.25 s, reaching a plateau of approximately 230 N (figure 1.9A). The 80 Hz stimulation produced a force that had a similar time constant, but reached a plateau that was larger, around 430 N. They also stimulated the gastrocnemius for four stimulation periods (figure 1.9B), and two stimulation frequencies. If we replot the data in figure 1.9B as a function of a force-time integral produced by the stimulations, we see that there is an approximately linear relationship between adenosine triphosphate (ATP) consumption and the force-time integral (figure 1.9C). Therefore, energy consumed during isometric force production of duration T may be approximated as follows:

$$e(T) = a_0 + a_1 \int_0^T f(t)\,dt \tag{1.16}$$

Using the above estimate of the energetic expenditure, we can write the utility of producing a constant force F for period T, in order to receive reward α:

$$J = \frac{\alpha - a_0 - a_1 FT}{1 + \gamma T} \tag{1.17}$$

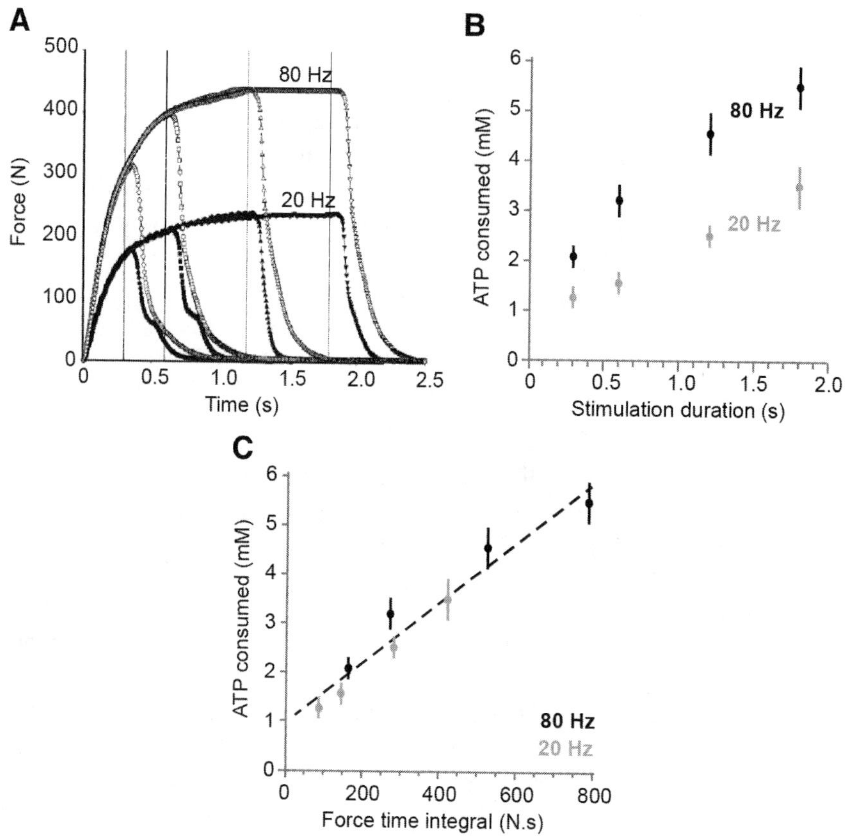

Figure 1.9
Metabolic cost of isometric force production. **A**. Force response of the gastrocnemius muscle to surface stimulation of various durations in a typical subject. **B**. Total ATP consumed as a function of stimulation duration. **C**. Using the force response curves in part A, we plotted the ATP consumed as a function of the force-time integral. Energy consumption is an approximately linear function of the force-time integral. (Data from Russ et al., 2002.)

The effort associated with the action is as follows:

$$U = -\frac{1}{1+\gamma T}(a_0 + a_1 FT) \tag{1.18}$$

The above equation makes a surprising prediction: as the duration of force production T increases, our measure of effort (i.e., temporally discounted metabolic cost) does not grow unbounded, but rather it reaches an asymptote. That is, subjects will become increasingly indifferent to duration. Interestingly, there is some evidence for this.

Konrad Kording and colleagues (Kording et al., 2004) asked volunteers to produce an isometric force F_1 for period T_1, and then an isometric force F_2 for period T_2. Subjects were then asked to choose which of the two forces they would like to experience again ("choose the force-time pair that you judge to be less effortful"). By increasing F_2, the authors determined the effort-indifference point. For example, they found that if $F_1 = 6$ N was held for duration $T_1 = 0.3$ s, this effort was subjectively equal to holding $F_2 = 9$ N for duration $T_2 = 0.1$ s. This makes sense, as it demonstrates that people find the effort associated with holding a large force for a short time equal to a smaller force held for a longer time.

However, the authors also found that as duration T_2 increased, the force that the subjects found equivalent did not go to 0 N, but rather reached a plateau (figure 1.10A). Therefore, the isoutility curves exhibited saturation as the duration of force production increased. This is a puzzling result, but one that we can account for if effort is represented as the temporally discounted metabolic cost.

The isoutility curve is the solution for the equality $J(F_1, T_1) = J(F_2, T_2)$. Using equation 1.17, if we solve this equality for F_2, we arrive at the following expression for the isoutility curve:

$$F_2 = \frac{\gamma(\alpha - a_0)(T_1 - T_2) + a_1 F_1 T_1 (1 + \gamma T_2)}{a_1 (1 + \gamma T_1) T_2} \tag{1.19}$$

As $T_2 \to \infty$, F_2 reaches an asymptote that depends on F_1, as shown in figure 1.10B. Manipulating a single parameter γ and setting all other parameters to 1 provides a reasonable fit to the measured data.

In contrast, let us consider an alternate model in which effort is represented as the undiscounted sum of squared forces (a common approach in the field of robotics as well as in computational motor control). In this case, production of force F for duration T results in $U = -aF^2T$ and a utility:

$$J = \frac{\alpha}{1 + \gamma T} - aF^2T \tag{1.20}$$

The isoutility curves become

$$F_2^2 = \frac{F_1^2 T_1}{T_2} - \frac{\alpha}{aT_2(1 + \gamma T_1)} + \frac{\alpha}{aT_2(1 + \gamma T_2)} \tag{1.21}$$

If effort is represented as sum of squared forces, then as $T_2 \to \infty$, $F_2 \to 0$ (left panel, figure 1.9C); this prediction is inconsistent with the measurements. A similar result is encountered if effort is represented as the undiscounted integral of force (right panel, figure 1.9C).

Our model of utility is not perfect. The measured isoutility curves show a curious dip (figure 1.10A), but our model cannot account for this observation.

In summary, in formulating a utility we made the assumption that the representation of effort depended on the temporally discounted energetic cost of that action. Our formula predicted that the perceived effort associated with generating force should not grow

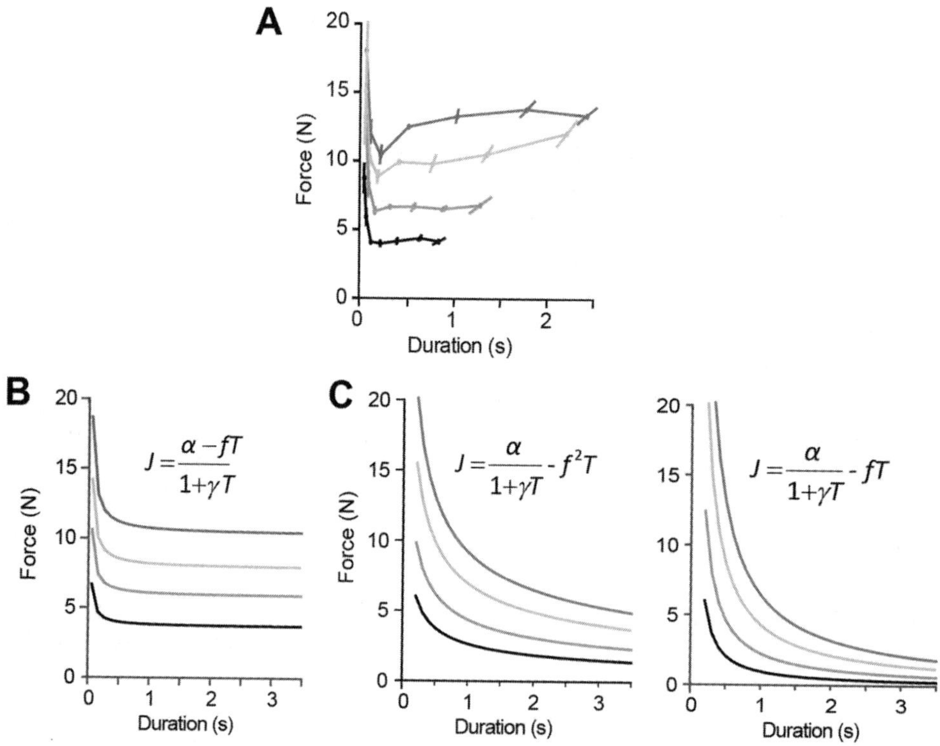

Figure 1.10
Constant utility curves that describe equal-effort lines as a function of force duration and force magnitude. **A**. Isoutility curves for a force production task; each curve connects the force-time pairs that were judged to require equal effort. For durations of greater than 0.5 seconds, force held for a short amount of time was judged to be approximately equal in effort to the same force magnitude held for a longer amount of time. **B**. The isoutility curves predicted by an effort representation in which the metabolic cost of force production (force-time integral) is discounted by a hyperbolic function of time. The function reaches a plateau as the duration of the force production increases. **C**. The isoutility curves (left panel) predicted by an effort cost that depends on the integral of the squared force (as is typical in optimal control models). The function goes to zero with increased duration. The isoutility curves (right panel) predicted by an effort cost that does not temporally discount the force-time integral. The function goes to zero with increased duration. (Data in part A from Kording et al., 2004.)

unbounded as a function of duration, but instead approach an asymptote. This prediction appears consistent with experimental data.

1.6 Alternative Representations of Effort

An objective measure of effort may be its energetic consumption, and for some actions like walking and running, this variable is relatively easy to measure. However, we have a harder time measuring this variable for other movements like saccades. In the field of robotics, which has had a strong influence on computational motor control, effort is often represented as the sum of squared motor commands (e.g., torques) over the course of the movement. This quadratic scaling of motor commands is not explicitly based on physiological principles, but rather a mathematical convenience (because it is easy to find the derivatives of quadratic variables). Nonetheless, models in which effort is represented as the sum of squared motor commands have had success in accounting for some features of movements.

Sum of squared forces approximates a measure of effort in reaching.
In 1998, Y. Uno, Mitsuo Kawato, and R. Suzuki examined the choice of movement trajectories during reaching (Uno et al., 1989). Subjects were asked to reach in a horizontal plane from point A to B, and they did so roughly along a straight line. However, when they were asked to reach between the same points with their hand attached to a spring that pulled it to the side, they reached in a curved trajectory. The authors found that by not moving straight, the curvature of the reach reduced the effort of the movement; effort was estimated as the sum of the squared time-derivative of joint torques.

More recently, Nicholas Schweighofer and colleagues examined the role of effort in determining the choice of which arm to use in performing a task (Schweighofer et al., 2015). The biomechanics of the arm is such that its effective mass varies on the basis of the direction of movement (see appendix A on effective mass of the arm). This is why your arm can feel heavier when you reach in one direction than when you reach in another. The authors asked whether these direction-dependent inertial effects would influence which arm was chosen to make a movement.

Subjects were asked to reach to an array of targets arranged on the perimeter of that circle, which had a radius of 10 cm (figure 1.11A). The subject would first reach for the target with the left arm and then reach for the same target with the right arm; finally, the subject would reach for that target with whichever arm he or she preferred. In this way, the authors examined the influence of movement direction on arm choice. The subjects' choices revealed that the direction of the target had a strong effect on arm choice. Results for the right arm are shown in the middle plot of figure 1.11A. The black line depicts the probability that a subject would choose the right arm to reach to a target in that direction. Subjects preferred to use the right arm when reaching for targets in the upper right and lower left quadrants.

To understand the reason for this preference, the authors estimated the effort required to reach in each direction by simulating dynamics of a planar two-link arm: they applied inverse dynamics by using the measured reach kinematics to estimate joint torques. The resulting effort costs, calculated as the sum of squared torques for a reach in a given direction, are shown in the right part of figure 1.11A. The effort curve for the right arm nicely mirrors the subjects' choices. The arm that people chose for making a movement was the one that produced that movement with the least amount of effort.

Ignasi Cos, Nicolas Belanger, and Paul Cisek designed a reaching experiment that leveraged the direction-dependent biomechanical properties of the arm to examine choices between movements with similar initial effort requirements but different final effort requirements upon reaching the target (Cos et al., 2011). Decoupling of the initial and final movement directions was accomplished by having subjects reach from a starting point to a target through a via point that induced curvature. Subjects were provided with two targets (T1 and T2) and selected which movement they preferred to make (figure 1.11B). Both movements involved initiating a reach in a direction that had low effective mass and thus required low effort (ellipses in figure 1.11B). However, as the reach went past the via point, the curvature made the reach to T2 conclude in a direction that had high mass. In comparison, the reach to T1 concluded in a direction that had low mass. The via points were positioned such that reaching to T2 required greater effort than reaching to T1, whereas reaching to T4 required lower effort than reaching to T3. For each pair of targets, the movement distance was varied from trial to trial. Thus, the results of the experiment showed both the effects of biomechanics-related effort and movement distance on preference.

The right part of figure 1.11B depicts the probability of choosing the low-effort movement over the high-effort movement (between T1 and T2, and between T3 and T4) as the difference in movement distance (log ratio) increases. As expected, when the distance to T1 or T3 increased, relative was greater than the alternative, and subjects were less likely to choose T1 and T3. The main point of interest is the preference when the two targets are equidistant (vertical line at 0). Although both T1 and T2 were 11 cm movements, subjects preferred T1, the low effort movement, over T2. Similarly, for the other target pair, T3 and T4, subjects preferred the low effort movement (T4) when the targets were equidistant. Taken together, the authors found that subjects consistently considered biomechanics of the arm (its mass) and preferred the movement that involved less mass.

A utility based upon minimizing the sum of squared forces is consistent with this behavior. However, there are alternative explanations. A utility in which effort is represented as the metabolic cost of the movement (equation 1.14) will also do well in accounting for direction-dependent movement preference, as shown in figure 1.11B (solid curves, right most figure). By setting $\gamma = 1$ in equation 1.14 and allowing only a single free parameter, α, we obtained a reasonable fit to the data.

In summary, our field is currently uncertain whether metabolic cost, sum of squared forces, or another variable best account for choices that subjects make in effort-based decision making. There are experiments that have begun to differentiate between these variables (Morel et al., 2017), but much remains to be explored. The answer likely lies in the neural basis of movement production, something that we will explore deeply in chapters 4 and beyond.

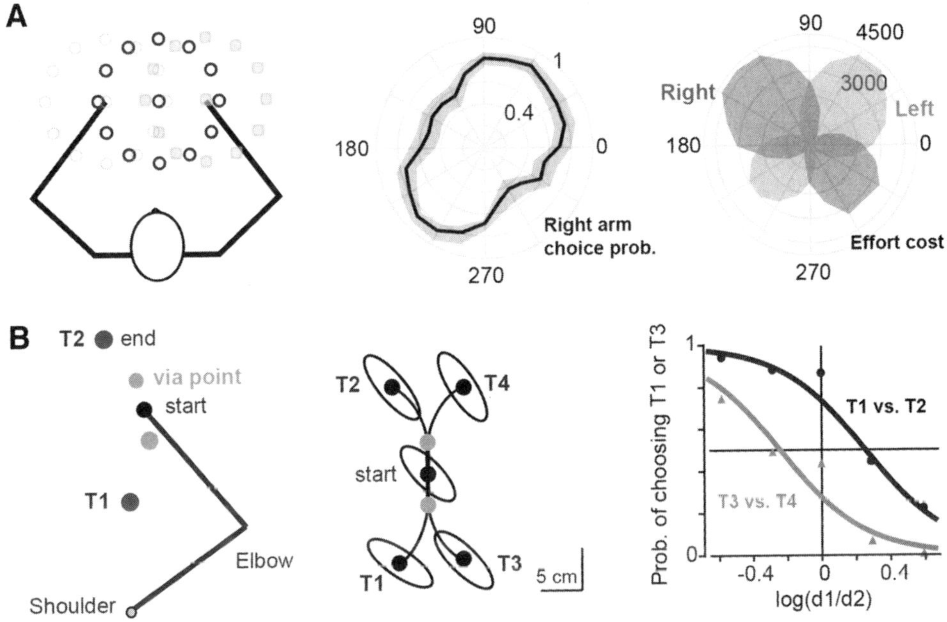

Figure 1.11
Minimization of sum of squared force can account for hand choice and movement preference. **A**. Subjects reached towards an array of targets equally distributed on the circumference of a 10 cm circle. Black circles are the visual feedback of the targets presented to the subjects in the center of the workspace. Gray circles indicate the actual location of the start and targets for each hand. Probability of choosing the right hand to reach to a target is shown in the middle figure. Effort costs for the right (black) and left (gray) arm for a reach in a given direction are shown on the right figure. Effort costs are calculated as the sum of muscle activations squared. (Data from Schweighofer et al., 2015.) **B**. Subjects were presented with the choice of reaching for one of two targets. Each reach had to pass through a via point. The configuration of the arm at the start point was such that movement initiation for all targets was in a low-effort direction. Final approaches towards T1 and T4 were in the low-effort direction (minor axis of the ellipse), while final approaches to T2 and T3 were in the high-effort direction (major axis of the ellipse). The right most figure shows the probability of choosing T1 over T2 (black circles) and T3 over T4 (gray triangles) as a function of the difference in movement distance. As the distance to T1 and T3 increased, subjects chose them less frequently. When targets were equidistant (vertical line), subjects preferred T1 to T2, and T4 to T3, which aligns with minimizing sum of forces squared. Solid lines depict predictions of a model of utility (equation 1.14) where utility is the sum of reward and effort costs discounted by time, and effort is represented as the metabolic cost of the reach. (Data from Cos et al., 2011.)

1.7 Cognitive Effort versus Movement Effort

The tasks that we will consider in this book often involve elementary movements like saccades and reaching, movements that may seem rather trivial and uninteresting. Our reason for this particular focus is that these movements lend themselves to computational and neurophysiological analysis. From a computational perspective, elementary movements are relatively easy to describe mathematically in terms of forces and states, and from a neurophysiological perspective they are tractable because they can be studied in nonhuman primates and other animals. What we call effort is referring to a metabolic or a neural representation of a measure of cost for performing these movements. Promise of reward affects how people and other animals perform movements, suggesting that attempts at forming a utility in terms of benefits and expenditures may be useful in the analysis of behavior. However, in limiting our scope to study of elementary movements, we are ignoring the literature on the effort involved in thinking, more formally known as cognitive effort. That effort, though currently poorly understood, does play an important role in decisions that people make.

While moving requires effort, so does thinking. Cognitive effort is part of most decisions that we make, yet an objective measure of cognitive effort remains elusive.

While there are no widely accepted definitions of cognitive effort, it is certainly a commonly experienced phenomenon. For example, if we were to read you a list of numbers and ask you to remember the last one that you heard (1-back task), you will likely find the task easy. However, you will find it much harder to perform a 3-back version of the same task. In a sense, the 3-back task requires greater cognitive effort than the 1-back task. Indeed, when young people are offered money to perform these tasks, they request 25% more money for the 3-back task than the 1-back task (Westbrook et al., 2013). In contrast, older adults request 75% more money for the same task differential (Westbrook and Braver, 2015). The 3-back task is harder than the 1-back task for the young adults, but it is even harder for the elderly.

One way to consider the concept of cognitive effort is to imagine that working memory is a limited and valuable resource; the 3-back task places greater demands on that resource. It is possible that greater demands on working memory require greater metabolic costs in terms of neural activity, but that relationship remains poorly understood (Westbrook and Braver, 2015). Whereas movement effort may be quantified objectively with metabolic costs, currently it is not obvious that energy expenditure is a relevant measure of cognitive effort.

An interesting comparison between these two forms of effort was in a study performed by Dan Ariely and colleagues (Ariely et al., 2009). They recruited a group of undergraduates and measured their performance in two tasks. In the key-pressing task (physical effort), the subjects were asked to alternate between pressing the "v" and the "n" keys on the keyboard. In the adding task (cognitive effort), they were given a set of 20 matrices, one at a time, each containing 12 real numbers, and were asked to find the two numbers in that matrix that would add to 10. They had 4 minutes to complete each task.

Each subject performed both tasks, but they were divided into two groups: low reward and high reward. In the low-reward group, for the adding task they received $0 if they solved nine or fewer matrices, $15 if they solved 10 matrices, and an additional $1 to $15 for each additional matrix (for a maximum of $30). In the high-reward group, the same task yielded 10 times the amount of reward ($0, $150, and $300). In the key-pressing task, the low-reward group received $0 if they pressed fewer than 600 keys, and an additional 10 cents for each additional key. In the high-reward group, the same task produced 10 times the amount of reward ($0 for fewer than 600 keys and $1 for each additional key).

The results showed that in the key pressing task, performance was much better in the high reward group: people pressed many more keys in the hope for higher pay. This makes sense in the framework in which reward (money) makes it worthwhile to spend effort. However, in the adding task, performance decreased with the increased reward: on average, people in the high-reward group solved 20% fewer matrices than in the low reward group. The authors performed other experiments that attempted to further test the effects of reward on various tasks that required cognitive effort. They generally found that while modest amounts of reward were occasionally effective in encouraging better performance, large amounts of reward were almost always counterproductive and often resulted in poorer performance.

There is an important caveat: the results in the cognitive tasks quantify performance, not effort expenditure. That is, whereas a higher number of movements in the key-pressing task is a reasonable indicator of greater effort expenditure, the lower number of matrices solved in the adding task does not imply that less effort was expended. Rather, the results indicate an incompatibility between performance measures in the cognitive task and effort expenditure in the same tasks.

In summary, cognitive effort is certainly part of most decisions that we make, yet an objective measure of its utility remains elusive. One idea is that tasks that require greater usage of our working memory also require greater cognitive effort. A greater reward can coincide with a poorer performance in a task requiring high cognitive effort (a phenomenon loosely called "choking"), but this poorer performance may not bear any relationship to the cognitive effort that was expended during that task. That is, it is likely that similar to physical effort, we do expend greater cognitive effort to get the greater reward, but for reasons potentially associated with emotional stress or anxiety, our performance may suffer, thereby eliminating the availability of behavior as an assay of effort expenditure in cognitive tasks.

Limitations

In trying to ask why we move more vigorously toward things we value more, we began with a utility in which the goodness of movements was defined as reward minus effort, divided by time. We imagined that effort could be objectively estimated as the energetic cost of the movement. For walking, energetics can be readily measured, but the problem is more difficult for reaching, and perhaps impossible for many other movements. For these movements, there are models that can relate muscle activation patterns with energetic expenditure (Tsianos et al., 2012). However, even if effort could be objectively measured, there are some broad limitations in our formulation of utility.

First, we made the assumption that reward is attained and consumed immediately at movement end. This is an oversimplification, as it usually takes time (and expenditure of further effort) beyond the completion of the movement to consume the reward. To illustrate the issue, consider the problem of walking to meet a friend. After the walk ends and you arrive where your friend is waiting, that reward is continuously enjoyed for the period that you are with him: the pleasure of being with your friend is not limited to the initial moments after the termination of your walk period. In our description of utility, we only considered the period of moving, not the period afterwards, and it is this period during which we harvest the benefits of our action. Harvesting takes time and effort. Currently, this post-movement period does not play a role in our determination of utility.

The second problem is that we did not consider the past history of the subject in our formulation of utility. For example, animals that are hungry tend to be jittery, moving with vigor in not just their approach toward food, but also in their grooming or other acts that have nothing to do with acquiring food. In an example that might be closer to home, consider what happens after you experience a bad meeting. Following that meeting, you might walk slower to your next appointment. Our history affects our movements, and this history is currently not represented in our formulation of utility.

These limitations hint that decisions that we make and the movements that follow should not be viewed in isolation (i.e., one-time events). Rather, the decisions and movements are influenced by our past experiences as well as our future expectations. For example, people who have experienced a high rate of success in their recent past generate a more vigorous movement in response to a novel stimulus than people who are given the same stimulus after a history of a low rate of success (Guitart-Masip et al., 2011). That is, the past rate of reward appears to influence current vigor (Yoon et al., 2018). In chapters 2 and 3 we will consider these limitations by expanding our time horizon to include the past and the future in the framework of utility. The result will be a normative approach, one in which our decisions and actions are evaluated based on their ability to maximize a long-term objective, a global capture rate.

Summary

We must spend energy in order to move, and the speed with which we typically move is near the one that minimizes this energetic cost. However, speed of movement also depends on expectation of reward at conclusion of that movement. To consider why animals move faster when there is greater reward, we introduced a concept of utility, defined as the expected reward minus the required effort, divided by duration of time. The effort required to move may be measured objectively via the metabolic cost of that movement. For walking, running, and reaching, as the value of the reward increases, moving faster requires greater effort but is justified because the time that it saves leads to an in increase in the rate of reward.

Our framework has important shortcomings. We considered each action in isolation, whereas in reality, our past history and future expectations affect both what we choose to do and how we choose to move. For example, after a good meeting, you might leave the room with a spring in your step. After a bad meeting, you might walk with your shoulders hunched and with slower steps. In our next chapter, we will consider why the history of reward, and not just the promise of reward, affects vigor.

2

Movements and Decisions in a Normative Framework

How shall I know, unless I go
To Cairo and Cathay,
Whether or not this blessed spot
Is blest in every way?
—Edna St. Vincent Millay

If you have ever visited New York, you will probably agree that people there seem to walk faster than people who live in other places. Between 1972 and 1974, Marc and Helen Bornstein visited cities around the world and measured how fast pedestrians walked over a distance of 15 m in functionally similar sites such as downtown or commercial areas (Bornstein and Bornstein, 1976). They found that on average, walking speed increased with the population of the city (figure 1.1C). To interpret their data, they suggested that "crowding has been thought to motivate … avoidance behaviors that reduce tension" and therefore "increased walking speeds serve to minimize environmental stimulation." That is, they suggested that people in big cities walked fast because of crowding. They conjectured that a faster speed of walking served as a form of avoidance.

Of course, there may be many other factors that influence walking speed: external factors, like temperature and humidity (factors that can affect the metabolic cost of walking), and internal factors, like fatigue. But if one could control for all of these factors, a critical element still remains: the subjective cost of time. That is, when your time is valuable, you will walk faster. This raises the possibility that some of the differences in the speed with which we move is due to differences in our subjective cost of time. Maybe people in New York walk quickly because they tend to be very busy.

To explain this idea, we will consider a thought experiment in which a child is reaching for a piece of candy. In one case she is reaching for this candy at home, where this is the only candy around, whereas in another case she is reaching for the same candy while trick-or-treating at a home during Halloween, during which other candies are available next door. The idea is that the utility of the action (in this case, reaching) depends not only on the available reward and expected effort, but also on the environment in which this action is being performed. Time is more valuable in some environments (Halloween) than in others. In a similar vein, sometimes you are busy and have many things to do, whereas other times you are on vacation. The decision regarding what to do, as well as the vigor of the movement that ensues, should depend on the value of your time.

We can define the value of time via a measure that considers actions over a long sequence: the rewards and efforts experienced in the recent past and expected for the near future, divided by time. In a rich environment with many possible rewarding actions, time is quite valuable. In this environment, doing any one action requires delaying and possibly missing out on the reward that could be attained by performing other actions. The opportunity cost of a rich environment is high. Behavior, as measured via decision-making and movement vigor, should be different in this environment from one in which there are few opportunities for rewarding actions.

In chapter 1, we considered a measure of movement utility in which effort was defined as the energetic cost of the action, and reward was akin to "energy" acquired at conclusion of that action. We defined utility as a measure of the capture rate: reward minus effort, divided by time. However, this was a local measure, as it described a currency to measure the goodness of the current act without considering other acts that we had done or planned to do. Here, we will expand our time horizon to incorporate events that preceded and will follow the specific decision/action that we are considering. By doing so, the policy that we are seeking maximizes a global capture rate, a variable that considers the current action as one of many among a sequence of actions that we plan to perform. The resulting framework will allow us to consider the effects that the cost of time has on movement vigor and hence provide a way to examine the question of why people walk faster in certain cities.

Our goal is to arrive at a normative function to describe utility. In economics, normative refers to "what ought to be," whereas positive economics refers to "what is." One measure of utility is global capture rate, which is defined as the sum of all rewards acquired minus the sum of all efforts expended, divided by total time. To arrive at a utility that is normative in the economic sense, we will employ a framework that has been advanced by ecologists: animals who have higher global capture rates tend to live longer and have more offspring. In this view, global capture rate is a good candidate for a normative utility because in principle, it is something that animals should strive to optimize.

Our discussion in this chapter will answer a basic question: if you prefer the doughnut to the apple, why in principle should you reach faster toward the doughnut? The answer that we will present is that our choices as well as our movements contribute to an important currency, the global capture rate. This currency is not arbitrary, but one that plays a significant role in longevity and fecundity, suggesting that living life in such a way that increases the global capture rate is evolutionarily advantageous.

Because movements require expenditure of effort, and utility-driven decisions affect reward accumulation, a policy that maximizes the global capture rate naturally coordinates motor control with decision-making. The link between decision-making and movements arises because both are elements of behavior that the brain must control in order to maximize a single currency: the global capture rate.

2.1 Optimal Foraging Theory

Behavior during foraging sometimes exhibits interesting patterns. A crow might dig up a small clam from a sandy beach but then abandon it in favor of trying to find a larger clam, even though the crow's decision entails leaving behind a perfectly good piece of food and requires spending additional time and energy in the hope of finding better food (Richardson and Verbeek, 1986). This is an example of a behavior that is locally suboptimal. Why not eat the small clam before moving on?

To understand this and other decisions that animals make, ecologist have considered a global form of utility called optimal foraging theory (Stephens and Krebs, 1986). In this theory, the hypothesis is that an animal decides what to eat and what to pass up on the basis of a single currency: the capture rate, defined as the energetic gain associated with the food minus the energetic loss associated with acquiring and consuming that food, divided by the total time it takes to find the food and consume it. Importantly, the gains and losses afforded by any action are measured in relation to other actions that are possible in the environment. Optimal foraging suggests that the decisions that animals make regarding how far to travel for food, what to eat when they arrive at the food patch, and how long to stay at that patch are based on an attempt to maximize the global capture rate.

However, optimal foraging is generally not concerned with the question of vigor; the theory is typically employed in order to understand decision-making, not motor control. This seems wrong to us because movements require effort, one of the key variables that affect the capture rate. If the objective is to live life in such a way that maximizes the global capture rate, then one should not only be concerned about one's decisions, but also careful about one's movements. In this chapter, our aim is to extend this theory so that it can do both. To achieve this goal, we will first summarize some of the experimental results that led to formation of optimal foraging theory, and then use the principles to recast our concept of utility in light of this normative approach.

Higher global capture rate results in improved fitness.

In optimal foraging theory, the hypothesis is that animals make choices based on an attempt to maximize their global capture rate. The rationale is that the pattern of decision-making that maximizes the global capture rate increases fitness (as measured via fecundity and longevity). To test this idea, William Lemon (1991) considered four populations of zebra finches that were housed in an indoor aviary. Each population received 200 g of seed each day, an amount that was more than they could consume. The control population (group A) received only the seed. The other groups received the same amount of seed but mixed with 400 g (group B), 600 g (group C), or 800 g (group D) of empty seed hulls (figure 2.1A). The zebra finches searched through the hull and seed mixture to find the seeds. As a result, groups B, C, and D had to spend greater time and effort in order to harvest the seeds. The feeding rate (number of seeds consumed per unit of time) was highest in group A and lowest in group D. While the total amount of food consumed was not different among the various groups, the energy expended to acquire that food was higher in the groups that had to rummage through the husks. As a result, the global capture rate, defined as the total food consumed minus energy expended, divided by time, was highest for group A and lowest for group D.

Lemon (1991) continued to feed the finches this way throughout their entire lives, tracking their fitness by quantifying patterns of reproduction and longevity. He found that female fecundity (figure 2.1A) and male and female probability of survival (figure 2.1B)

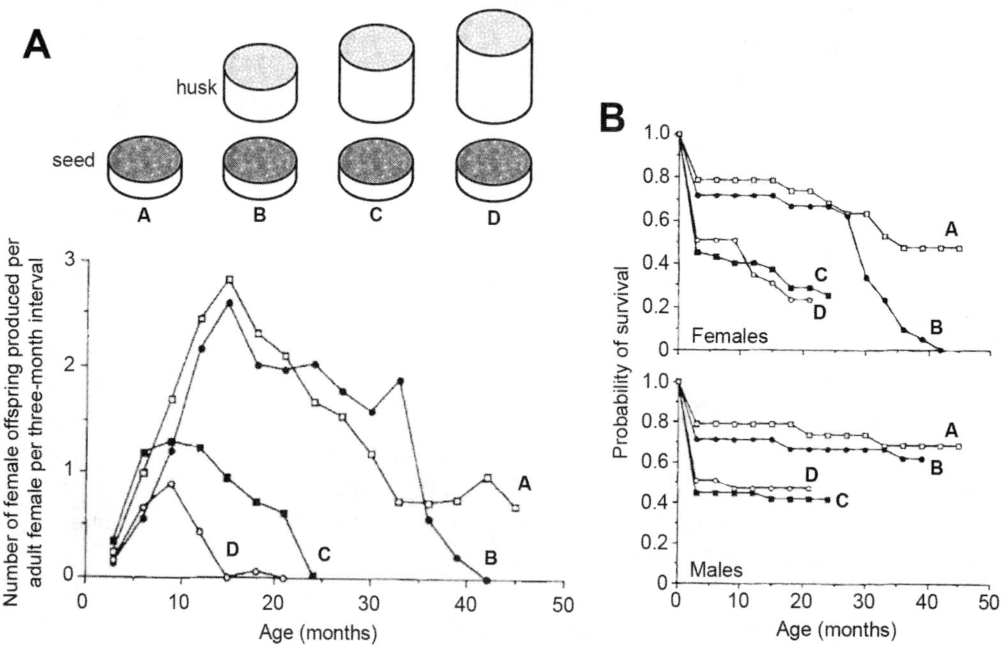

Figure 2.1
Higher global capture rate results in greater fecundity and longevity. Four groups of zebra finches were fed a daily diet of 200g of seed mixed with 0, 400, 600, or 800 g of seed husk. **A**. Fecundity as a function of age in the female subgroup of the finches, measured as the number of female offspring produced per adult female per three-month interval adjusted for the probability of surviving through the interval. **B**. Probability of survival from birth to a given age for female and male finches. (Data from Lemon, 1991.)

were highest for group A and generally lowest for group D. Groups with higher global capture rate demonstrated greater fitness, suggesting that capture rate is a currency with which one could compare the utilities of various actions.

Crows are selective in what food they eat and what they reject. Their policy can be evaluated in the framework of optimal foraging.

Although it takes time and effort to find food, it usually takes additional time and effort to harvest that food. For example, an animal may spend time and energy finding its prey, but killing that prey and extracting the meat will take additional time and energy. All of these costs affect the capture rate, which in turn can determine whether it is worthwhile to spend the time and effort to eat an inferior piece of food, or better to abandon that food in favor of finding something superior. For example, if one expects to find better food nearby, then one should abandon the inferior piece of food and start searching. Maximizing the global capture rate allows us to make predictions regarding foods that animals harvest, foods that they leave behind, and their choices regarding mode of transport (e.g., walking or running).

Crows that live in the Pacific Northwest of the United States rely on clams as an important part of their diet. Clams hide in sandy beaches, and crows search for them, dig them up, carry them by flying to a rocky area, drop them a few times (until they break open), and then finally consume the meat. However, the crows do no attempt to open all clams after they have dug them up; sometimes they simply leave the clam they have dug up and move on to search for another clam. Although they have already spent energy finding the clam, they abandon it and decide not to spend any further energy trying to open it. Whether they try to open the clam or discard it is generally a function of the size of the clam: they leave the smaller clams but attempt to open the larger ones.

Howard Richardson and Nicolaas Verbeek (1986) considered this behavior and asked whether the decision-making of crows (i.e., which clam to eat, which to pass up) could be understood in terms the global capture rate. In their formulation, they considered both the time and the energy it took to search and find the clam on the beach (t_s and e_s), to fly to a rocky area with the clam and drop it (t_f and e_f), to retrieve the clam and handle it (t_h and e_h), to extract the meat from the clam (t_e and e_e), and to return back to the beach and restart the search process (t_r and e_r). All of these terms were constant, except for the time and energy it took to extract the meat, which depended on size of the clam, represented by variable x: $t_e = 0.13x^{1.58}$, and $e_e = 7.2t_e$, where x is in millimeters, time is in seconds, and energy is in Joules. The energy gained from eating a clam also depended on its size: $\alpha = 0.012x^{3.47}$. The average rate of energy intake (i.e., local capture rate) for finding and consuming a clam of a given size is as follows:

$$J(x) = \frac{\alpha(x) - e_s - e_f - e_h - e_e(x) - e_r}{t_s + t_f + t_h + t_e(x) + t_r} \tag{2.1}$$

The clams had a length distribution that ranged from about 10 mm to 40 mm: the distribution was roughly flat between 15–30 mm, and then declined at the margins. The crows did not know the size of the clam until they had dug it up. Interestingly, despite having spent time and effort in the process of digging up the clam, they abandoned it if it was smaller than a certain size. The size that was rejected 50% of the time was 29 mm (they ate the larger clams, and abandoned the smaller ones). Was this policy of rejecting the smaller clams reasonable?

In the framework of optimal foraging, the currency that we use to gauge the goodness of a policy is the global capture rate (i.e., the net energetic intake divided by total time). Suppose that a crow's policy is to eat any clam that is equal to or bigger than a minimum size y. Let us use the random variable $z^{(n)}$ to indicate whether on encounter n with a clam, the crow decides to eat it: $z=1$ if $x^{(n)} \geq y$ (the clam is eaten), and $z=0$ if $x^{(n)} < y$ (the clam is abandoned). After N encounters on the basis of this policy, we have the following global capture rate:

$$\bar{J} = \frac{\sum_{n=1}^{N} -e_s + (\alpha(x^{(n)}) - e_e(x^{(n)}) - e_f - e_h - e_r) z^{(n)}}{\sum_{n=1}^{N} t_s + (t_e(x^{(n)}) + t_f + t_h + t_r) z^{(n)}} \tag{2.2}$$

In the above formulation, the crow spends energy e_s and time t_s to find each clam, and then it experiences additional expenditures if it decides to consume that clam. The expected value of the random variable z is $E[z] = \sum_{x=y}^{x_{\max}} p(x)$, where $p(x)$ is the probability of finding a clam of size x. After a large number of encounters N, the global capture rate $\bar{J}$ incorporates both the energetic gain associated with clam size x, as well as the probability of finding such a clam $p(x)$:

$$\bar{J} = \frac{-e_s + \sum_{x=y}^{x_{\max}} (\alpha(x) - e_e(x) - e_f - e_h - e_r) p(x)}{t_s + \sum_{x=y}^{x_{\max}} (t_e(x) + t_f + t_h + t_r) p(x)} \tag{2.3}$$

The above quantity is plotted in figure 2.2. The global capture rate is positive when the minimum clam size (variable y)is small, which implies that a crow that eats every clam is likely to survive. The global capture rate becomes quite small (or negative) when the minimum clam size is large. This implies that a picky crow, waiting to find only the largest clams, will likely starve. However, the highest global capture rate is found when the crow eats only clams that are 27 mm or larger, which is what crows actually do. Thus, crows have a decision-making process that is consistent with maximizing their global capture rate.

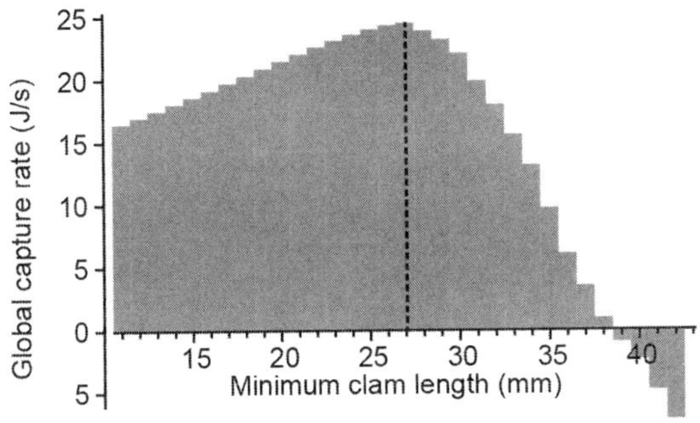

Figure 2.2
Clam diet of crows maximizes the global capture rate. Distribution of clams on the beach was roughly uniform between 15–30 mm. After digging up a clam, crows rejected the smaller ones but chose to expend further effort to open the larger ones. The figure shows the global capture rate as a function of minimum clam size in the diet. The optimum diet rejects clams that are smaller than around 28 mm.

Starlings can choose between walking or flying to the food site. Their choice can be evaluated in the framework of optimal foraging.

The currency of capture rate might not only explain diet selection, but also decisions that animals make with regard to their mode of movement during foraging. For example, if the animal has the option of walking or flying, walking may be less costly in terms of energetic expenditure, but flying gets the animal to the food site sooner. The choice of whether to walk or to fly may also be predicted on the basis of the policy of maximizing the capture rate.

Alejandro Kacelnik and colleagues (Bautista et al., 2001) investigated this question by asking whether maximization of the capture rate could reliably explain an animal's choice of transportation. Specifically, they asked whether the capture rate could predict whether a starling would walk or fly to obtain food. The investigators trained four starlings to make a number of walking or flying trips to obtain a food reward. Each starling was housed in its own aviary, and in that aviary were three elevated perches: a base perch, a near perch (placed 0.35 m from the base), and a far perch (placed 4.65 m away). From the base perch, the birds could walk over an elevated platform to the near perch, but flying was the only way to reach the far perch. During training, there was a light cue next to each perch that indicated to the starling where it had to go. For example, by sequentially lighting the far and base perch lights, each starling learned to fly a number of times to the far perch and back. Then after a given number of flights, the starling received the food reward next to the far perch. A similar method was used to train the birds to walk a number of times to the near perch and back in order to receive the food reward.

Once the starlings were trained, on each day they experienced a sequence of forced and free trials. In the forced trial, only one of the perches was active, so the animal had to walk (near perch) or fly (far perch). Once there, the cue light switched back to base. After a preset number of repetitions, food was delivered. For example, on day 1 of the experiment in the forced trials, the animals experienced 6 flights to the far perch and then 35 walks to the near perch. In the free trial that followed, both perches were lit, giving the bird the option of flying or walking. The first two sets of trials in each cycle were forced, hence exposing the birds to the number of walks and flights that had to be done to get the reward. The third and fourth trials in the cycle were free; the birds chose whether to walk or fly. Suppose that in this case the bird chose to fly. During the next cycle, the researchers reduced the required number of walks until the bird was indifferent between walking and flying. (Indifference implies that the animal would choose roughly equally between the two options.)

Each day, the number of flights required was set to a constant integer ranging from 1 to 11, and on each day the animals converged on a specific number of walks for which the walk and fly options were deemed to be equivalent (figure 2.3A). For example, one flight

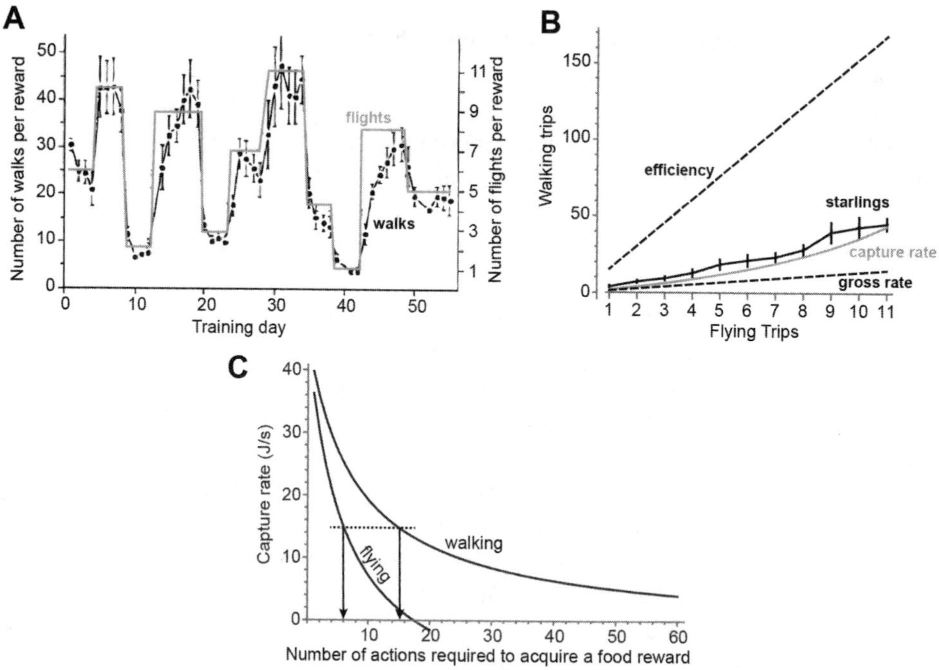

Figure 2.3
Foraging choices made by starlings regarding modes of transportation. Starlings chose between walking and flying in order to acquire a food reward. **A**. On each training day, the number of flights was fixed and the number of walks was adjusted until the animals were indifferent between the walking and flying options. The error bars are the standard errors of mean. **B**. Observed number of walks per number of flights for the birds (solid dark line, mean and standard deviation), as well as the predicted values from three models (dashed and gray lines). **C**. The capture rate (rate of net energetic gain) as a function of number of walks or flights required to acquire reward. For example, 6 flights have the same capture rate as 15 walks. (Data in parts A and B from Bautista et al., 2001.)

was equivalent to about 4 walks, and 10 flights was equivalent to about 42 walks, resulting in 11 indifference points measured for each of the four starlings (data labeled starlings, figure 2.3B).

To account for these observations, the authors considered the possibility that the animals were making choices in order to maximize a utility. They considered three candidates. The first utility was the capture rate, defined as energy α gained from the food minus energy E expended to acquire the food, divided by total time T spent acquiring that food: $(\alpha - E)/T$. The second utility was the gross rate of energetic gain, defined as α/T. The final utility was efficiency, defined as the ratio of reward acquired to energy expended: α/E. In these models, the reward was the energetic content of the food, and effort was the energetic expenditure of the action. As a result, there were no free parameters (that is, nothing to fit). Each utility was used to make a prediction regarding behavior. The authors asked whether any of these currencies could be used to predict a behavior that resembled their starlings.

Let us begin with the capture rate. For this model, the decision making process proceeds with a comparison of the capture rates of the options. For walking, the capture rate is as follows:

$$J_w = \frac{\alpha - E_w}{T_w} \tag{2.4}$$

To compute energy expended, the authors considered that the walks consisted of a period of walking and perching, followed by a period of handling the food. For example, if the task required walking n_w times, the time and energetic expenditure would define the following utility:

$$J_w = \frac{\alpha - n_w e_w - e_h}{n_w t_w + t_h} \tag{2.5}$$

In the above equation, e_w is the energy expended while walking for duration t_w, and e_h is the energy expended while handling the food for duration t_h. The energetic expenditure during walking and handling the food are related to the durations of each action:

$$e_w = c_w t_w \qquad e_h = c_h t_h \tag{2.6}$$

A similar capture rate can be written for flying:

$$J_f = \frac{\alpha - n_f e_f - e_h}{n_f t_f + t_h} \tag{2.7}$$

In these equations, all variables are known from empirical observation (duration of flight, walking, or handling food) and measurement (energetic cost of flying, walking, or handling the food). For example, the energetic cost of flying was roughly 15 times that of walking.

Given the option of flying n_f times or walking n_w times, we compute $J_f(n_f)$ and $J_w(n_w)$, and then select the option that has the greater utility. Figure 2.3C shows the capture rate for each mode of transportation and demonstrates that as the number of walks or flights increase, the capture rate decreases. The indifference point occurs when $J_f(n_f)=J_w(n_w)$ (illustrated by the dashed line in figure 2.3C). If the option of flying is chosen n_f times, the energetic value of the food is α, and the time and energetic costs of flying and walking are as previously defined, the equivalent number of walks can be calculated:

$$n_w = \frac{n_f(\alpha t_f - e_h t_f + e_f t_h)}{e_w(n_f t_f + t_h) + t_w(\alpha - e_h - e_f n_f)} \approx \frac{34 n_f}{20 - n_f} \tag{2.8}$$

The above equation implies that 10 flights should be equal to 34 walks. We have plotted the above equation in figure 2.3B (labeled capture rate). The results are close to choices made by the starlings.

In contrast, if the utility is the rate of gross energetic gain, then for walking we have $J_w = \alpha/(n_w t_w + t_h)$. We have a similar equation for flying. As a result, for n_f flights, the equivalent number of walks is

$$n_w = \frac{t_f}{t_w} n_f \approx 1.3 n_f \tag{2.9}$$

This model predicts that 10 flights should be equal to 13 walks, a value that is much smaller than the number of walks that the starlings were willing to do (gross rate, figure 2.3B).

Finally, if the utility is efficiency (energetic gain over energetic loss), then for walking we have $J_w = \alpha/(n_w e_w + e_h)$. As a result, given the option of flying n_f times, the equivalent number of walks is

$$n_w = \frac{e_f}{e_w} n_f \approx 15 n_f \tag{2.10}$$

This model predicts that 10 flights should be equal to 150 walks, a value that is much greater than the number of walks that the birds were willing to do (figure 2.3B). Therefore, with no parameters to fit, the authors demonstrated that the choices that the animals made were well accounted for via a utility that described goodness of walking or flying as the capture rate.[1]

However, the authors noted that another aspect of the starlings' behavior remained unaccounted for: the choice between foraging and resting. In a given day, the animals could

1. The model we used here is a simplified version of the one implemented by Bautista et al. (2001). They considered changes in energetic cost due to changes in the daily weight of the animal. Their model gave a more accurate prediction than our simplified model, placing the model results within one standard deviation of the measured data.

forage for food by performing the requisite walking and flying trips, or they could choose not to forage and simply rest instead. The capture rate was always positive for the number of walks and flights the birds actually completed (figure 2.3C), so in accordance with the model, it was always better to forage than to rest. (For resting, the capture rate is negative because no explicit rewards are available.) Yet, the starlings chose to spend the majority of the day resting, passing up the opportunity to forage. As a result, they lost weight, especially on days on which the requisite number of walks or flights was large. The authors suggested that perhaps there were other costs not accounted for in the utility model. For example, foraging may place the birds in greater danger of being preyed upon than would resting. Furthermore, walking and flying could be equally dangerous and thus explain the birds' choice to rest.

In summary, policies that increase the rate of net energy intake, what we termed the global capture rate, increase the fitness of the animal as measured via longevity and fecundity. Optimal foraging theory posits that choices regarding what to eat and what to pass up, as well as the mode of transportation during foraging, may be predicted by assigning a value to each option on the basis of its capture rate. Choices that crows make regarding which clams to eat and which to pass up, and choices that starlings make regarding walking or flying during foraging, appear generally consistent with predictions of a utility that depends on the capture rate.

2.2 Marginal Value Theorem

Global capture rate describes a currency with which we can measure the goodness of a policy. This rate depends on two kinds of time periods: the time required to find food, and the time required to harvest it. Suppose food is distributed in patches. There may be a few bushes that are distributed throughout a field, with each bush containing a number of berries. In this scenario, $t_s^{(i)}$ describes the time it takes to travel to patch i (search time), and $t_h^{(i)}$ describes the time it takes to harvest the food at that patch. Imagine that in one environment the food patches are placed far from each other, and in another the patches are closer. If the number of patches and the number of berries available at each patch are the same in the two environments, maximizing the global capture rate would require that when patches are far part, the animal should stay longer and eat more of the berries at a given patch before traveling to another patch. In this environment, because the patches are far apart, time is not very valuable. However, in the environment where patches are close, time is valuable—after eating the berries that are easily found, the animal should abandon the remaining berries in the current patch and move on to the next one.

Marginal value theorem describes a policy that maximizes the global capture rate. This line of thinking suggests that if our objective is to maximize the global capture rate, then we should be able to make concrete predictions regarding the harvest duration at each patch. Eric Charnov (1976) considered this question and provided an important insight. He assumed that there is a finite amount of food α in a patch (e.g., berries on a bush). While at a given patch, the animal eats the available food over the harvest period t_h. As the animal eats, the energy intake initially rises rapidly but then saturates as the food at that patch runs out (reward in figure 2.4A). As the animal is eating this food, it also expends energy, which is represented as $k_h t_h$ (where k_h is metabolic rate during harvesting). For patch i, the net energy intake over the time period t_h is expressed via the reward harvest function $f^{(i)}$:

$$f^{(i)}(t_h) = \alpha^{(i)}\left(1 - \frac{1}{1+\beta^{(i)}t_h}\right) - k_h^{(i)}t_h \tag{2.11}$$

The first term in the above expression represents energy gained by eating the available food, and the second term is the energy expended during the act of harvesting. As the harvest period t_h increases, the amount of food consumed also increases, but with diminishing returns. The term β expresses the rate at which the food is consumed. For example, when β is small, it takes a great deal of time to find and consume the food at the patch. A large β implies that the food can be found and consumed quickly. (The exact form of the harvest function is not critical for our discussion. What is important is that the function is concave downward.)

In addition to the energy spent to harvest the food at the patch, there is energy spent traveling to the patch, which is represented by $k_s t_s$, in which t_s is time spent searching for the patch, and k_s is the metabolic rate during searching. The local capture rate for traveling to and harvesting patch i is as follows:

$$J^{(i)} = \frac{f^{(i)}(t_h^{(i)}) - k_s t_s^{(i)}}{t_h^{(i)} + t_s^{(i)}} \tag{2.12}$$

In an environment with N patches of food, this is the global capture rate:

$$\overline{J} = \frac{\sum_{i=1}^{N} f^{(i)}(t_h^{(i)}) - k_s t_s^{(i)}}{\sum_{i=1}^{N} t_h^{(i)} + t_s^{(i)}} \tag{2.13}$$

Note that the global capture rate $\overline{J}$ is not the average value of $J^{(i)}$. Indeed, as an alternative hypothesis, we could posit that the utility that is being optimized is this average value: $J = N^{-1}\sum_{i=1}^{N} J^{(i)}$. If we were to define utility as the average value of $J^{(i)}$, then behavior at any particular reward site would be independent of the history of the subject, something that we will see is inconsistent with experimental data. For more information on the topic of alternative utilities, see appendix B.

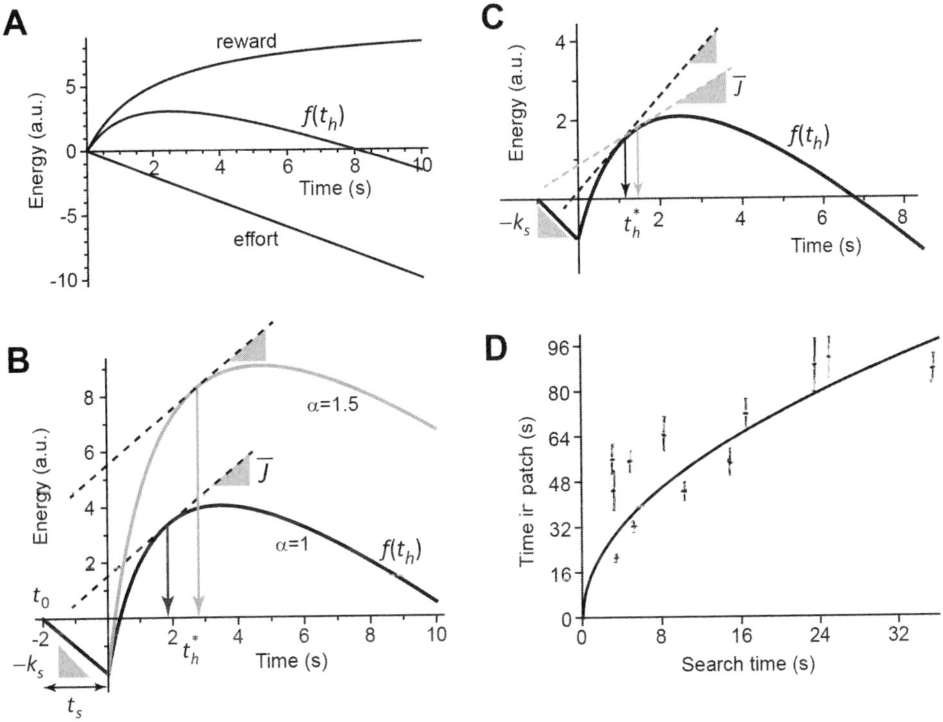

Figure 2.4
Graphical representation of marginal value theorem. **A**. At the food patch, the energy intake associated with harvesting of reward is an increasing function of time spent. However, harvesting requires expenditure of energy, which is depicted by an effort function that grows with duration of harvest. The net energy, sum of energetic intake due to reward and loss due to effort, is depicted by the harvest function $f(t_h)$. **B**. The curve $f(t_h)$ represents the net energy gain as a function of time spent in a food patch. The slope of the line represents $\bar{J}$, the global capture rate. The animal should leave the patch (time t_h^*) and abandon the remaining food at the current patch when the marginal capture rate (derivate of $f(t_h)$) becomes smaller than the global capture rate. Time t_0 indicates start of the search process. Time $t_0 + t_s$ indicates when the food patch was found. If the current patch has a larger amount of food, indicated by α, the animal should stay longer at this patch (gray vertical arrow indicating t_h^*). **C**. Effect of global capture rate. When the global capture rate is low (gray dashed line, representing a poor environment), the animal should stay longer at a given food patch and consume more food than when the global capture rate is high (rich environment, solid line). **D**. Time spent harvesting should increase roughly with the square root of time spent searching. Data from six birds that spent time foraging in two environments: one which required a brief travel time between opportunities to eat (easy environment), and one that required a long travel time (hard environment). The figure shows the relationship between the mean search time and the mean time spent eating at each patch for each bird in each environment. (Data from Cowie, 1977.)

Charnov (1976) asked how long the animal should stay at the current patch (i.e., the optimum value for $t_h^{(i)}$). He noted that the animal's actions in patch n had no consequence on availability of food in any other patch. That is, the above equation could be written as follows:

$$\bar{J} = \frac{f^{(n)}(t_h^{(n)}) + A}{t_h^{(n)} + B} \tag{2.14}$$

Here, terms A and B do not depend on the harvest period in the current patch n; instead, they depend on behavior in other patches. The optimum value for $t_h^{(n)}$ is found by maximizing the above expression. He did so by finding its derivative with respect to $t_h^{(n)}$:

$$\frac{d\bar{J}}{dt_h^{(n)}} = \frac{df^{(n)}}{dt_h^{(n)}} \frac{1}{t_h^{(n)} + B} - \frac{f^{(n)}(t_h^{(n)}) + A}{(t_h^{(n)} + B)^2} \tag{2.15}$$

At the optimum harvest period $t_h^{(n)*}$, the above expression is equal to 0. He therefore arrived at the following equation:

$$\left.\frac{df^{(n)}}{dt_h^{(n)}}\right|_{t_h^{(n)*}} = \bar{J}\,\Big|_{t_h^{(n)*}} \tag{2.16}$$

The left side of the equation is the instantaneous capture rate in the current patch, termed *marginal capture rate*, evaluated as $t_h^{(n)*}$. (Marginal is the term used in economics to refer to derivatives in mathematics.) The interesting prediction is that the optimum time to leave the patch, $t_h^{(n)*}$, occurs when the marginal capture rate is equal to the global capture rate. If at every patch this policy is carried out, then the global capture rate will be maximized. As a result, Charnov (1976) stated that "the predator should leave the patch it is presently in when the marginal capture rate in the patch drops to the average capture rate for the habitat."

A useful way to express the above condition is graphically (figures 2.4B and 2.4C). Suppose that the animal begins its search for food at time t_0. The energy expended during the search period is represented by the line with negative slope ($-k_s$). At time $t_0 + t_s$, the animal arrives at the patch and begins harvesting, at which time the energetic intake rises, as described by equation 2.11. It is up to the animal to decide when to leave the patch. If the objective is to maximize the global capture rate, then equation 2.16 predicts that the animal should abandon the patch when the marginal capture rate falls below the global value for the environment.

We can represent the right side of equation 2.16 with a line that has the slope $\bar{J}$. The optimum harvest duration t_h^* occurs when the slope of a line tangent to $f(t_n)$ is equal to the slope of the line that has slope $\bar{J}$. As a result, the decision of when to abandon the food at the current patch and restart the search process comes down to a simple calculation: is the current rate of capture greater than the global rate? If it is, the animal should stay and continue harvesting. If it is not, the animal should abandon the patch and restart its search for food.

How does harvest duration depend on the reward available at that patch? The marginal capture rate is the derivative of the harvest function with respect to harvest duration:

$$\frac{df^{(n)}}{dt_h^{(n)}} = \frac{\alpha^{(n)}\beta^{(n)}}{(1+\beta^{(n)}t_h^{(n)})^2} - k_h \tag{2.17}$$

Because of the concave shape of the reward harvest function (equation 2.11), the marginal capture rate is large when animal arrives at the patch and begins harvesting food. When the animal has just arrived at the patch, t_h is small, it is easy to find food; hence, the marginal capture rate is high. As the harvesting continues at the patch, the marginal capture rate decreases and eventually becomes negative; now the rate of energy expended to harvest the food has become larger than the rate of energy acquired from feeding. Solving for t_h in equation 2.16 and equation 2.17, we find that the harvest duration grows as the square root of the amount of reward available at the patch (figure 2.4B):

$$t_h^{(n)*} = \left(\frac{\alpha^{(n)}}{\beta^{(n)}(k_h^{(n)} + \bar{J})}\right)^{1/2} - \frac{1}{\beta^{(n)}} \tag{2.18}$$

Therefore, as the reward value increases, the animal should stay longer. However, the theory leads to a prediction we would not have expected: in a rich environment with a large amount of reward, the animal should stay only a short period of time at any given patch, abandoning more of the food. We can infer this from equation 2.18. the harvest duration should *decrease* with the global capture rate $\bar{J}$.

The global capture rate is akin to cost of time. In an environment where $\bar{J}$ is high, indicating that this is a rich environment where reward is easily obtained, the slope of the line is high. Time is quite valuable in this environment. In this rich environment, t_h^* will be small. That is, the animal should stay only a brief time in each patch, capture the easily harvested reward, and then leave (black dashed line, figure 2.4C). In contrast, in an environment where the global capture rate is low, indicating that this is an environment in which reward is hard to find, the cost of time is low (the slope of our line is small, gray dashed line figure 2.4C). In such an environment the animal should stay a long time in each patch and capture more of the reward before restarting the search for another patch. This means that in a rich environment, more of the reward will be left behind than in a poor environment.

In addition, harvest duration decreases with the energetic cost of harvesting k_h and with rate of harvest β. As a result, if it takes a great deal of effort to acquire reward in a patch, the animal should stay only briefly in that patch.

Note that in figure 2.4B, t_h^* occurs before the peak of the $f(t_h)$ function. That is, despite the fact that staying longer than t_h^* continues to provide a positive rate of energy, it is not worthwhile to stay. The reason for this is that according to $\bar{J}$, there is better harvest available elsewhere.

Marginal value theorem predicts that duration of harvest should depend on past history of rewards and efforts.

To test for some of the predictions of this theory, Richard Cowie (1977) placed six birds in their own large cages containing five artificial trees. Each tree supported six plastic containers filled with worms. We can think of each container as a patch of food. The search time to access the worms was manipulated not by putting the trees farther apart (which was not possible because of the limitations of the cage size), but by covering each container with a lid that the birds had to remove before they began eating. The birds were repeatedly exposed to two environments. In one environment, all the lids were easy to remove. In the other environment, all the lids were difficult to remove. In the easy environment, the average time between opportunities to eat, labeled as $E[t_s]$, was short (around 5 s), whereas in the hard environment, this time was long (around 20 s). We can write the global capture rate in terms of the average time between opportunities to eat:

$$\bar{J} = \frac{-Nk_s E[t_s] + \sum_{i=1}^{N} f^{(i)}(t_h^{(i)})}{NE[t_s] + \sum_{i=1}^{N} t_h^{(i)}} \tag{2.19}$$

This expression implies that as the mean time between eating opportunities increases, $\bar{J}$ decreases. Equation 2.18 predicts that if behavior is guided by a desire to maximize global capture rate, then harvest duration (time spent eating worms) should grow as the global capture rate decreases. That is, the animals should stay longer (and eat more of the food) if they had to work harder to open the food container. Figure 2.4D shows the empirical data, quantifying the relationship between the mean search time (time expended to open the container) and the mean time spent eating at each patch. The time spent eating increased roughly as the square root of time spent searching. (The curve is proportional to the square root of search time.)

Sara Constantino and Nathaniel Daw (2015) tested some of the theory-based predictions on humans. They presented volunteers with an image of an apple tree. The subject could stay and harvest the apples, incurring a short harvest delay t_h, or abandon the tree and move (figuratively) to another one, incurring a longer travel delay t_s. If the subject stayed, the number of apples received at each harvest decreased incrementally: on the first harvest of the tree, the number of apples was α_0. At each subsequent harvest, the number of apples declined at a rate specified by a depletion rate $b < 1$ such that $\alpha_{n+1} = b\alpha_n$. At some point, the subject chose to abandon the tree, causing it to disappear and another to appear after a "travel" delay period t_s.

The total experiment duration was fixed and the parameters of the environment were varied in a block design. In some blocks, travel delay time t_s was manipulated, and in other blocks, apple depletion rate b was manipulated. The authors measured the number of harvests that the subjects performed at each tree in each environment and asked whether this number reflected a policy aiming to maximize the global capture rate.

At a given tree, the subject spent time t_h and acquired a certain number of apples. To estimate effort, let us assume that passage of time incurred an effort-like cost specified by rate k. After N harvests from the same tree, the sum of apples acquired and effort spent is as follows:

$$f(N)=\sum_{n=1}^{N}(\alpha_o b^{n-1}-kt_h)=\frac{\alpha_o(1-b^N)}{(1-b)}-Nkt_h \tag{2.20}$$

The marginal value theorem states that the subject should stop harvesting the current tree when the marginal return is less than the global capture rate. The marginal capture rate is the derivative of the previous expression. We have the following.

$$\frac{df}{dN}=-\frac{\alpha_o b^N \log(b)}{1-b}-kt_h \tag{2.21}$$

Figure 2.5A shows equation 2.21 via a dashed line. (To make this plot, all parameters were known, except for effort term k, which we assumed was negligible, an assumption that we will return to.) As expected, as the number of harvests increased, the marginal capture rate decreased, implying that initial harvests had greater benefits than later ones.

To compute the global capture rate, suppose that the subject encountered m trees in an environment and on average harvested each tree N times. The global capture rate for that environment is

$$\bar{J}(N)=\frac{-mkt_s+m\sum_{n=1}^{N}(\alpha_o b^{n-1}-kt_h)}{m(Nt_h+t_s)}=\frac{\alpha_o(1-b^N)}{(1-b)(Nt_h+t_s)}-k \tag{2.22}$$

The term $\bar{J}$ expresses the global capture rate in units of apples per period of time. The marginal capture rate is expressed in terms of apples per harvest. To make the two measures consistent, we multiply $\bar{J}$ by time t_h per harvest. Optimum number of harvests N^* is found when the following equality holds:

$$\left.\frac{df}{dN}\right|_{N^*}=t_h\bar{J}(N)\Big|_{N^*} \tag{2.23}$$

Note that in this expression, when we substitute equation 2.22 and equation 2.21, the term kt_h appears on both sides of the equation; this allows us to ignore effort costs, thus eliminating the only unknown in our formulation.

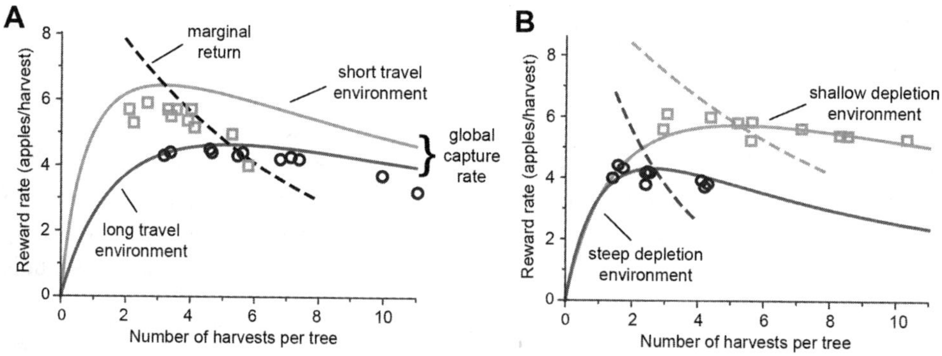

Figure 2.5
Behavior of humans in a virtual apple harvesting task. **A**. The global capture rate is plotted for two environments as a function of number of harvests per tree. The marginal capture rate indicates the local return rate as a function of number of harvests. Note that the marginal capture rate intersects the global capture rate at its maximum. Therefore, the optimal policy is to continue harvesting until the marginal capture rate falls below the global capture rate. In the environment that requires a shorter travel time between trees, the optimum number of harvests per tree is less than in the other environment. The data points represent performance of human subjects in each environment. People harvested a smaller number of times at each tree when the travel time decreased. **B**. The two environments differed in the rate of apple depletion per harvest. People harvested a larger number of times at each tree when the depletion rate became shallow. (Data adapted from Constantino and Daw, 2015.)

Our policy is expressed as follows: stop harvesting when the marginal capture rate for the next harvest is expected to be less than the global capture rate. Figure 2.5A shows the global capture rate $t_h\bar{J}$ as a function of N for two environments in which travel duration was different. (When making this plot, all values were known from the experiment design except k, which here we assumed to be negligible). When travel duration was short (a rich environment), the optimum number of harvests was $N^* \approx 3$, and when travel duration was long (a poor environment), $N^* \approx 5$. The dashed line, representing the marginal capture rate, precisely intersects the global capture rate at its peak, demonstrating the optimality of the policy. Each data point represents a subject's performance in the associated environment. Constantino and Daw found that all the subjects reduced their average number of harvests when the travel duration decreased. In their next experiment, the authors manipulated the depletion rate b and kept other variables constant. Figure 2.5B shows the marginal capture rate (dashed line) and the global capture rate as a function of number of harvests in each environment. The volunteers increased their number of harvests when the depletion rate decreased.

Because the model has no free parameters, comparison of its predictions with behavior are noteworthy. The changes in behavior from one environment to another are all in the right direction, but individual subjects are not behaving optimally (as evidenced by the fact that few subjects performed near the peak of each global capture rate). However, in environments where the global capture rate has a sharp terrain about its peak, the variance across subjects is small and centered on the peak. When the peak is broad, the variance across subjects is large. This highlights that in order to be optimal, the subjects must compute the global capture rate from their trial-by-trial experience, perhaps via a running average. How might this computation be performed, and what might be the neural correlates of it, are questions that are currently being investigated.

A limitation of the marginal value theorem is that it can only be applied to problems in which the choice of action in one patch has no bearing on states of other patches. For example, if the problem involves choosing which patch to travel to, then future choices depend on the present choice. In that case, the marginal value theorem cannot be applied. Rather, optimizing the global capture rate requires solving a state-dependent optimal control problem. Tools for solving this more general class of problems are discussed in many texts (including Shadmehr and Mussa-Ivaldi, 2012).

In summary, optimal foraging theory suggests that animals make choices so that they maximize the global capture rate $\bar{J}$. Policies that maximize the global capture rate may result in greater fitness, as measured via longevity and fecundity. In order to maximize the global capture rate, the animal's decision regarding what to eat and what to pass up (harvest duration) should be based on both the local energetic landscape (food availability and effort expenditure at the current patch) and on the global landscape (history of reward and effort as well as future expectations). The marginal value theorem is a solution to the optimal foraging problem. It suggests that the harvest duration should depend on a critical comparison: stay and harvest reward only as long as the local capture rate is larger than the global capture rate. In other words, the theorem states that you should leave when your current rate of capture falls below the average value that you can get elsewhere.

2.3 Optimizing Movement Vigor

Until recently, foraging theory had not been applied to the problem of movement control. This seemed incorrect to us, because our movements require energetic expenditure; therefore, how we move affects the effort that we have spent to arrive at the reward site. If our objective is to maximize the global capture rate, then we should care not just about our decisions, but also about our movements.

In his formulation of capture rate, Charnov (1976) did not consider energetic cost of travel between patches in a realistic fashion. If he had done so, he may have been able to predict not just how long the animals would stay at a patch and harvest food, but also how fast they would travel to that patch. Here, we will consider this problem. Our goal is to cast movements in the framework of optimal foraging theory so that we can make predictions regarding vigor (how fast to move to the reward site), and decision-making (how long to stay and harvest the reward).

A unified approach to decision-making and movement vigor was first described by Yael Niv, Peter Dayan, and colleagues (Niv et al., 2007). Their aim was to describe a framework in which one could select which action to perform as well as dictate the vigor with which to perform that action. In their framework, they assumed that effort expenditure declined hyperbolically with the duration of the action. (A more vigorous movement would require more effort). This encouraged sloth: move slowly to conserve effort. However, any time spent performing an action took time away from other potential actions that could have been done. Those other actions represented an opportunity cost, a cost that could be quantified via the capture rate that the animal had experienced in the past multiplied by the duration of the current action. This encouraged vigor: move quickly to minimize the opportunity cost. Together, maximizing the global capture rate could be achieved by performing actions in such a way that balanced the cost of effort (assumed to reduce hyperbolically with duration of the movement) and the cost of time (opportunity cost, assumed to increase linearly with duration of the movement).

The basic prediction of this framework was that as the past capture rate increased (rich environment), the vigor should also increase. That is, if in the recent past the rewards had been acquired easily, resulting in an increase in the capture rate, then movements of the animal should tend to become more vigorous. To test this prediction, Guitart-Masip and colleagues (Guitart-Masip et al., 2011) asked human subjects to perform a perceptual discrimination task. Before the trial began, subjects were told the reward at stake. This reward value varied gradually from trial to trial, from high to low, and then back to high, and finally back down to low. This changing reward, and the success patterns of each subject as a function of trial number, resulted in a reward rate that varied with the progression of the trials. The authors asked whether reaction time of the decisions that the subjects made also changed as a function of this reward rate. They found that as the reward rate (averaged over some length of history) increased, the reaction time (a proxy for vigor) for the current decisions also decreased. That is, the past reward rate appeared to modulate the reaction time of the current movement.

At its core, the framework of Guitart-Masip and colleagues is the same as that of optimal foraging theory: perform decision and actions in such a way as to maximize the global capture rate. Here, we will extend this framework in three ways. First, we consider effort to be objectively related to the energetic cost of performing the action. Second, we consider reward to be acquired not instantaneously, but through harvesting, an act that consumes time and energy. Finally, we apply the marginal value theorem and describe the optimal policy that maximizes the global capture rate.

Marginal value theorem can predict the effects of reward and effort on movement vigor. Marginal value theorem is concerned with the length of time the animal chooses to stay at the harvest site. It predicts that the animal should leave the harvest site when the local capture rate drops to the global capture rate. However, the theorem is based on the assumption that "the length of time between patches should be independent of the length of time the predator hunts within any one" (Charnov, 1976). In other words, the assumption is that factors that influence harvest duration do not affect movement vigor. Contrary to this assumption, experiments have repeatedly found that increased reward at the destination shortens travel duration to that destination (Takikawa et al., 2002; Xu-Wilson et al., 2009; Opris et al., 2011; Manohar et al., 2015; Summerside et al., 2018). In anticipation of greater reward, animals move faster to the harvest site. Because reward magnitude also modulates harvest duration, it seems likely that speed of movement and duration of harvest are not independent, as was assumed by Charnov (1976), but influenced by the reward and effort characteristics of the environment. Let us resolve this theoretical problem.

Suppose that the total energy spent performing a movement that covers distance d and duration t_m is specified by the function $g(d,t_m)$. This function is typically concave upward as a function of t_m: moving fast results in large energy expenditure, but moving slow can also be costly. For example, as we saw in chapter 1, the total energy that a person of mass m will spend to walk a distance d in duration t_m can be expressed this way:

$$g_w(d,t_m)=(a_0+a_1)mt_m+\frac{bmd^2}{t_m} \tag{2.24}$$

The term a_0 reflects the energetic cost of being alive for duration t_m, and the term a_1 is a bias specific to walking. Energetic cost of reaching also has a similar concave form (Shadmehr et al., 2015). Now suppose we consider our action not in isolation, but as part of a sequence of actions, such as walking to a sequence of locations. At each location n, we spend an amount of time $t_h^{(n)}$ and an amount of energy $k_h^{(n)}t_h$ harvesting the reward $\alpha^{(n)}$. The harvest function $f^{(n)}(t_h^{(n)})$, as shown in equation 2.11, describes the time course of gaining reward and expending effort. In addition, we spend energy moving to this reward site, which we denote via the function $g(d^{(n)},t_m^{(n)})$. The capture rate for traveling to and harvesting reward at location n is as follows:

$$J^{(n)}=\frac{f^{(n)}(t_h^{(n)})-g(d^{(n)},t_m^{(n)})}{t_h^{(n)}+t_m^{(n)}} \tag{2.25}$$

This expression describes the local capture rate, referring to the events that transpired during our visit to location n. In an environment with N such locations of reward, the following is the global capture rate:

$$\bar{J}=\frac{\sum_{n=1}^{N} f^{(n)}(t_h^{(n)})-g(d^{(n)},t_m^{(n)})}{\sum_{n=1}^{N} t_h^{(n)}+t_m^{(n)}} \tag{2.26}$$

Based on the objective of maximizing the global capture rate, our goal is to find the optimum movement duration $t_m^{(n)*}$ that brings us to reward site n, and the optimum harvest duration $t_h^{(n)*}$ at that site. If we assume that our behavior at location n has no effect on rewards or efforts that come before or after, we can rewrite the above expression as follows:

$$\bar{J} = \frac{f^{(n)}(t_h^{(n)}) - g(d^{(n)}, t_m^{(n)}) + A}{t_m^{(n)} + t_h^{(n)} + B} \tag{2.27}$$

In this equation, A and B indicate terms in $\bar{J}$ that do not depend on behavior associated with patch n. To predict behavior in patch n, we seek two unknowns, $t_m^{(n)*}$ and $t_h^{(n)*}$, variables that maximize the value of equation 2.27. We note that the derivative of $\bar{J}$ with respect to the unknowns can be written in terms of $\bar{J}$:

$$\begin{aligned} \frac{d\bar{J}}{dt_m^{(n)}} &= \frac{-1}{t_m^{(n)} + t_h^{(n)} + B}\left(\bar{J} + \frac{dg(d^{(n)}, t_m^{(n)})}{dt_m^{(n)}}\right) \\ \frac{d\bar{J}}{dt_h^{(n)}} &= \frac{1}{t_m^{(n)} + t_h^{(n)} + B}\left(\frac{df^{(n)}(t_h^{(n)})}{dt_h^{(n)}} - \bar{J}\right) \end{aligned} \tag{2.28}$$

The optimum vigor and harvest durations are found when these two equalities are simultaneously equal to 0. As a result, the optimum vigor is specified by a relationship between the effort expenditure during the movement and the global capture rate:

$$\begin{aligned} \left.\frac{dg}{dt_m^{(n)}}\right|_{t_m^{(n)*}} &= -\bar{J}\Big|_{t_m^{(n)*},\, t_h^{(n)*}} \\ \left.\frac{df}{dt_h^{(n)}}\right|_{t_h^{(n)*}} &= \bar{J}\Big|_{t_m^{(n)*},\, t_h^{(n)*}} \end{aligned} \tag{2.29}$$

Therefore, the global capture rate affects both the duration of the movement period (vigor of the movement) and the duration of the harvest period.

To illustrate the meaning of these equations, suppose that one is harvesting reward at patch n. How long should one stay at this patch? Harvesting commences when the subject arrives at the patch (equation 2.11), resulting in intake that rises (figure 2.6A). Harvesting requires effort, which produces a peak in this function. If the objective was to maximize harvest intake at the current patch, the subject should stay until the time specified by the peak of this curve (when $df^{(n)}/dt_h^{(n)} = 0$). However, equation 2.29 states that the harvest should end when the rate of intake equals the global capture rate in the environment, specified by $\bar{J}$. That is, the subject should abandon the patch when the rate of intake falls below the rate indicated by her past experience and future expectations. As a result, local factors (current reward and the effort required to harvest that reward) interact with global factors (past rewards and future efforts) to determine how long the subject should stay at the patch.

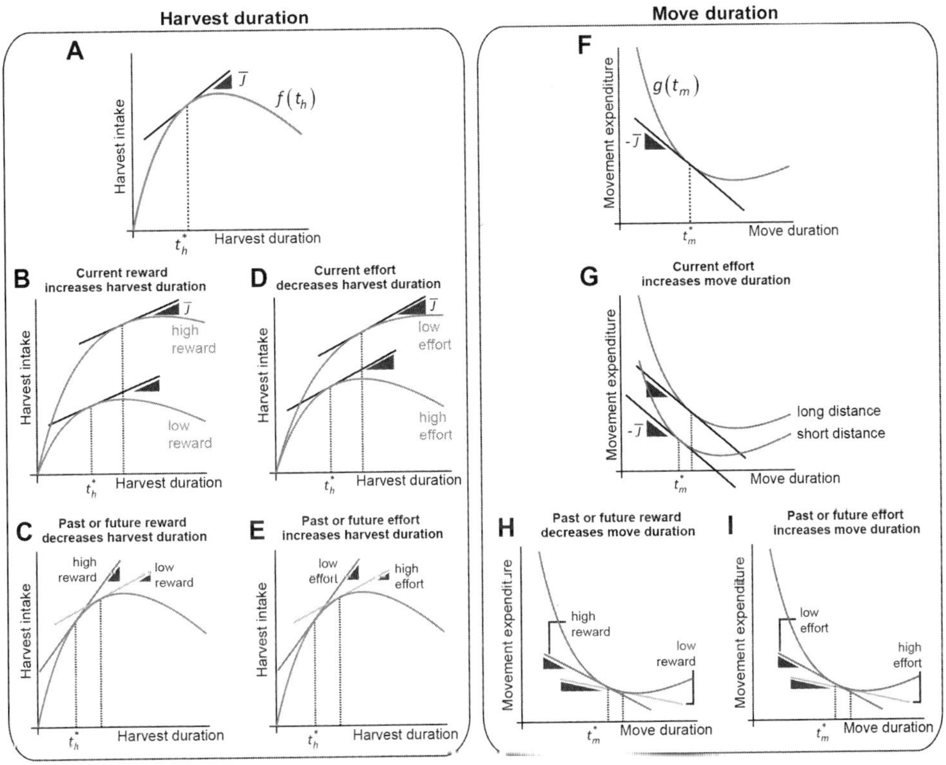

Figure 2.6
The theoretical link between harvest duration and movement vigor during foraging. **A**. During harvest, reward is accumulated while effort is expended, resulting in harvest intake $f(t_h)$, a function that is concave downward. The patch should be abandoned not when the intake is maximized, but when the rate of intake (local capture rate) is equal to the average capture rate $\bar{J}$ (at time t_h^*). **B**. Higher reward at the current location extends the harvest duration. Subject should stay longer when there is greater reward. **C**. After experience of high reward, indicating a rich environment, subject should stay a shorter period in the current patch. Similarly, in anticipation of high reward in a future harvest, subject should shorten harvest duration at the current patch. **D**. Harvest duration should be shortened when the current harvest requires a large amount of effort. **E**. Harvest duration should be lengthened if the past history has included high effort. Similarly, if future harvests are expected to encounter high effort, current harvest should be extended in time. **F**. During movement, energy expenditure is a concave upward function of duration. Movement duration is optimum not when the effort cost of travel g is minimized, but when the rate of energetic loss during travel is equal to the negative of the average capture rate. **G**. Increased effort requirements of travel result in increased movement duration. **H**. After experience of high reward (larger $\bar{J}$), or in anticipation of high reward in the future, movement duration t_m^* should be decreased. That is, vigor should be increased after experience of high reward or moving toward a high-reward patch. **I**. After experience of high effort (lower $\bar{J}$), or in anticipation of high effort, movement duration t_m^* should be increased. That is, vigor should be reduced after experience of high effort.

The theory leads to the prediction that the duration of stay at patch n should increase with the reward available at that patch (effect of $\alpha^{(n)}$, figure 2.6B). If past actions have been rewarding (increasing $\bar{J}$), the subject should exhibit impatience at the current patch and thus leave the patch early (figure 2.6C). In contrast, the duration of stay should decrease with the effort required to capture the reward (effect of $k_h^{(n)}$, figure 2.6D). If past actions have required large effort (decreasing $\bar{J}$), the subject should stay longer at the current patch (figure 2.6E). Finally, if future actions are anticipated to require high effort (decreasing $\bar{J}$), or if future actions are expected to encounter low reward, one should stay longer at the current patch.

More interestingly, the theorem can be used to predict patterns of movement vigor. Movement requires energetic expenditure (function g). If we consider only the local conditions, then we would move in such a way as to minimize the energetic cost of travel; that is, we should find $t_m^{(n)}$ such that $dg/dt_m^{(n)} = 0$. This provides a ballpark estimate of how fast a movement takes place, as demonstrated by those who study speed of locomotion in humans and other animals (Ralston, 1958; Hoyt and Taylor, 1981; Selinger et al., 2015). However, our objective is to maximize the global utility $\bar{J}$. In that case, duration of movement will be modulated by both the local contingencies and the history of the subject. In particular, the duration of the movement toward patch n is optimal when the derivative of the energetic loss during the movement $dg/dt_m^{(n)}$ is equal to $-\bar{J}$ (figure 2.6F). Therefore, the global capture rate affects both the duration of the harvest and the vigor of the movement.

Another prediction based on the theory is that when moving toward patch n, a longer distance will result in a longer walk duration (figure 2.6G). As the reward at the patch increases, the travel duration will decrease, implying a faster walk (figure 2.6H, increasing future reward $\alpha^{(n)}$ increases $\bar{J}$). As the effort associated with harvesting the reward at the patch decreases, walking speed toward the patch increases (figure 2.6I, decreasing future $k_h^{(n)}$ decreases $\bar{J}$).

Surprisingly, the theory also predicts that when the subject is moving away from patch n, her movement will be faster if she has just harvested a large amount of reward (figure 2.6H, increasing past reward $\alpha^{(n-1)}$ increases $\bar{J}$), or has just spent a small amount of effort harvesting that reward (figure 2.6I, decreasing $k_h^{(n-1)}$ decreases $\bar{J}$). We see that the theory provides a rationale for why you might walk fast after a reunion with a dear friend, and walk slow after a bad business meeting.

2.4 Example: Walking Speed

Suppose that an individual has distance d to walk to location n, where he expects to harvest reward $\alpha^{(n)}$. His action is not performed in isolation, but as a part of a sequence of many actions. We would like to predict two things: (1) how fast the subject will walk, and (2) how long the subject will stay at that location harvesting reward.

Consider the convex function that defines the energetic cost of walking, g_w. This function has a minimum that describes the duration of the walk under the assumption of minimizing the energetic cost. However, if our global capture rate is a positive number (our past rewards have been greater than our efforts), we should walk faster than energetically optimal (spend the extra energy so we arrive sooner).

At the reward site n, the harvest function is as follows:

$$f^{(n)}(t_h) = \alpha^{(n)}\left(1 - \frac{1}{1+\beta^{(n)}t_h}\right) - a_0 t_h \tag{2.30}$$

Based on the energetic cost of walking g_w, as shown in equation 2.24, the local capture rate is

$$J^{(n)} = \frac{f^{(n)}(t_h^{(n)}) - g_w(d^{(n)}, t_m^{(n)})}{t_h^{(n)} + t_m^{(n)}} \tag{2.31}$$

Suppose that our objective was to maximize the local capture rate. In that case, the optimum walk duration is achieved when $dg_w/dt_w = -J^{(n)}$, and the optimum harvest duration is achieved when $df^{(n)}/dt_h = J^{(n)}$. We therefore have two equations and two unknowns:

$$\begin{aligned} \left.\frac{df^{(n)}}{dt_h^{(n)}}\right|_{t_h^{(n)*}} &= \frac{\alpha^{(n)}\beta^{(n)}}{(1+\beta^{(n)}t_h^{(n)*})^2} - a_0 m = \left.J^{(n)}\right|_{t_h^{(n)*}, t_m^{(n)*}} \\ \left.\frac{dg}{dt_m^{(n)}}\right|_{t_m^{(n)*}} &- (a_0 + a_1)m \quad \frac{bd^2m}{t_m^{(n)*}} - -\left.J^{(n)}\right|_{t_h^{(n)*}, t_m^{(n)*}} \end{aligned} \tag{2.32}$$

Figure 2.7A provides a graphical representation of the solution: the time when the line with the slope $J^{(n)}$ becomes tangent to g_w specifies the walk duration, and the time when the same line becomes tangent to $f^{(n)}$ specifies the harvest duration. In this case, the travel distance is 30 m and the reward $\alpha^{(n)} = 10$ kJ, the combination of which produces an average walk speed of 92 m/min.

However, our objective is to maximize the global capture rate. To do so, we must integrate the capture rate for the action that we are planning (walk to location n) with the capture rate that we have experienced in the past (global capture rate until now). To see the trial by trial influence of each action on the global rate, suppose that we have performed $n-1$ actions in the past that have consumed total time $T^{(n-1)}$ and produced a global capture rate of $\bar{J}^{(n-1)}$. For the action that we are planning, the local capture rate is $J^{(n)}$, and its expected duration is $t_s^{(n)} + t_h^{(n)}$. The global capture rate after completion of the n-th action will be as follows:

$$\bar{J}^{(n)} = \bar{J}^{(n-1)} \frac{T^{(n-1)}}{T^{(n-1)} + t_m^{(n)} + t_h^{(n)}} + J^{(n)} \frac{t_m^{(n)} + t_h^{(n)}}{T^{(n-1)} + t_m^{(n)} + t_h^{(n)}} \tag{2.33}$$

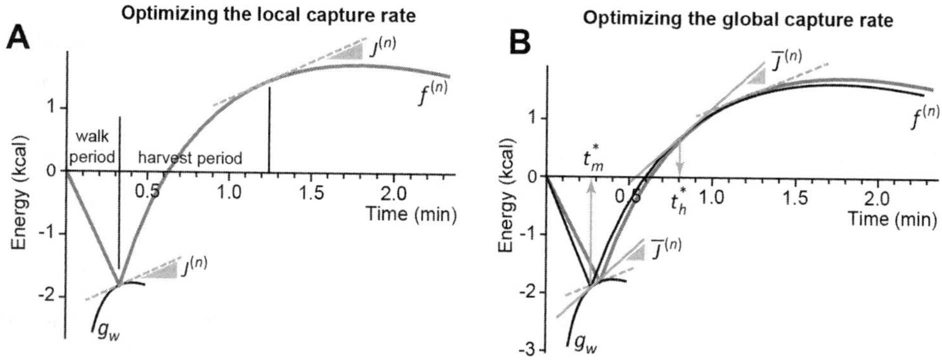

Figure 2.7
Speed of walking and the duration of harvest that follows the walk should depend on the capture rate. **A**. Walking in order to maximize the local capture rate $J^{(n)}$. The task is for 75 kg subject to walk 30 m in order to harvest $\alpha = 10$ kJ reward. The function g_w specifies energetic cost of walking as a function of walk duration, and the function f specifies energy acquired during harvesting following arrival at the destination. The walk and harvest durations that maximize the local capture rate are specified by the time points at which $J^{(n)}$ is tangent to g_w and f, respectively. **B**. Two movements that are made for the purpose of acquiring the same reward will be performed differently for different global capture rates. The local conditions are the same as in part A (i.e., the reward at stake and distance to walk are the same). However, if the previous movements have produced a high global capture rate, then the integration of the local rate with the previously experienced global rate will produce a faster movement and a shorter harvest period.

To maximize the global capture rate, we replace the local capture rate $J^{(n)}$ in equation 2.31 with the global capture rate $\overline{J}^{(n)}$ in equation 2.33, and again we have two equations and two unknowns (the optimal harvest and travel periods).

On the basis of the theory, if our past actions have been successful, producing a high global capture rate $\overline{J}^{(n-1)}$, the next action will be performed with greater vigor. Thus, now we have a rationale for why in principle after a good meeting, you should walk faster.

To illustrate this idea, suppose that we begin with $\overline{J} = 0$. At trial $n = 1$, we walk a distance of 30 m in order to acquire reward $\alpha^{(1)} = 10$ kJ. This produces an average walking speed of 92 m/min. The next walk requires the same distance, but has a higher reward, $\alpha^{(1)} = 20$ kJ. As a result, our global capture rate following the completion of the second movement will be larger than that the rate following the first movement, $\overline{J}^{(2)} > \overline{J}^{(1)}$. Now suppose that for the third trial, we have the same local condition that we had in the first trial (same distance and reward). That is, the local capture rates are equal: $J^{(1)} = J^{(3)}$. However, because our global capture rate has risen since that trial, our walk will be faster, and our harvest duration will be shorter, as illustrated in figure 2.7B.

When there are many past actions ($n - 1$ is a large number), equation 2.33 implies that the global capture rate following completion of the current movement, noted by $\overline{J}^{(n)}$, will be dominated by the past global capture rate $\overline{J}^{(n-1)}$. In that case, using the second equality in equation 2.32, we can describe a relationship between movement duration (for walking) and the global capture rate:

$$t_m^{(n)*} = \frac{d\sqrt{bm}}{\sqrt{(a_0 + a_1)m + \bar{J}^{(n-1)}}} \tag{2.34}$$

This equation implies that given the goal of walking a distance d, the average speed of walking d/t_m^* should increase with the square root of the global capture rate $\bar{J}$. That is, we should walk faster in environments in which we have experienced a high rate of reward.

It is useful to compare this result to traditional approaches in biomechanics, in which the objective is to minimize energetic cost of the current action; that is, the objective is to minimize $g_w(d, t_m)$. In that case, the optimum duration of the walk is

$$t_m^{(n)*} = \frac{d\sqrt{b}}{\sqrt{a_0 + a_1}} \tag{2.35}$$

Assuming that the global capture rate $\bar{J}$ is a positive quantity, equation 2.35 implies that the effect of the global capture rate is to make us walk faster than the speed that minimizes energetic costs. We have illustrated this idea by simulating energetic cost of walking a distance of 100 m (figure 2.8A). If our objective is to minimize the energetic expenditure of this specific walk, then then based on the parameter values from chapter 1, the optimum speed of walking is around 74 m/min (equation 2.35). However, if our objective is to maximize the global capture rate, then this walk is only one of the actions that we will be performing. According to equation 2.34, the optimum speed depends on our history, as represented by the global capture rate $\bar{J}$: we should arrive at our destination at the time when the slope of our energetic loss is equal to the global capture rate.

In summary, when we view movements in the framework of optimal foraging, we find that movement vigor depends on two variables: (1) the local capture rate that the subject expects for the upcoming action, and (2) the global capture rate experienced in the past. The local capture rate depends on the immediate conditions, such as how much reward is expected at the destination (α), how far away it is (d), and how readily it can be harvested (β). The global capture rate depends on the history of the subject. Integration of the local and the global capture rates determines three variables: (1) the utility of the act, specified by the maximum capture rate that can be achieved for the action, (2) the vigor of the ensuing movement, specified by the optimal time to arrive at destination, and (3) the harvest duration, specified by the duration of time that the subject should stay at the reward site before abandoning it and moving on to the next action.

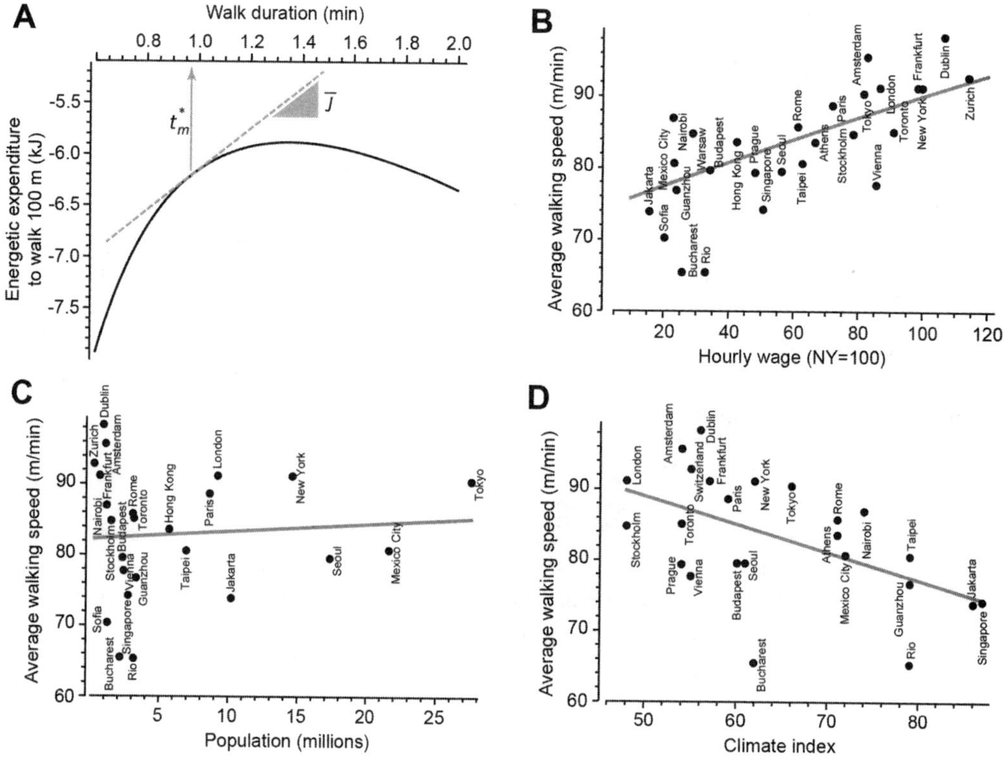

Figure 2.8
Effect of global capture rate on speed of walking: average speed of walking in various cities. **A**. Energy expended in order to walk 100 m as a function of walk duration. If the objective is to minimize the energetic expenditure of the current walking episode, then the optimum walk duration is around 1.3 min. However, if the objective is to maximize the global capture rate, then $\bar{J}$ specifies the optimal duration. A high global capture rate (rich environment) predicts a faster walking speed. **B**. Average walking speed, measured by Levine and Norenzayan (1999), as a function of average hourly wage, measured by Hoefort and Hofer (2006). The curve is the prediction of equation 2.36. Goodness of fit: $R=0.73$, $F(1,25)=29.2$, $p=0.000013$. **C**. Effect of population on walking speed. Goodness of fit: $R=0.08$, $F(1,22)=0.16$, $p=0.69$. **D**. Effect of climate on walking speed. Climate was measured as the average monthly high temperature during the month in which the walk data was gathered. Goodness of fit: $R=0.53$, $F(1,22)=8.6$, $p=0.0077$.

2.5 Walking Speed at Various Cities

We are in position now to tackle the puzzle that we started this chapter with: the fact that the average speed of walking is affected by where people live. Recall that Bornstein and Bornstein (1976) visited cities around the world and measured the natural walking speed of pedestrians. They found that the walking speed increased roughly with the population of the city (figure 1.1C). To interpret their data, they suggested that "crowding has been thought to motivate … avoidance behaviors that reduce tension", and therefore "increased walking speeds serve to minimize environmental stimulation."

Our theory suggests a very different possibility: the variability in walking speed may be partly related to the differences in the global capture rate of the people who live in the various cities. We assume that between individuals, the energetic cost associated with walking, the distance to the destination, and the reward expectations are similar. (The first assumption can be debated because of various factors, including temperature, humidity, and body mass.) With these assumptions, the remaining variable is the global capture rate. An objective measure of reward rate is the average hourly wage of the individuals, normalized to the cost of living in that city.

We computed hourly wage and cost of living data for 27 cities from previously published reports (Hoefort and Hofer, 2006), and average speed of walking data from observations in the same cities (Levine and Norenzayan, 1999). We represented this hourly wage with the variable z, and assumed that $\bar{J} = kz$. That is, in this model, the global capture rate was proportional to the average wage with respect to cost of living in each city. As a result, on the basis of the marginal value theorem, walking speed should be related to the wages in the following way:

$$v^* = \sqrt{\frac{(a_0 + a_1)m + kz}{bm}} \tag{2.36}$$

This equation had one unknown parameter, k, and provided a very good fit to the measured data (figure 2.8B, $R = 0.73$, $F[1,25] = 29.2$, $p = 0.000013$). That is, estimation of vigor via foraging theory provided a reasonable fit to the average walking speed of people in various cities.

Of course, walking speed likely depends on other parameters. We considered two other potential variables: population and climate of the city. Using data from Robert Levine and Ara Norenzayan (1999), we found that population was a poor predictor of walking speed (figure 2.8C, $R = 0.08$, $F[1,22] = 0.16$, $p = 0.69$), which implied that these data do not agree with the observations reported by Bornstein and Bornstein (1976). However, climate (reported by Levine and Norenzayan [1999] as the average monthly high temperate in each city) was a reasonable predictor of walking speed: people who lived in colder climates tended to walk faster (figure 2.8D, $R = -0.53$, $F[1,22] = 8.6$, $p = 0.0077$). The best predictor, however, was income: the average income of the inhabitants, along with the metabolic cost of walking in the context of optimal foraging, accounted for roughly twice the walking speed variance as our next best variable, climate. This result is noteworthy because our foraging model had only one free parameter k, whereas models for climate and population data had two free parameters (offset and slope).

In summary, according to the marginal value theorem, the speed of walking should increase with the square root of the global capture rate. Using income as a proxy for the global capture rate, the theory accounted for some of the variability that had been observed in the average walking speeds of people across various cities. Perhaps the global capture rate of the inhabitants was one of the factors that contributed to differences in walking speed.

2.6 Foraging with Our Eyes

Locomotion is an action that has a well-defined metabolic cost. However, it is difficult to perform foraging experiments that use locomotion because such experiments require long amounts of time to complete and large amounts of space. Instead, in a laboratory there are other, simpler kinds of movements that can be used to explore behavior during foraging. One such movement is the saccade (eye movement).

When we move our eyes to a particular location, we place a part of an image on our fovea. As we maintain gaze, holding the eyes still, we acquire information from that part of the image, akin to harvesting of reward. This harvesting is not without effort: as we move our eyes, the eye muscles are activated, requiring energy that depends on distance of travel as well as speed of the movement. When we hold our eyes still, further energy is expended that depends on eccentricity of gaze (the farther the eyes are away from center, the more force one has to produce with the extra-ocular muscles). That is, the rate of energy expenditure grows with eccentricity of the eyes.

For example, we might consider behavior in a foraging paradigm in which reward patches consist of small images. People harvest the reward by gazing at the image, and once satisfied, move their eyes to view another image. Gaze duration is a proxy for harvest duration, and saccade velocity is a proxy for vigor of movement. When people perform such a task, do they control their harvest duration and vigor in a way that we might understand in the framework of optimal foraging?

Gaze duration and saccade velocity are modulated by history of reward.
In an experiment performed by Tehrim Yoon and Robert Geary in our lab (Yoon et al., 2018), subjects were presented with a center fixation that was followed by two simultaneous images, each selected randomly from various categories: noise, simple shapes, inanimate objects, animate objects, and faces (figure 2.9A). The images were always 20° apart, but their positions varied with respect to the midline. Subjects had 2 s to freely gaze. During each trial, we measured the time they spent on each image and their saccade velocity as they moved their eyes from one image to another. Image category served to modulate reward magnitude α (value increasing from noise to face), and image position served to modulate effort expenditure rate k (k increasing with eccentricity of the image).

To establish the relative value of each image, we quantified the probability of choosing that image after removal of the center fixation point. This probability increased with the value of the image type (figure 2.9B, left panel), suggesting that α increased from noise to face. To establish the effort associated with gazing at each image, we quantified choice probability as a function of image eccentricity. The probability decreased with image eccentricity (figure 2.9B, right panel).

Once the eyes landed on an image, gaze duration increased with image value (figure 2.10A, left panel). That is, as reward amount at the current patch increased, so did harvest

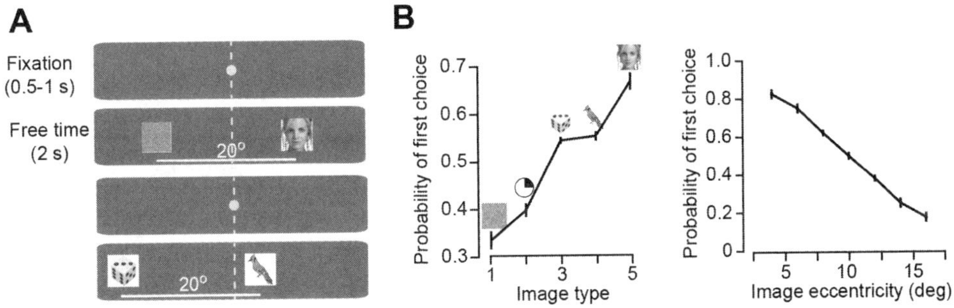

Figure 2.9
Effect of image content and eccentricity of probability of being selected. **A**. Trials began with presentation of two simultaneous images, each selected randomly from five categories. The two images were 20° apart, but their positions varied with respect to the midline. Subjects had 2 seconds to freely gaze. **B**. Probability of choosing an image after removal of fixation dot. Face images had greater utility, resulting in higher probability of being selected, but image eccentricity reduced that utility, resulting in smaller probability of selecting that image. (From Yoon et al., 2018.)

duration, confirming a prediction of the theory (figure 2.6B, effect of $\alpha^{(n)}$ on $f^{(n)}$). More interestingly, gaze duration decreased as the value of the competing image increased (figure 2.10A, middle panel). As the environment became richer (total value of the images on the screen), the amount of time devoted to harvesting any given image decreased.

We expected that as reward magnitude at the destination increased, saccade velocity toward that destination would also increase (effect of $\alpha^{(n+1)}$ on $\bar{J}$, figure 2.6G). Indeed, saccade peak velocity increased with the value of the destination image (figure 2.10B, left panel).

More interestingly, the theory predicted that past history of reward should modulate movement vigor (figure 2.6H): after a rewarding event, $\bar{J}$ increases, and therefore the eyes should move faster toward the next image. To test for this, we quantified the effect of the recently viewed image on the saccade that took the gaze away from that image toward the competing image. We found that after viewing of an image of high value, the saccade peak velocity was greater as the eyes moved away to the next image (figure 2.10B, middle panel). As a result, saccade velocities were highest in the richest environment (figure 2.10B, right panel).

The theory predicted that as the effort expenditure rate at the current patch increases, the harvest duration should decrease (effect of $k^{(n)}$ on $f^{(n)}$, figure 2.6D). Image eccentricity increased effort expenditure associated with gazing. Indeed, time spent at the current image depended on the effort requirements for that image; as the image eccentricity increased, gaze duration mildly decreased (figure 2.10C).

In summary, when subjects were presented with two reward patches, the duration of harvest at a given patch depended on both the magnitude of reward (image type) and the magnitude of effort required for harvesting that reward (eccentricity). People spent greater time gazing at more valuable images, but their gaze duration decreased as the effort required

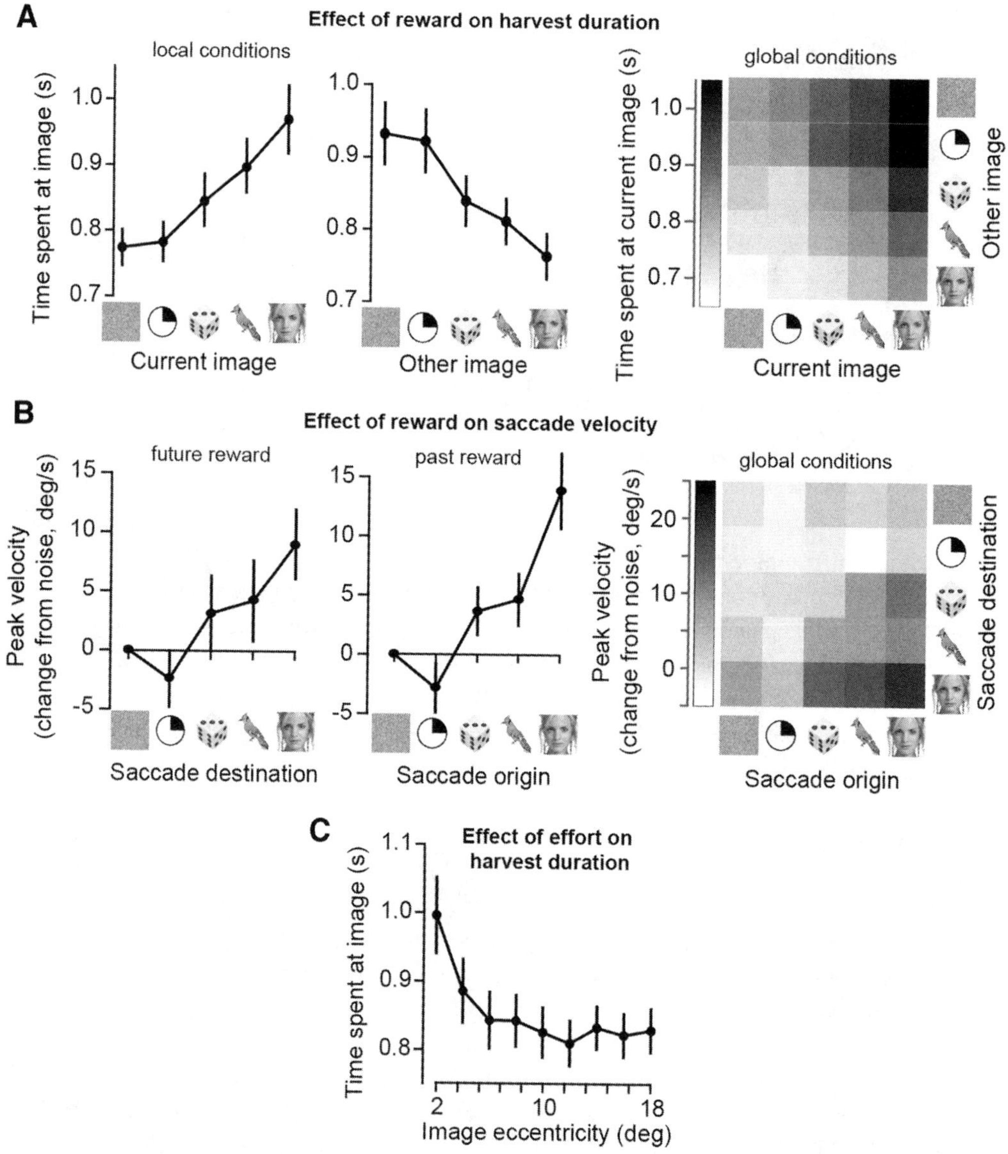

Figure 2.10
Richness of the environment affected harvest duration and saccade vigor. **A**. Effect of image content (proxy for reward value) on harvest duration. Time spent gazing increased with current image value and decreased with competing image value. As the environment became richer (right figure, from top row to bottom row of each column of the matrix), the amount of time spent at any given image decreased. **B**. Effect of image content on saccade velocity. Saccade velocity was measured with respect to the noise condition. Velocity increased with image value at destination, and also increased with value of the image at previous harvest. As the environment became richer (right figure, from top row to bottom row of each column of the matrix), saccade peak velocity increased. **C**. As image eccentricity (proxy for effort requirements during harvest) increased, gaze duration tended to decrease. (From Yoon et al., 2018.)

to harvest (view those images) increased. The duration of the harvest depended on global conditions. As the value of the competing image increased, people spent less time gazing at the current image. Reward available at the destination affected saccade vigor. Increasing the image value at the destination increased vigor. However, the history of reward also affected vigor: saccade velocities were higher when subjects were leaving an image of high value. As a result, the richness of the environment affected both the duration of harvest and the vigor of movement. In a rich environment, people spent less time at each reward site and moved faster between the sites.

History of effort modulates harvest duration as well as saccade vigor.

An important limitation in the experiment in figure 2.10 is that saccades were made toward images, raising the possibility that low-level differences in the images, such as luminosity and contrast, might have affected saccade vigor. Furthermore, this experiment imposed an explicit time limit on harvesting (2 s); gazing at one image reduced the time available for gazing at the other image, thereby imposing an opportunity cost. For our next set of experiments, we considered a foraging environment in which there were no explicit time limits on harvesting, and all saccades were toward a single control stimulus.

Yoon et al. (2018) presented one image at a time, allowing subjects to choose how long to gaze at that image without a time limit (figure 2.11A). We employed three image types: simple shapes, realistic objects, and faces. While subjects gazed at the image, they were presented with a dot that identified the location of a future image (located randomly at a distance of 10°, 15°, or 20°). Once they made a saccade to the dot, they were presented with a randomly selected image at that location.

The theory predicted that harvest duration should increase with the reward available at the current patch (effect of $\alpha^{(n)}$ on $f^{(n)}$). To test for this, we measured each subject's average gaze duration $\overline{t}_h$ across all images (633.8±25 ms), and then represented gaze duration at the current image with respect to their $\overline{t}_h$. As expected, gaze duration increased with the amount of reward (figure 2.11B). People lingered longer on images of faces than on those of simple shapes. To test whether the history of past rewards altered the duration of the current harvest, we computed gaze duration at the current image as a function of the previously viewed image. Our findings were consistent with the theory: as the value of the previous image increased, gaze duration at the current image decreased (figure 2.11C). Therefore, after a good harvest (increased $\overline{J}$), people lingered for a shorter period of time at the next reward site.

In accordance with the theory, if effort expenditure at the current patch is high, subjects should shorten their harvest (effect of $k^{(n)}$ on $f^{(n)}$). We compared gaze duration at images of varying eccentricity while keeping the next option constant and found that as current image eccentricity increased, gaze duration decreased (figure 2.11D).

On the basis of the theory, we predicted that if the previous harvest required great effort, reducing $\overline{J}$, then one should linger longer at the next harvest. Indeed, we found that if

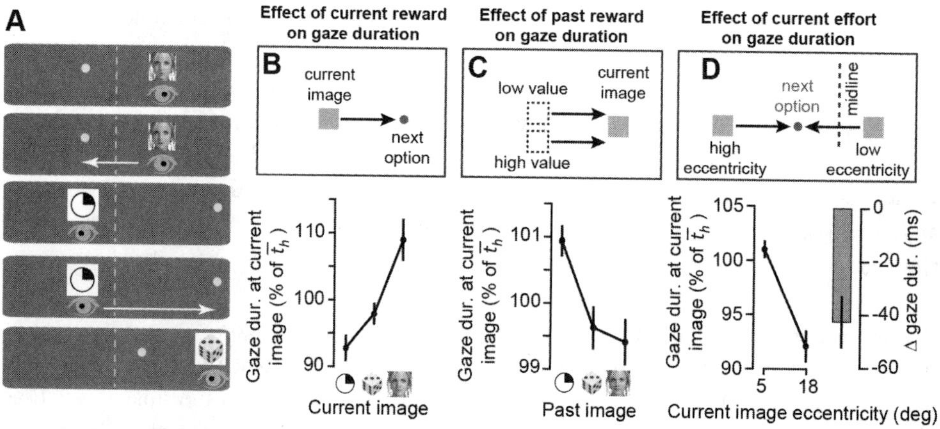

Figure 2.11
Effect of reward and effort history on gaze duration. **A.** Subjects were presented with a single image, thereby allowing them to gaze without competition from another image and without a time limit. While they gazed, they were presented with a dot that identified the location of the future image. The image was presented at the identified location only after the subjects made a saccade to the dot. Images were selected randomly. **B.** Gaze duration at current image, normalized within subject to their average gaze duration $\bar{t}_h$ across all images. **C.** Gaze duration at the current image was longer if the past image was less valuable. **D.** Gaze duration at current image was shorter if that image had high eccentricity. Right subplot shows within subject change in gaze duration. In this comparison, the next option was at the same eccentricity regardless of the current image eccentricity. (From Yoon et al., 2018.)

the travel distance to the current image was large, greater time was spent at the current image (figure 2.12A). If one expected to exert great effort in the subsequent patch, one should increase harvest duration at the current patch. As the eccentricity of the future image increased, gaze duration at the current image increased (figure 2.12B). Therefore, the past, current, and future effort expenditures all modulated harvest duration in the direction predicted by the theory.

We next asked whether the effort required to harvest affected the vigor of movements. Because in this experiment the saccades were of various amplitudes, and peak saccade velocity varies with amplitude, we normalized the data for each saccade by using procedures that took into account the direction and amplitude of each movement (Choi et al., 2014; Reppert et al., 2015; Reppert et al., 2018). The theory predicted that when high effort was expected at a future location, one should make a low velocity movement toward that location (effect of $k^{(n+1)}$ on $\bar{J}$). Indeed, saccade peak velocity toward the dot decreased as dot eccentricity increased (figure 2.12C).

However, contrary to the theory's prediction, after the expenditure of high effort, the saccade to the next target was of high vigor (figure 2.12D). This finding was repeatedly observed in various experiments by Yoon et al. (2018): past expenditure of effort was consistently followed by high saccade vigor. We do not know what the reason might be for this discrepancy between the theory and the experiments. One possibility is that there is

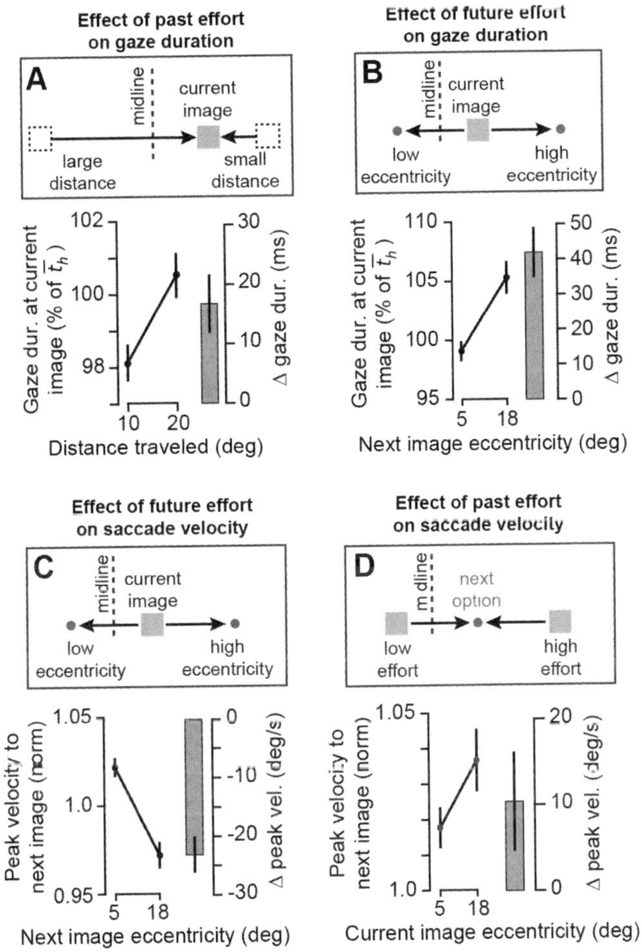

Figure 2.12
Effect of reward and effort history on gaze duration and movement velocity. **A**. Gaze duration was longer if the effort required to travel to that image (distance) was greater. Note that previous image eccentricity was kept constant in this comparison. **B**. Gaze duration at current image increased if future effort requirements (dot eccentricity) were high. **C**. Saccade velocity was lower if the future harvest required large effort. Saccade vigor refers to amplitude normalized peak velocity. **D**. Saccades that followed a high effort harvest were faster than saccades that followed a low effort harvest. Data are mean±SEM. (From Yoon et al., 2018.)

an elevation in the subjective value of a reward after the expenditure of great effort, generally termed "justification of effort" or the "IKEA effect" (Aronson and Mills, 1959; Neuringer, 1970; Kacelnik and Marsh, 2002; Clement et al., 2000; Klein et al., 2005; Zentall, 2013). That is, the subjective value of reward α is elevated after the expenditure of high effort to acquire it. We will return to this puzzle in chapter 6 when we examine dopamine and its relationship to past effort expenditure.

In summary, in this experiment, subjects had unlimited time to stay and harvest reward. They were unaware of reward magnitude (image type) at the future harvest site, but they were provided with information regarding its effort requirements (eccentricity). The results confirmed all of the theory-based predictions regarding harvest duration. Harvest duration increased with the value of the current reward but decreased with the value of the past reward. Harvest duration decreased with the amount of the current effort but increased with the amount of past and expected future effort. Saccade vigor decreased with the amount of effort expected in the future. However, contrary to our expectations, past expenditure of high effort was followed by high vigor.

Experiments that rely on eye movements in a foraging environment produce thousands of decisions and movements in a short period of time, thereby allowing for examination of a rich body of behavioral and neurophysiological data. However, our focus on simple movements like saccades has a number of disadvantages. First, we do not know the harvest function associated with gazing at images, nor do we know the energetic costs of moving the eyes or holding them still. Second, in a typical foraging paradigm, travel is via locomotion, which takes minutes to complete. In an eye movement experiment, travel is measured by saccades, which take only milliseconds. Should the brain care about saving a few milliseconds of travel time?

These problems are not insurmountable. To infer the shape of the harvest function during image gazing, one can control gaze duration and then measure its effect on saccade vigor. If the function is concave downward, presenting two images during a given period should produce a higher capture rate than presenting only one. In one experiment (Yoon et al., 2018), this manipulation was considered, and gaze duration was found to modulate saccade vigor; thus we can infer that during gazing, the harvest function is indeed concave downward. (We will review that work in chapter 3.)

Regarding the shape of effort-related functions, we assumed that during gazing, effort expenditure increased with the eccentricity of the image. This may be justified because holding the eyes requires activity in the oculomotor neurons, and the firing rates of these neurons during gazing increase with eccentricity of the eyes (Robinson, 1970; Sylvestre and Cullen, 1999; Shadmehr, 2017). It takes energy to produce an action potential, to release neurotransmitters, reuptake them at the postsynaptic site, and then produce a postsynaptic current. The energetic cost of generating spikes increases with the number of spikes (Attwell and Laughlin, 2001). Therefore, the energetic cost of holding still may be inferred from the neuronal activity associated with the period of holding.

We assumed that the effort expenditure associated with moving the eyes during a saccade was a concave upward function of movement duration. The firing rate of oculomotor neurons is proportional to saccade velocity (Sylvestre and Cullen, 1999): as velocity decreases, the rate of action potentials decreases. However, we do not know whether an increase in saccade duration results in a reduction in the total number of action potentials that act on muscles that move the eyes. Indeed, the metabolic cost of making a saccade remains undefined at this stage.

Finally, do a few milliseconds matter? We make about two saccades per second, resulting in about 100,000 saccades during our waking hours each day. During each saccade we are effectively blind. Remarkably, a 5 ms reduction in the average duration of a saccade results in the elimination of 10 minutes of blindness during each day of our life. If there is a utility that guides the vigor of our movements, saccades should be among the most exquisitely optimized voluntary movements with respect to that utility. This makes the measurement of saccades an excellent method to study the neural basis of utility and vigor, something that we will explore in chapter 4.

2.7 Salience versus Utility

We like to believe that our choices are determined by the computation of a utility, a logical process of deliberation that results in the maximization of the gain ascribed to the utility. But in fact, we cannot help but be influenced by the physical properties of the stimuli that motivate or symbolize those choices, properties that attract movements toward them, thereby swaying the decision-making process simply because their sensory properties make them stand out among their competition.

For example, at the grocery store, while you are looking at the display of jams and attempting to decide which one to buy, your visual system is affected by a low-level process that attracts your attention on the basis of the elementary properties of the various items with respect to their neighborhood; these properties include color, luminance, and orientation. If the color of the label on the jam or the shape of its jar makes that jam stand out with respect to its neighbors, it will attract your attention and draw your gaze. This, in turn, will affect the decision-making process, especially if you are in a hurry. Indeed, when you are distracted, you are more likely to choose the item that has greater salience, despite the fact that it may have a lower utility (as measured when you make the same decision absent the pressure of time). Salience is a fundamental aspect of decision-making because it influences movements (gaze) through the capture of attention, which in turn influences choice.

When there is a time limit, choice relies more on salience rather than utility.
Visual salience refers to the low-level properties of an item that bias the viewer, thus making that item easier to detect. That is, visual items that are more salient are defined as being those that are easier to find. Absent of a goal, the eyes are automatically drawn to

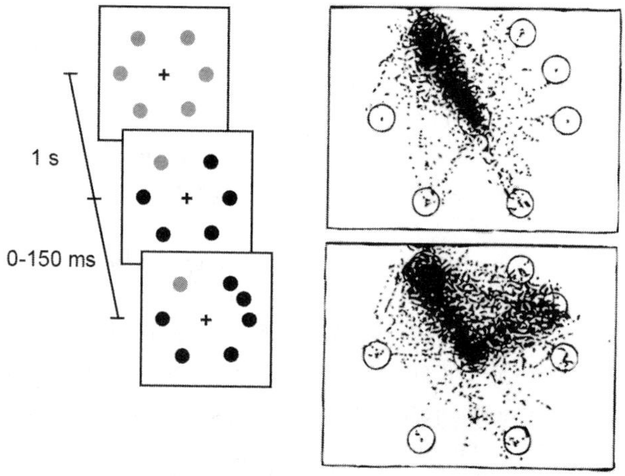

Figure 2.13
Sudden appearance of an irrelevant visual stimulus attracts gaze. **A**. Subjects were instructed to saccade to the odd stimulus (gray among black, in this depiction). However, during some trials, after various delays, a distractor stimulus appeared at another location. In the control trials, this distractor was present at trial onset. **B**. Eye position traces for the control (top plot), and the distractor (bottom plot, the distractor stimulus appeared at 0 ms with respect to onset of the target). The target is at top left part of the image, and the distractor appears at top right. (Theeuwes et al., 1999.)

regions of a scene where there is high contrast. This contrast refers to a differential property with respect to the neighborhood, where property can be luminance, orientation, or motion. Image locations that are salient are detected faster and more accurately among distractors.

For example, the sudden appearance of a stimulus often results in a saccade toward it. Jan Theeuwes, Gregory Zelinsky, and colleagues (1999) investigated the influence of salience on gaze by considering a task in which subjects were instructed to search for an odd-colored circle (figure 2.13A). The task began with a fixation cross and six gray circles. After 1 s, all gray circles except one turned red. Coincident with this event, another red circle appeared at a random location. Inside each circle, there was a letter; the subject had to identify the letter inside the odd-colored (gray) circle. In the control trials, the additional circle was present at trial onset (at fixation). Therefore, the onset of the additional circle at the moment when the circles changed color acted as a distractor: the salience of the distractor competed with the salience and utility of the odd target.

The authors found that the abrupt onset of the distractor attracted the eyes; the first saccade was often toward it rather than the target (figure 2.13B, bottom plot). In contrast, during the control trials, there were very few saccades toward the distractor (figure 2.13B, top plot). Importantly, reaction times were shorter for saccades that went to the distractor than for those that went to the target. This suggests that in order to direct gaze toward the target that had high utility, one had to inhibit the tendency to make movements toward

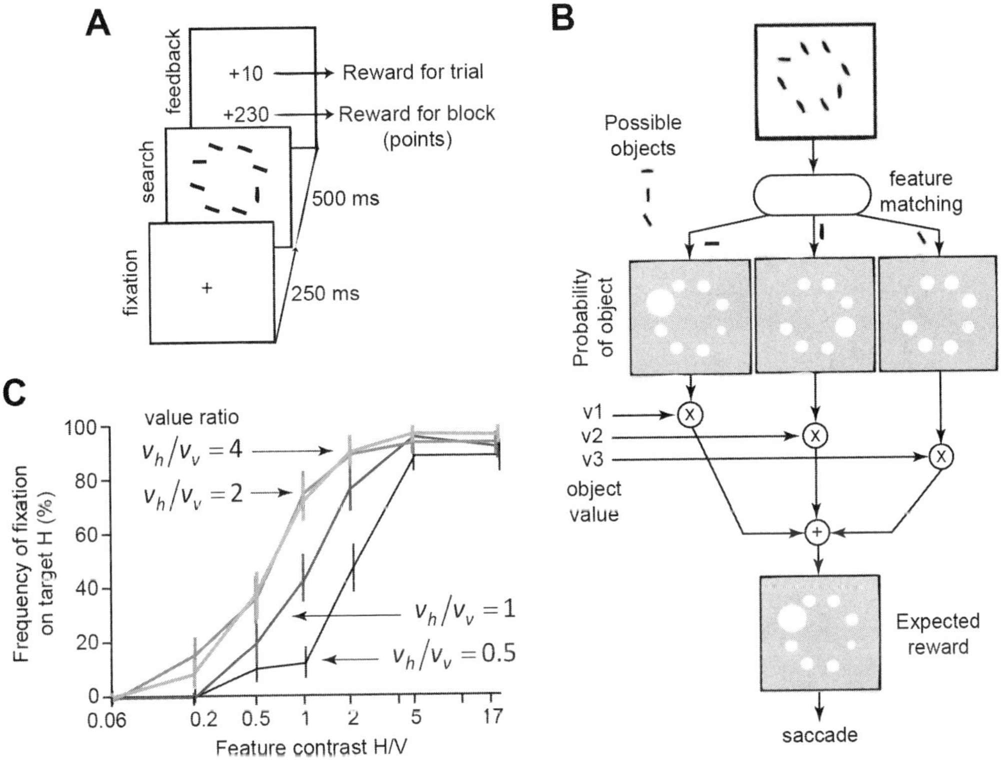

Figure 2.14
Salience of the stimulus affects the choice of the target in a reward maximization task. **A**. Horizontal and vertical stimuli were rewarded, but other stimuli were distractors. **B**. Model of decision-making. For an observation, the model assigns a posterior probability of the identity for each stimulus location. This probability is shown here via the radius of the white circle in the probability of object layer. This probability is then weighed by the value of each possible object, resulting in an estimate of reward for each location. Saccade is made to the location where this estimate is highest. **C**. A value ratio is the ratio of the value of the horizontal target to the value of the vertical target. For a value ratio, as the feature contrast for the horizontal target increases, subjects choose that target more frequently. Feature contrast refers to the ratio of the absolute value of the difference between the horizontal target and the distractors to the absolute value of the difference between the vertical target and the distractors. In our example of distractors at 5°, this ratio for the horizontal target to the vertical target is $|(0° - 5°)|/|(90° - 5°)| = 5°/85°$, or roughly 0.06, corresponding to very low salience for the horizontal target. (From Navalpakkam et al., 2010.)

the salient but irrelevant stimulus (distractor). The two properties, salience and utility, appeared to compete.

Does salience play a role in decision-making? To investigate this question, Vidhya Navalpakkam, Pietro Perona, and colleagues (2010) asked subjects to fixate on a central location and then displayed eight stimuli for 500 ms (figure 2.14A). One stimulus was horizontal, and one was vertical. These two stimuli were the task-relevant targets. The other stimuli were distractors: they were all diagonal with variable orientations. Subjects had 500 ms to choose by making a saccade to the vertical or horizontal bars. (That was enough time to make one saccade followed by fixation.) Across blocks, the value and salience of the targets were varied. For example, at the beginning of the block, the subjects were instructed that the horizontal bar was valued at 20 points and the vertical bar was valued at 10 points, so $v_h/v_v = 2$. The distractors were always valued at 0 points. To vary the salience of each target, the researchers manipulated the orientation of the distractors, and they reduced the salience of one of the targets by making the angle of the distractors closer to the angle of that target (and simultaneously farther from the other target). For example, when the distractors were at 5°, the salience of the horizontal target was very low, while salience of the vertical target was near maximum. Feature contrast in figure 2.14C refers to the ratio of the absolute value of the difference between the horizontal target and the distractors to the absolute value of the difference between the vertical target and the distractors. In our example of distractors at 5°, this ratio for the horizontal target to the vertical target is $|(0° - 5°)|/|(90° - 5°)| = 5°/85°$, or roughly 0.06, corresponding to very low salience for the horizontal target.

To optimize payoff, the naïve policy is to find the stimulus that has the highest value. However, a better policy is to find the expected reward at each stimulus location, and then choose the location that has the largest expected reward. To explain this, the first step is to find the probability that a given stimulus is horizontal, vertical, or a distractor. For example, in a field of nearly vertical distractors (figure 2.14B), the posterior probability of a stimulus at location x being labeled as the vertical target, given the measured angles at all locations, noted by $\Pr(T_x = V \mid \hat{\theta})$, is lower than when the same stimulus is placed in a field of nearly horizontal distractors. To optimize payoff, subjects should choose the stimulus location x that has the highest expected value of reward α_x:

$$E[\alpha_x] = \sum_{i=\{V,H,D\}} v_i \Pr(T_x = i \mid \hat{\theta}) \tag{2.37}$$

In this expression, v_i is the value of the vertical, horizontal, and distractor stimuli. When the subject is looking at the fixation dot and the stimuli appear, the brain might process the image and assign a label to each stimulus location, quantifying the probability that the stimulus is horizontal, vertical, or distractor. For the stimuli presented in figure 2.14B, this probability for each of the three possible stimuli is represented with a circle whose radius is proportional to its probability. The authors found that when the salience values of the horizontal and vertical targets were equal (i.e., the distractors were at 45°), and the value of the horizontal and vertical targets were also equal ($v_h/v_v = 1$), the subjects chose the horizontal target at around 50% (figure 2.14C). However, as the salience of the horizontal target increased, so did the likelihood that the first saccade would be toward the horizontal target. As the value ratio increased (horizontal target became more valuable), the probability of the first saccade being toward the horizontal target further increased. Later experiments showed that the choice was influenced by salience regardless of the type of motor response used to indicate that choice. (For example, the results were similar when the choice was indicated with the arm rather than the eyes.)

Therefore, when the values of two stimuli were kept constant, the increase in the salience of one of those stimuli led to the increase in the frequency of its selection. A model that accounted for the data was one in which people chose a visual stimulus location that maximized the expected value of the reward. In this model, the expected value was defined as a combination of its bottom-up (salience) and its top-down features (value).

Do the low-level properties of a stimulus influence how people make choices in more realistic scenarios? Milica Milosavljevic, Antonio Rangel, and colleagues (2012) asked volunteers to rank their preference among 15 snack food item (chips and candy bars) without any time pressure. This ranking served to determine their baseline preferences. After a fixation period, they were shown two different target items, one on the left and the other on the right side of the screen, each surrounded by eight other distractor snack items. All distractors and one target item were low salience (brightness set at 60%). The other target was set at high salience (brightness at 100%). As a result, in some trials, there was a conflict between the value and the salience of the two items: the preferred item was the one that was harder to find because it had low salience. After 70–500 ms of presentation (block design), the display was masked, and subjects indicated their choice via an eye movement toward their preferred snack item.

Under the easiest set of conditions (strong preference, high saliency of the preferred item, and longest exposure time), subjects chose their baseline preferred items in 90% of the trials. However, when salience of the baseline preferred item was lower than that of the chosen item, this rate dropped to around 60%. To quantitatively assess the weights assigned to salience and value, the researchers fitted the probability of choosing an item to a logistic regression model at each exposure duration. They found that at low exposure durations, salience had very strong weight. However, as exposure duration increased, the weight of the stimulus value became around 4 times the weight of salience.

Therefore, during short durations of exposure, decisions were primarily based on salience. As the length of exposure duration increased, decisions were primarily based on value. However, even during long periods of exposure, salience continued to play a small role in the decision.

In summary, choices are affected both by their economic value (as measured via subjective preference ranking without time pressure) and visual salience. The effect of salience is strongest when there is time pressure, which limits exposure to the available choices.

Limitations

In a number of experiments, the actual behavior of animals disagrees with the exact formulation of optimal foraging. Optimal foraging assumes that animal's attempt to maximize an objective function defined as the ratio of two sums: sum of all gains over sum of all times.

$$\bar{J} = \sum_{n=1}^{N} e^{(n)} \Big/ \sum_{n=1}^{N} t^{(n)} \tag{2.38}$$

In contrast, consider a different objective function, one in which the animal attempts to maximize the expected value of the capture rate $J^{(n)}$. In this case, if a single event has the utility $J^{(n)} = e^{(n)}/t^{(n)}$ and occurs with probability $p^{(n)}$, then its expected value is as follows:

$$E[J] = \sum_{n=1}^{N} J^{(n)} p^{(n)} \tag{2.39}$$

In general, $E[J] \neq \bar{J}$, except for the special condition where $p^{(n)} = t^{(n)} / \sum_{j=1}^{N} t^{(j)}$. Is the animal's objective function to optimize the ratio of sums (equation 2.38) or to optimize the sum of ratios (equation 2.39)?

Suppose that our utility is defined as ratio of sums (equation 2.38). We are given a choice of 5 units of food after a 20 s delay with certainty (fixed option) or a delay of either 2.5 s or 60.5 s with equal probability (variable option). For the fixed option, the utility is 5/20. For the variable option, the utility is (5 + 5)/(2.5 + 60.5), or 10/63. The fixed option has a higher utility.

Now suppose that our utility is defined as sum of ratios (equation 2.39). For the fixed option, the utility is 5/20. For the variable option, the utility is 0.5(5/2.5) + 0.5(5/60.5), or 62.5/60.5. The variable option has a higher utility. Therefore, the ratio of sums results in a different evaluation than the sum of ratios.

To test which is a better predictor of behavior, for each objective one can calculate the amount of reward needed or the amount of delay needed so that the two ratios are equal. This indifference point will be very different for the two theories. Melissa Bateson and Alejandro Kacelnik (Bateson and Kacelnik, 1996) presented starlings with various amounts of food and delay options and measured indifference points, then compared the predictions of the two hypotheses. They found that the choices that the birds made were better explained by the sum of ratios (i.e., $E[J]$, rather than $\bar{J}$). In another experiment, Bateson and Kacelnik (1995) found that if time to reward was variable, $E[J]$ was the better predictor. However, if the amount of food was the variable, then $\bar{J}$ was as good a predictor as $E[J]$.

Kacelnik and Bateson (Kacelnik and Bateson, 1996) have suggested that animals appear to be risk-averse to variability in reward, but risk-prone to variability in the magnitude of delay to reward. If bees are given an option between two types of flowers, one of which provides 0.1 μL of nectar and the other of which provides 1.0 μL on 10% of the trials and 0 on 90% of trial, bees are risk-averse, preferring the sure option (Waddington et al., 1981). If pigeons are presented a key that once pecked would release water after a 15 s delay and another key that variably delays the release of water but averages a 15 s delay, the birds prefer the variable delay (Case et al., 1995).

In general, when faced with variability in the amount of food, animals are risk-averse. However, when faced with variability in the delay to acquiring food, they are risk-prone. This is inconsistent with optimal foraging theory because the theory only considers the average value of the variables. What has emerged is risk-sensitive foraging theory, which assumes that the subjective value of reward is a nonlinear function of its objective value. If this function is positively accelerating, it is better to be risk-prone. If the function is negatively accelerating, then it is better to be risk-averse. This complexity in representation of risk, arising from the subjective values of reward, effort, and time, currently limits the predictive power of the theory.

There are other limitations to the theory. In a patchy environment, predictions based on the marginal value theorem generally underestimate the amount of time the animal actually stays at a food patch (Nonacs, 2001). One possibility is that there are other costs besides reward and effort. For example, a prey animal may stay at a patch in order to avoid predators, or may leave the patch in order to find a mate. Indeed, risk of predation acts like an energetic cost, offsetting reward. To test for this, Mark Abrahams and Lawrence Dill (1989) placed guppies in a fish tank and provided food at two locations, one in each half of the tank with variable distances. The two sides were separated by a screen, but on one side they placed one or two predators. (Cichlid and gourami fish feed on guppies.) Guppies could swim from one side to the other, but not the predator.

The authors measured how the guppies distributed themselves in response to amount of food that was being released at each location, and how close the location was to the screen (risk). When there was no predator, the guppies distributed themselves equally at the two locations. When the predator was introduced, if the food available at each location was equal, then a smaller number of guppies used the risky feeder. As risk increased (feeder was taken farther from the screen), the number of guppies using the risky feeder dropped. They then computed how much more food would need to be made available at the risky feeder to offset the risk: double the food for every one level increase in risk. The results demonstrated that risk of predation acted as an effort cost in units of energy in the optimal foraging equation.

Another example of a behavior that currently is hard to explain with optimal foraging is food wasting, a trait that is particularly prevalent in parrots (Sebastian-Gonzalez et al., 2019). In the wild, parrots have been observed to pick fruits, flowers, leaves, twigs, bark, and so on, from a tree, but then simply discard it. On average, they waste about 21% of the food that they pick. One idea is that this behavior is akin to a gardener who prunes the orchard by picking the unripe, smaller fruits so that the remaining fruits grow bigger. Perhaps food wasting is a form of horticultural cultivation, but that remains a speculation.

Finally, as we have noted, our choices are influenced by visual salience as well utility. In chapters 4 and 5, when we consider the neural basis of movement utility, the question of salience versus utility will come into focus, affecting our discussion of how the cerebral cortex and brainstem structures cooperate to direct our movements. We will see that salience initially plays a significant role in composing neural activity in the decision-making process, but if time is allowed for deliberation, utility becomes the dominant factor in guiding the decision of where and how to move. However, whereas the effect of utility on vigor is somewhat well studied, the effect of salience is less understood.

Summary

Life is a sequence of decisions and movements. The goodness of these behaviors can be measured via the global capture rate, defined as sum of all rewards acquired minus all efforts expended, divided by total time. Animals that make choices and produce movements that increase this rate tend to live longer. Marginal value theorem provides predictions regarding how animals should make choices in order to maximize the capture rate. An extension of this theory also predicts how animals should control their movement vigor. On the basis of the theory, when there is greater reward at stake, the animal should move faster: expenditure of energy saves time in the process of acquiring reward, and that increases the capture rate. The link between decision-making and movements arises because both are elements of behavior that the brain must control in order to maximize a single currency: the global capture rate.

In our next chapter, we will consider the observation that certain people seem to consistently move with high vigor, while others seem to move more slowly. To explore this observation, we will focus on the period before movement onset—the reaction time—as this is the period when the decision to move is taking place. We will explore whether there is a link between utility of the movement and reaction time. Because reaction time is akin to a period of deliberation, the link between reaction time and movement speed will serve as a bridge between individual preferences in decision-making and movement vigor. If we were to observe how people move toward an option, can we infer how much they value that option?

3

Reaction Time and Deliberation

The most ordinary movement in the world,
Such as sitting down at a table and pulling the inkstand towards one,
May agitate a thousand odd, disconnected fragments [of memory],
Now bright, now dim, hanging and bobbing and dipping and flaunting,
Like the underlinen of a family of fourteen on a line in a gale of wind.
—Virginia Woolf

So she was considering in her own mind … whether the pleasure of making a daisy-chain would be worth the trouble of getting up and picking the daisies.
—Lewis Carroll, *Alice's Adventures in Wonderland*

It takes time to start a movement, and this period is often much longer than the time that neurons need to detect the presence of a stimulus and trigger a movement. For example, when a visual stimulus appears to one side of the fovea, it produces activity in the neurons of the retina, which then convey signals to the neurons on the superficial layers of the superior colliculus. These neurons register the visual stimulus at a latency of 70 ms (Mays and Sparks, 1980). When the saccade-related neurons in deeper layers of the superior colliculus burst, they generate a saccade within 20 ms (Munoz and Wurtz, 1995). However, between the time that the cells in the superficial layers detect the stimulus and the time that the cells in the deeper layers burst to generate a saccade, often another 100 ms have passed. As a result, the latency (or reaction time) to initiate a saccade is around 190 ms. Why does it take so long to start a movement?

From the time the stimulus leaves its impression on the nervous system to the time a movement is made in response to it, the brain appears to accumulate a decision variable that describes the merit of performing that movement. This decision variable rises toward a threshold with a rate that depends on the stimulus properties. In one analogy, the stimulus is "sampled" at fixed intervals, and the value that is sampled is added to previous values, resulting in a rising decision variable. The movement is initiated when the decision variable reaches a threshold.

For example, when people are sitting in a dark room and see a bright light, they produce a movement in response to it earlier than if they had been prompted by a dim light. The rate of rise of the decision variable is greater for the bright light because at each moment of time, the value that is sampled from the stimulus is larger than if the stimulus had been a dim light. This difference in the rate of rise would in principle account for the observation that movements triggered by more salient stimuli (in this case, luminance) occur after shorter reaction times. Therefore, in these models, the delay in the start of the movement is related to the time it takes for the decision variable to reach threshold, and in turn, this stretch of time depends on the properties of the stimulus.

To illustrate the relationship between stimulus intensity and reaction time, consider an experiment performed by James Cattell (1886). He produced light by applying high voltage to two metal electrodes, each placed at the ends of a tube filled with mercury vapor (called a Geissler tube). The high voltage caused an electric current to flow through the gas and thus made it emit light (the basis of neon lighting). Cattell then altered the intensity of light by putting a smoked glass in front of it. He had his subjects (himself and his assistant) sit in darkness, look at the light through a telescope lens, and raise their hand as soon as they saw the light. He found that as light intensity increased, reaction time decreased (figure 3.1A). However, the data had another important feature: for each subject, at each light intensity, the distribution of reaction times had a mean that was usually larger than the median (shown for a data set gathered by Grice et al. [1979] in figure 3.1B). This implied that for a given stimulus intensity, the reaction time distribution was asymmetric, skewed toward long reaction times. (An example of the reaction times is shown in figure 3.2A).

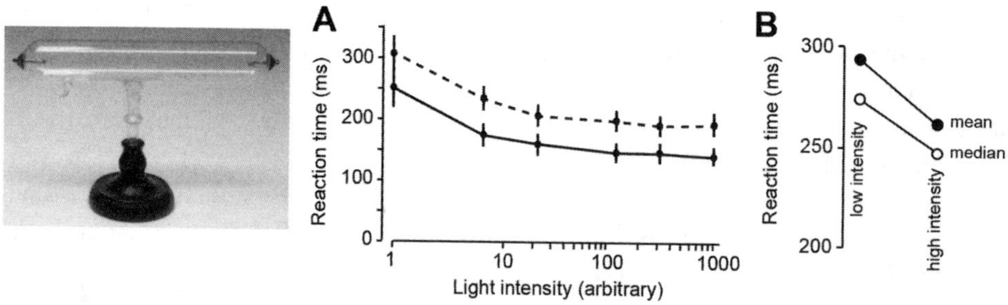

Figure 3.1
Reaction time as a function of stimulus intensity. **A**. Light intensity was varied by placing smoked glass in front of a lamp. Mean reaction times are shown for two subjects (dash and solid lines). Error bars are within subject standard deviation. The plot was made from data tabulated by Cattell (1886). **B**. Twenty subjects took part in trials in which after a warning signal, they responded as fast as possible to presentation of a light. The light had one of two intensities randomly selected during each trial. Mean and median of reaction times for each subject were calculated. The data show the mean of the means and the mean of the medians. The medians were generally smaller than the means, indicating that the distribution of reaction times was skewed. (Plot made from data from Grice et al. 1979.)

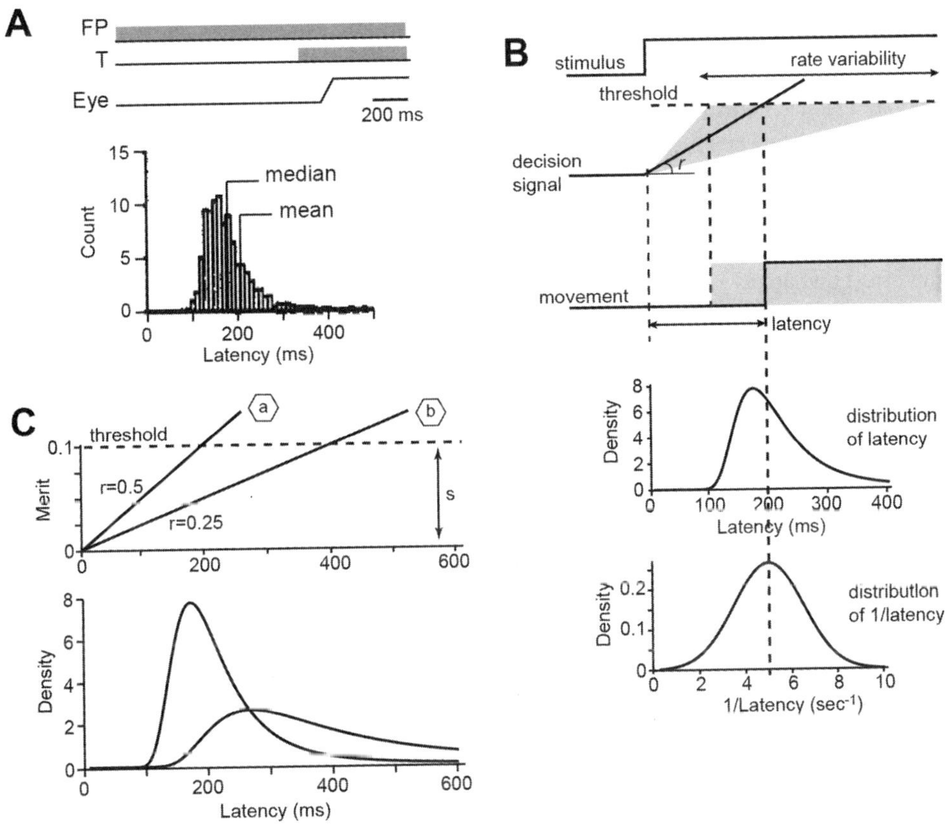

Figure 3.2
The LATER model of reaction time. **A**. Following presentation of a fixation point (FP), a target (T) was displayed in one of two locations (randomly selected) within a block of trials. Distribution of reaction times is shown for a single monkey. (From Everling et al., 1999.) **B**. The appearance of a stimulus is associated with a decision signal that increases with a constant rate. A movement is performed once this decision signal, indicating the merit of the action, reaches a threshold. If the rate is a random variable (indicating trial-to-trial variability) and normally distributed, then the latency of the resulting movement will have a skewed distribution. In contrast, the inverse of the latency will be normally distributed. **C**. Comparison of reaction times to stimulus A (high intensity) and stimulus B (low intensity). The rise rate r of the decision signal for stimulus A rises is twice as large as the rate for the signal for stimulus B; however, the variances are equal. The mean of the reaction time for stimulus B is twice as large as that for stimulus A.

To account for some of these behaviors, William McGill (1961) proposed a model. In his work, McGill was concerned with understanding the distribution of the latency associated with detecting a sound. He varied loudness of a tone that played for 0.5 s and asked volunteers to move their finger as soon as they heard the sound. Like Cattell, McGill found that the mean latency (i.e., the time needed to detect the auditory signal) was inversely related to its loudness. He then hypothesized that "the detection mechanism works by integrating a constant number of impulses coming through the auditory nervous system and that the time distribution of impulses is random with an average rate determined by the intensity of the stimulus." That is, McGill suggested that the detection of a stimulus required the integration of a rate that had a mean proportional to the strength of the stimulus. The subject reacted to the stimulus only when this integrated rate reached a threshold. This event occurred sooner for the stimulus with the greater intensity because the rate of rise was faster. The variability in reaction time came from the stochastic nature of the rate parameter.

However, reacting to a stimulus depends not only on its intensity, but also the reward and effort that are associated with the movement: people and other animals react faster to stimuli that promise greater reward and react more slowly to stimuli that demand greater effort. That is, reaction time depends not only on the sensory properties of the stimulus, but also the value of the act that the stimulus invites.

In this chapter, we will review data that demonstrate the effect of reward on the resulting movement. Researchers have performed experiments in which animals are trained to associate moving toward a visual stimulus with acquisition of reward (for example, juice). Animals saccade toward a given stimulus after a shorter delay period and with a higher peak velocity when the reward is large amounts of juice than when the reward is only a few drops of juice. When people are asked to push on a force transducer, their reaction times are longer when they are asked to exert greater forces. Thus, we move earlier and with greater velocity toward things that we value more.

Starting with a mathematical framework that accounts for the influence of stimulus intensity on reaction time, we will ask why reaction time should vary as a function of reward and effort. As before, we will measure utility of movements in terms of the reward that is at stake minus the effort that it demands, divided by the duration of time that is required. We imagine that the decision variable, the variable that is integrated to threshold, is specified by the expected capture rate. This framework will allow us to make predictions regarding how reward and effort should affect reaction time as well as movement velocity. The key idea of here is that utility not only describes how the movement should be performed (fast or slow), but also when it should be performed (sooner or later).

3.1 Latency to Start a Movement: The LATER Model

Building on models initially described by McGill (1961), Vickers (1970), and Ratcliff (1978), Carpenter (1999) described the process of reacting toward a stimulus in a framework called the LATER model (Linear Approach to Threshold with Ergodic Rate). In the model, the merit of performing an action is represented by variable x, which rises and falls through accumulation of evidence in favor of or against performing that action. If we imagine that the threshold for performing an action is set by x^*, once the merit of moving to a particular location reaches and exceeds that threshold value, the associated movement is performed (figure 3.2B).

Imagine that the evidence for performing an action rises with a rate specified by variable r. The larger this rate, the sooner the merit of this action will reach threshold, triggering the movement. Therefore, in this model, the reaction time is the latency from the presentation of the stimulus to the time t that the merit signal reaches threshold.

The model provides a way to understand the distribution of reaction times. Reaction times are not normally distributed; they have a skewed distribution, with many movements that take an unusually long time to start (figure 3.2A). Hence, the distribution of reaction times exhibits a mean that is larger than its median.

For example, consider a simple task in which the animal is holding still, looking at a central fixation point. After a period of time, a target appears at one of two locations. (The position is selected randomly from the two fixed locations on each trial.) Data in figure 3.2A display the reaction times exhibited by the subject: the distribution is skewed, with a median that is smaller than the mean.

The LATER model accounts for the skew in the distribution of the reaction times by representing the rate of rise r as a random variable with a normal distribution. Because of this, the reaction time t will have a skewed distribution with a long tail. However, the reciprocal of reaction time $1/t$ will have a normal distribution (figure 3.2B).

During the reaction time, the merit of each possible action accumulates on the basis of rate parameter r. However, there is trial-to-trial variability in r. Let us assume that r is a normally distributed random variable. To explain the relationship between rate of rise r and reaction time t, let us label the probability density of r as p_r and the probability density of reaction time t as p_t. If at start of the trial, the baseline merit for performing an action is $x(0)$ and the threshold needed to perform that action is x^*, then the reaction time is specified by the following ratio:

$$t = \frac{x^* - x(0)}{r} \tag{3.1}$$

If variables x^* and $x(0)$ are constants (and not random variables), then the probability density function for the random variable t (i.e., reaction time) can be written as follows (see appendix C on Algebra of Random Variables):

$$p_t(t) = \frac{x^* - x(0)}{t^2} p_r\left(\frac{x^* - x(0)}{t}\right) \tag{3.2}$$

If we assume r is normally distributed with mean μ_r and variance σ_r^2, we have the following:

$$p_t(t) = \frac{x^* - x(0)}{t^2 \sigma_r \sqrt{2\pi}} \exp\left(-\frac{1}{2\sigma_r^2}\left(\frac{x^* - x(0)}{t} - \mu_r\right)^2\right) \tag{3.3}$$

Let us define the threshold $s = x^* - x(0)$. We have $\mu_0 = \mu_r/s$, and $\sigma_0 = \sigma_r/s$. We can rewrite equation 3.3 as follows:

$$p_t(t) = \frac{1}{t^2\sigma_0\sqrt{2\pi}}\exp\left(-\frac{1}{2\sigma_0^2}\left(\frac{1}{t}-\mu_0\right)^2\right) \tag{3.4}$$

This equation describes the distribution of movement latency t for a rate of rise r, where r is a normally distributed random variable. The distribution of latency is skewed with a long tail (figure 3.2B). Therefore, starting from a normal distribution of rate parameter r and a constant threshold, we arrive at a skewed distribution of reaction times.

To compute the mean of the resulting distribution of reaction times, it is useful to consider the random variable $z = 1/t$ (i.e., the inverse of the reaction time). This random variable has the following distribution:

$$p_z(z) = \frac{1}{\sigma_0\sqrt{2\pi}}\exp\left(-\frac{1}{2\sigma_0^2}(z-\mu_0)^2\right) \tag{3.5}$$

Equation 3.5 implies that the inverse of the reaction time is normally distributed with a mean $\mu_0 = \mu_r/s$ and standard deviation $\sigma_0 = \sigma_r/s$, where $s = \mathrm{x}^* - x(0)$. This last result is useful because it tells us the mean of reaction time:

$$\mathrm{Mean}[p_t(t)] = \frac{x^* - x(0)}{\mu_r} \tag{3.6}$$

As the rate μ_r increases, mean reaction time falls hyperbolically. We can compute the mode of the reaction time by finding the time that maximizes the distribution $p_t(t)$. We find the derivative of equation 3.3 with respect to t and then set the result equal to 0. We find this:

$$\mathrm{Mode}[p_t(t)] = (x^* - x(0))\frac{\sqrt{\mu_r^2 + 8\sigma_r^2} - \mu_r}{4\sigma_r^2} \tag{3.7}$$

In accordance with equation 3.7, an increase in the mean rate μ_r produces a reduction in the mode of the reaction time. (The peak of the distribution shifts earlier.) Surprisingly, an increase in the variance σ_r^2 also produces a reduction in the mode.

Using this formula, we can describe the distribution of latencies for reacting to stimuli of different intensities. Suppose the intensity of stimulus A produces a rate with the following distribution, as shown in figure 3.2C: $r_A \sim N(\mu_A, \sigma_A^2)$. (The variance in the rate parameter refers to trial-to-trial variability, not within-trial variability, which is 0.) Stimulus B produces a rate $r_B \sim N(\mu_B, \sigma_B^2)$, where $\mu_A = 2\mu_B$, and $\sigma_A = \sigma_B$. The resulting distributions for the two stimuli are plotted in figure 3.2C. The subject should respond earlier to stimulus A than to stimulus B. As the rate r increases, the merit for that movement reaches threshold sooner, resulting in an earlier execution of that movement (i.e., a smaller latency). According to the LATER model, doubling the rate r should half the mean latency.

3.2 The Variable Threshold Model

The LATER model suggests that latency of a movement is due to a process in which the merit of an action (the decision variable) rises with a variable rate to a constant threshold. The roots of this idea can be traced back to the work of McGill (1961), who had hypothesized that the detection of an auditory cue depended on integrating the impulses coming through the auditory nerve until it reached a threshold.

Robert Grice (1968) was also concerned with understanding the relationship between latency and stimulus intensity. In his experiments, like those of McGill, he noted that two stimuli of different intensities produced latencies with distinct means. A stimulus of high intensity typically elicited a movement after a short reaction time, whereas a lower intensity stimulus elicited movements after a longer reaction time. He considered a block of trials in which the stimulus had a constant intensity, and he found that the reaction times were variable. He considered the variable rate model of McGill, in which the variability came from trial-to-trial randomness of the rate with which the decision variable increased. Because this rate was associated with the stimulus intensity and that intensity was constant across trials, he had difficulty imagining why the rate should be variable across trials. Instead, he considered an alternate model. He wrote, "It is relatively easy to think of a sensory input as a rather stable process, determined by the stimulus energy, temporally dense and uniform on the rather gross time scale in which reaction latency displays its variability. In other words, it is being suggested that we may alternatively regard all of the variability as residing in fluctuations of the criterion rather than in the input rate." He proposed that the variability in reaction time came not from variability in the rate of rise of the decision variable, but from variability in the threshold.

An example of the variable threshold model is plotted in figure 3.3. The illustration suggests that if the threshold is variable but normally distributed and the rate is constant, then the variability in the latency will also be normally distributed and not skewed. To see this result, let us label the threshold with random variable s and the rate with parameter r, then write the reaction time as follows:

$$t = \frac{s}{r} \tag{3.8}$$

In the variable threshold model, r is a constant, and s is a random variable with probability density p_s. The probability density for the random variable t (i.e., reaction time) becomes the following:

$$p_t(t) = rp_s(rt) \tag{3.9}$$

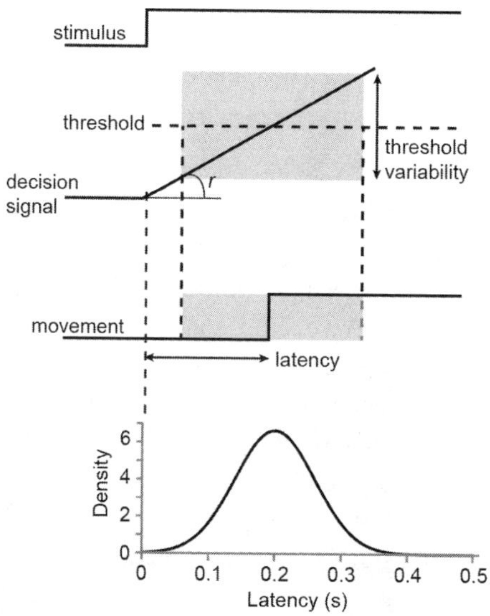

Figure 3.3
The variable threshold model of decision-making. The variability in latency is associated with a decision variable that rises with a constant rate r, but must pass a threshold that has trial-to-trial variability. In this simulation, the threshold s has mean 0.1, standard deviation of 0.03, and the rate r is 0.5. The variable threshold model produces latencies that have equal mean and median.

(See appendix C on the Algebra of Random Variables.) If we assume that the threshold s is normally distributed with mean μ_s and variance σ_s^2, we have the following:

$$p_t(t) = \frac{r}{\sigma_s\sqrt{2\pi}} \exp\left(-\frac{1}{2\sigma_s^2}\left(rt - \mu_s\right)^2\right) \tag{3.10}$$

The resulting reactions time are normally distributed.

In summary, whereas a variable rate model with a constant threshold predicts a skewed distribution of reaction times (figure 3.2B), a constant rate model with a variable threshold predicts a normal distribution of reaction times (figure 3.3). The skewed distribution of reaction times is a better match to the observed data.

3.3 Linking Latency and Utility

In chapter 2 we considered merits of each action via a quantity called utility, which we defined as the capture rate. In our framework, utility is a rate, carrying the units of energy per time, defined as the reward that is gained during the harvest period (food) minus the effort expended to acquire that reward (movement), divided by the total time required for moving and harvesting. When the movement and harvest durations are optimum, the utility achieves a maximum value at the conclusion of the harvest period. Increasing reward at the harvest site makes the utility larger, reduces the optimum movement duration, and increases the optimum harvest duration. Hence, on the basis of the framework, increasing reward should result in faster movements and longer harvests.

Here, in order to consider the period of time before movement onset, we posit that the rate parameter r in the LATER model is proportional to the capture rate that is expected for the upcoming action. That is, we will assume that before a movement is initiated, there is an expectation regarding its utility: the capture rate at the conclusion of the harvest period. During the reaction time period, this expected capture rate is integrated until it reaches threshold, at which time the movement is initiated. As a result, variables that affect the capture rate, such as the reward at the harvest site or the effort needed to move to the harvest site, will not only affect movement vigor and harvest duration, but also movement latency. Furthermore, when one is considering multiple acts, such as whether to reach for an apple or a doughnut, the utility of each act is integrated, and the utility that reaches the threshold first specifies the action that is chosen and performed.

A theoretical relationship between capture rate and reaction time.
Consider an experiment in which in trial n, a reach target is placed at distance $d^{(n)}$. After the completion of the movement, the subject can harvest reward $\alpha^{(n)}$. Our objective is to specify the capture rate that is expected at the conclusion of this action and predict how the reaction time should vary as a function of distance and reward magnitude.

From the energetic model of reaching described in chapter 1, the energy expended during the reach depends on the resting rate a_0, the duration of the reach t_r, and the reach-specific rates a_1 and b:

$$g(d,t_r) = (a_0 + a_1)t_r + b\frac{d}{t_r^2} \tag{3.11}$$

As before, we assume that once the reach is completed, the reward is harvested during period t_h with diminishing returns, incurring energetic costs. The harvest function is concave downward:

$$f^{(n)}(t_h) = \alpha^{(n)}\left(1 - \frac{1}{1+\beta^{(n)}t_h}\right) - a_0 t_h \tag{3.12}$$

For the act of reaching for and harvesting reward, the local capture rate is the reward accumulated over the harvest period minus the energy expended to arrive at the harvest site, divided by total time:

$$J^{(n)} = \frac{f^{(n)}(t_h^{(n)}) - g(d^{(n)},t_r^{(n)})}{t_h^{(n)} + t_r^{(n)}} \tag{3.13}$$

Our main idea here is that during the reaction time period, the brain integrates the utility of the act, namely its capture rate. As a result, acts that have high capture rates (for example, high reward or low effort) will be performed with a shorter latency.

To illustrate the effects of reward and effort on reaction time, we need to compute the expected capture rate at the conclusion of the reach and harvest periods. For now, let us ignore the history of the subject (i.e., the global capture rate) and focus only on this specific act (i.e., maximize the local capture rate). The reach and harvest periods ($t_r^{(n)*}$ and $t_h^{(n)*}$) are specified when the following conditions are met:

$$\left.\frac{dg}{dt_r}\right|_{t_r^{(n)*}} = -\left.J^{(n)}\right|_{t_r^{(n)*},t_h^{(n)*}}$$
$$\left.\frac{df^{(n)}}{dt_h}\right|_{t_h^{(n)*}} = \left.J^{(n)}\right|_{t_r^{(n)*},t_h^{(n)*}} \tag{3.14}$$

Figure 3.4 provides a graphical representation of the solution: the time when the line with the slope $J^{(n)}$ (dashed line) becomes tangent to g specifies the reach duration, and the time when the same line becomes tangent to $f^{(n)}$ specifies the harvest duration. The middle row in figure 3.4 describes the time course of the local capture rate: during the movement, the capture rate is negative; one is spending energy, but has not yet received any reward yet. When one arrives at the target and begins harvesting, the capture rate rises, eventually becoming positive. As reward at the destination increases, reach duration decreases (figure 3.4, left column, top plot); hence, one is willing to spend greater energy to reach the target. Importantly, as reward increases, the maximum local capture rate ($J^{(n)}$ evaluated at $t_r^{(n)*}$ and $t_h^{(n)*}$) increases (figure 3.4, left column, middle plot). That is, despite having spent greater effort to reach the target, the greater reward there allows for a higher capture rate as measured during the entire process, from the start of the reaction time through the harvest of the final reward.

Now suppose that in trial n, during the period before movement onset, the decision regarding when to start the movement depends on the amount of time it takes to integrate to threshold a random variable that rises with a rate that has a mean that is proportional to the expected capture rate $J^{(n)}$ of that movement:

$$r^{(n)} \sim N\left(k\left.J^{(n)}\right|_{t_r^{(n)*},t_h^{(n)*}},\sigma^2\right) \tag{3.15}$$

As reward increases, the expected capture rate increases, increasing the mean value of r. As a result, with increasing reward, the distribution of movement latency shifts earlier (figure 3.4, left column, bottom figure).

Therefore, if we increase the reward associated with an action, we can expect four outcomes: (1) the movement latency (i.e., time to start the movement) should decrease; (2) the movement should be made with greater velocity; (3) the harvest period that follows the

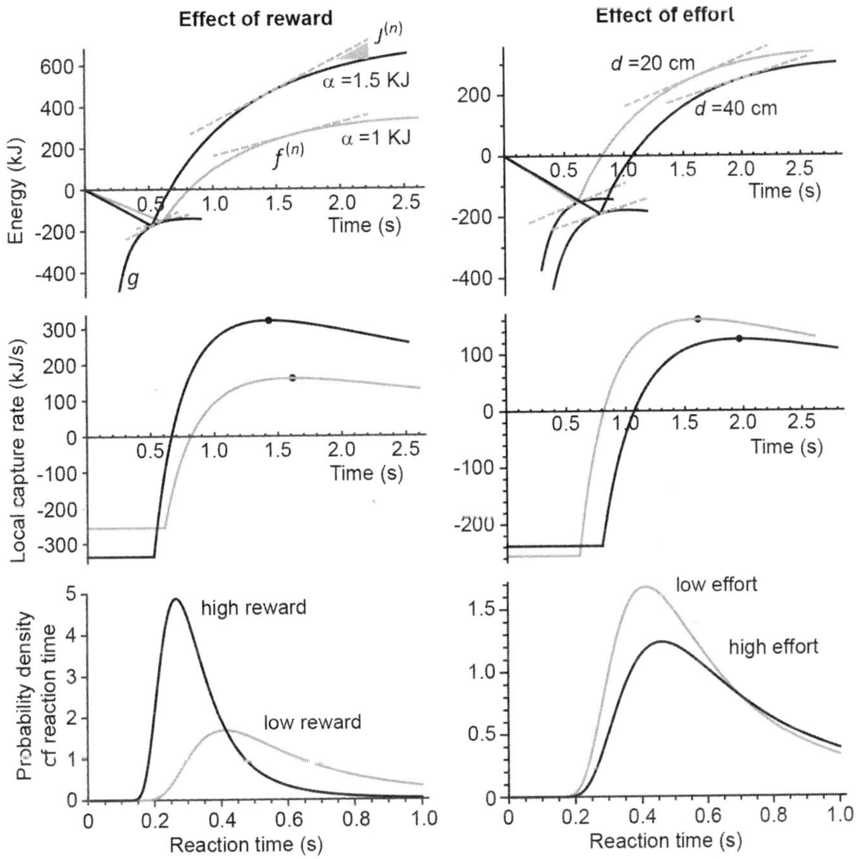

Figure 3.4
A theoretical relationship between movement latency and movement utility. In these simulations, the rate of increase of the decision signal for performing a task is assumed to be proportional to the utility of that action. Utility in these simulations reflects metabolic cost of reaching. **A**. When reward increases, the optimum duration of the movement is reduced, and the maximum utility (filled circle) is increased, which reduces the latency of the movement. **B**. When distance of the movement increases, the optimum duration of the movement increases, and the maximum utility decreases; as a result, the latency of the movement increases.

movement should be longer in duration; and (4) the utility of the movement (i.e., its expected capture rate) should increase, making it more likely that this action will be selected over other options.

As the distance to the reward site (reach distance) increases, reach and harvest durations both increase (figure 3.4, right column, top plot). That is, the increased effort required to arrive at the harvest site implies that the subject should stay longer there to harvest a greater proportion of the reward. Importantly, because of the increased effort needed to reach the reward site, the expected capture rate decreases (figure 3.4, right column, middle plot). This decrease in $J^{(n)}$ decreases the mean value of r. As a result, the reaction time will increase as distance to reward increases. That is, it will take longer to initiate a movement when the effort required for performing that movement has increased.

The history of the subject will often influence how a movement is performed: if the history is one of high capture rates, then the current act is performed with more vigor. To incorporate the influence of history on reaction time, we proceed by integrating the local capture rate with the history of those rates. As we described in the previous chapter, suppose that we have performed $n-1$ actions in the past, which have consumed total time $T^{(n-1)}$, and produced a global capture rate of $\bar{J}^{(n-1)}$. For the action that we are planning, the local capture rate is $J^{(n)}$, and its expected duration is $t_r^{(n)} + t_h^{(n)}$. The global capture rate following completion of the n-th action will be as follows:

$$\bar{J}^{(n)} = \bar{J}^{(n-1)} \frac{T^{(n-1)}}{T^{(n-1)} + t_r^{(n)} + t_h^{(n)}} + J^{(n)} \frac{t_r^{(n)} + t_h^{(n)}}{T^{(n-1)} + t_r^{(n)} + t_h^{(n)}} \tag{3.16}$$

To maximize the global capture rate, we replace the local capture rate in equation 3.14 with the global capture rate above, and again we have two equations and two unknowns (the optimal harvest and travel periods). The theory predicts that if our past actions have been successful, producing a high global capture rate $\bar{J}^{(n-1)}$, the next action will be performed with a shorter reaction time, exhibit greater vigor, and conclude with a shorter harvest period.

3.4 Reward Modulates Vigor

Saccades and reaching movements are faster toward more valuable stimuli.

Humans and other animals tend to move faster toward things that they value more. An example of this is an experiment that Reiko Kawagoe, Okihide Hikosaka, and colleagues (1998) performed on monkeys. In that work, a target was flashed at one of four possible locations; after a delay, the subject (a monkey) was instructed to move its eyes to the remembered location of the target. (The cue to move was removal of the fixation point.) The result was a memory-guided saccade.

In the first block of trials, one of the four locations was consistently paired with reward (juice), so the animal knew which target was paired with reward. During the subsequent block, the rewarding target was changed. (The animals made the saccade to the non-rewarding target because by doing so, they could proceed to a future rewarding trial.) The peak speed of the saccadic eye movement was higher when that action was paired with a reward than when it was not (figure 3.5A). In addition, saccade latency was shorter when the target was associated with a reward (figure 3.5B). We can conclude that in monkeys, expectation of reward reduced the latency of the saccade and increased its velocity. If we define vigor as the reciprocal of the sum of reaction time and movement time, we see that expectation of reward increased vigor.

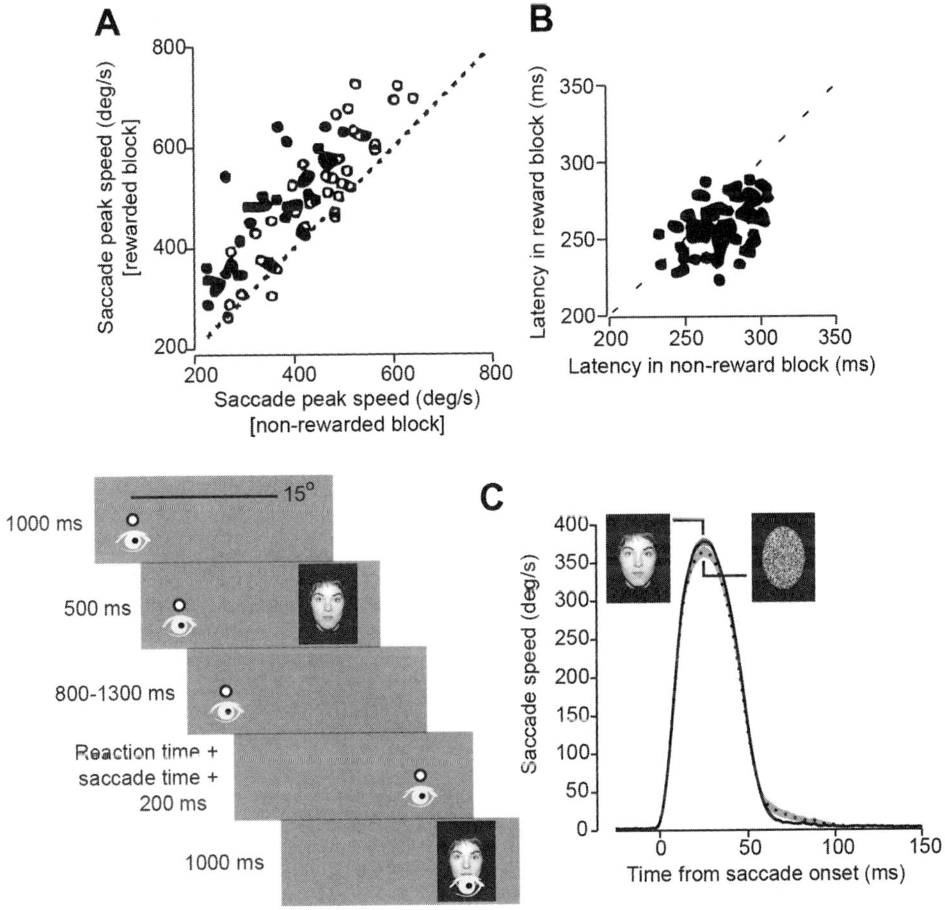

Figure 3.5
Effect of reward on saccade velocity and latency. **A**. Four target locations were presented during a block of trials, but in a given block, only one location was paired with a reward. Saccades were memory guided: target was flashed, and then after a delay, the fixation spot was removed, instructing the monkey to make a saccade to the remembered location of the target. Data illustrate peak velocity of saccadic eye movements in blocks of trials for which the target of the movement was associated with reward (juice) versus blocks of trials for which the same target was not associated with reward. Data are shown for two monkeys (Kawagoe et al., 1998). **B**. Average movement latency for saccades made to rewarded targets versus non-rewarded targets. Each dot is for a single experiment. (From Takikawa et al., 2002.) **C**. Saccadic eye movements were made to a spot of light. After the completion of the movement, the subjects (humans) were rewarded with either a picture of a face, an upside down face, an inanimate object, or noise. Saccades that were made in anticipation of seeing a face were fastest. (Data from Xu-Wilson et al., 2009b.)

In the real world, we do not make saccadic eye movements to get juice. Rather, we move our eyes to place the part of the visual scene that we are interested in on our fovea, thereby allowing us to harvest information. Do we make faster saccades to things that we value more? We explored this idea with an experiment (Xu-Wilson et al., 2009b) in which humans maintained gaze at a fixation point while a small image appeared for 500 ms at a distance of 15°. That image could be a face, an inverted face, an object, or simply noise. Subjects were not allowed to look at the image. Rather, they maintained gaze at the fixation point until it disappeared, and then reappeared where the image had been. This was the cue to move their eyes to the new fixation point. Once they completed their eye movement, they were rewarded with the image, at which they were allowed to gaze for 1 s. Therefore, the saccades were made to single points of light, but with the promise that after the saccade and a delay, the subjects would view a face or another image. In this way, like the monkeys in our previous example, the humans made saccades, but sometimes were rewarded with the image of a face, while other times were presented with the unrewarding image of noise. The saccades that were made in anticipation of viewing a face were fastest (figure 3.5C).

Increased reward also affects vigor of reaching movements. Erik Summerside in our lab (Summerside et al., 2018) asked human subjects to make out-and-back reaching movements to one of four quadrants (figure 3.6A). Each trial began with the presentation of a large circle and a marker on one of its quadrants that indicated the goal. The sole criterion of success was that the hand should cross the ring within a 100° arc centered on the marker. As the reach began, visual feedback of the hand was blanked. Once the hand crossed the outer ring within the quadrant, the outer ring changed color, indicating that the trial was complete and that the hand should be brought back to center. There was no time limit to complete the reach, and no instructions were provided regarding a desired reach velocity. Importantly, during various periods of the experiment, only one of the four quadrants was paired with reward, and the transition from reward to non-reward status of a quadrant occurred randomly. The reward was not money or food, but a pleasing sound, a visual animation of the ring at the moment that the invisible hand passed through it, and points that accumulated but were not associated with money. Thus, the experiment provided only an abstract reward, with effectively no dependence of the reward on accuracy of the reaching movement (because the target was so wide). This allowed the subjects considerable latitude in self-selecting their movement speed, amplitude, and variability.

We found that when a quadrant was paired with reward, the reaching movement toward it occurred after a shorter reaction time, had a higher peak velocity, and had a greater amplitude (figure 3.6B). That is, people started their reach earlier and moved with greater vigor in anticipation of receiving the rather modest reward of a pleasing sound. Furthermore, movements toward the rewarded quadrant were performed with less variability: both reaction times and reach kinematics (hand cross-point on the outer circle) were less variable when there was expectation of reward.

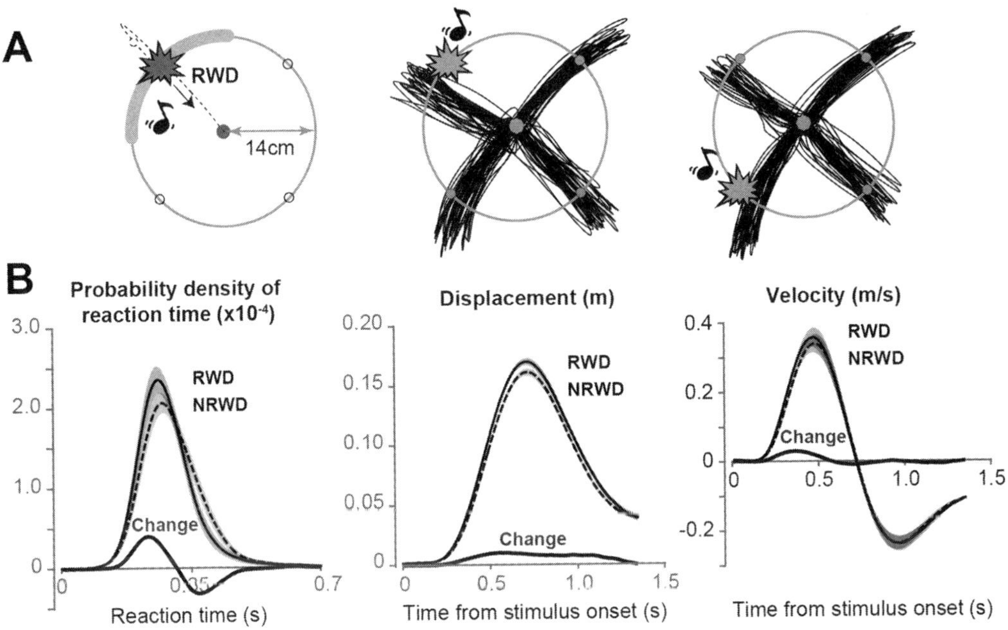

Figure 3.6
Effects of reward on reach velocity and reaction time. **A**. Participants completed out-and-back reaches to one of four alternating targets located 14 cm from the home circle. Each block of trials had one target paired with reward (indicated by quadrant with shaded gray region). The reward (RWD) consisted of an exploding target, auditory stimulus, and four points. The curve labeled change is the average of within-subject difference in the reward and non-reward (NRWD) reaction time distributions, displacement, and velocity. **B**. Probability density of reaction time shifted earlier for rewarded trials. In addition, the participants reached farther (displacement plot) and with higher velocity in the rewarded trials. (From Summerside et al., 2018.)

There are more ecologically relevant versions of this experiment in which reward has objective value. Kareisha Sackaloo and colleagues (2015) asked human volunteers to reach for a candy bar, pick it up, and then bring it back to a start location. Various candy bars were presented in random order, and after the reach trials, the volunteers completed a survey describing their preference for the bars. The authors found that reach duration was shortest for the most preferred candy bar and longest for the least preferred bar.

Reaction times are shorter toward more valuable stimuli.

In our framework, as the expected capture rate increases, the reaction time for initiating that movement should become shorter. One way to increase this capture rate is through the expected value of the stimulus: the probability of the stimulus occurring at a given location, multiplied by its reward value.

David Milstein and Michael Dorris (2007) designed a task to assess the influence of the expected value of a stimulus on the latency of the ensuing movement (figure 3.7A).

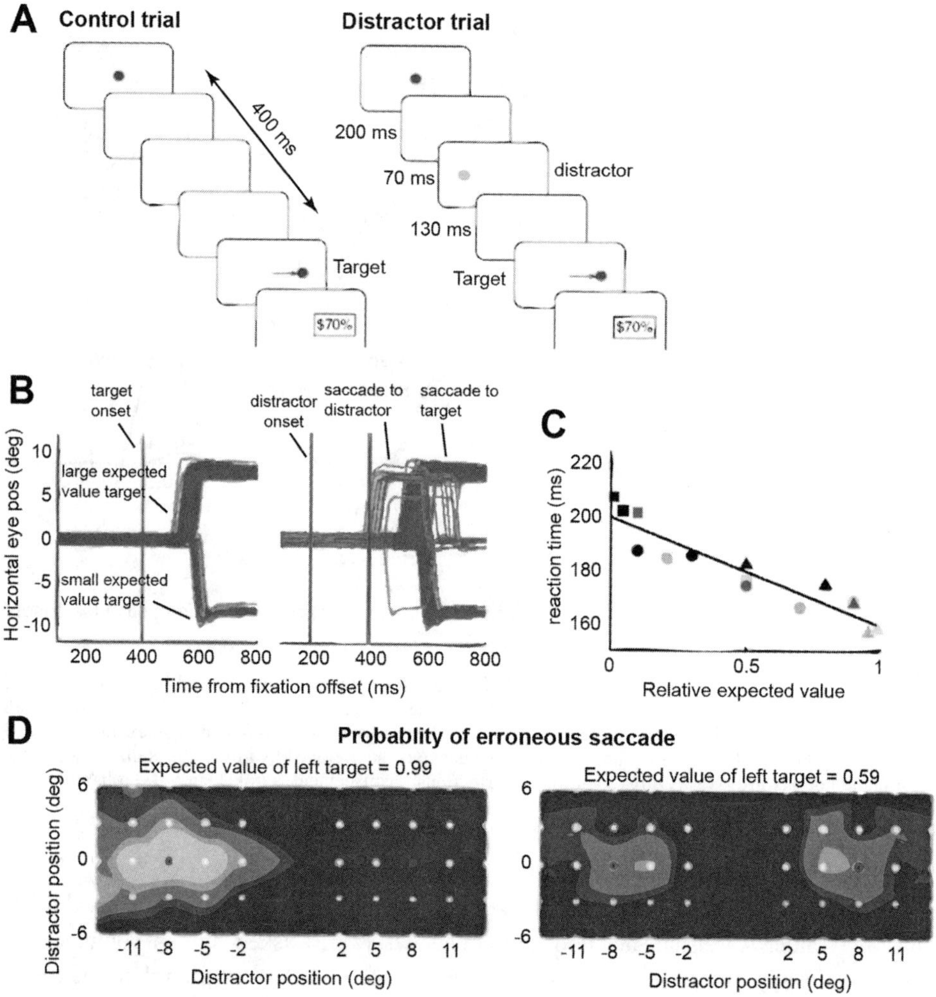

Figure 3.7
Effects of expected value of a stimulus on latency of the saccade. **A**. People were asked to produce a saccade in response to a red stimulus that appeared in one of two locations. The probability of stimulus location (left or right) and its reward value were manipulated in blocks to produce variation in the expected value of the action. In some trials, a distractor green stimulus appeared to increase the probability of an erroneous movement. **B**. The left figure shows trials that only had a red stimulus (presented at 400 ms). When the stimulus had a larger expected value, the latency of the movement was shorter. The right figure shows trials that had a green stimulus (presented at 200 ms) followed by a red stimulus (presented a 400 ms). Presentation of the green stimulus occasionally induced an erroneous saccade to the distractor target, but this error was much more likely to occur if the green stimulus was located at a place where a high-valued target was likely to appear. **C**. Saccade latency in correct trials as a function of stimulus expected value. Larger expected value coincided with smaller movement latency. **D**. A heat map representing the probability that the distractor stimulus produced an erroneous saccade. The location of the distractor stimulus is indicated by the white dot. (From Milstein and Dorris, 2007.)

Subjects viewed a fixation point that disappeared. Then after a 400 ms delay period, a red stimulus appeared either to the right or left. Subjects made a saccade to the red target. The expected value of the target was manipulated across 15 blocks of trials by varying two parameters: the probability of the target presentation in the left portion of the workspace (this probability was varied over 3 levels), and the magnitude of monetary reward (this magnitude was varied over 5 levels), resulting in 15 conditions. In this way, the experiment controlled the expected value of each stimulus (probability of the stimulus appearing at a particular location, multiplied by its reward magnitude). The authors found that when a target location had a large expected value, the saccade to that location had a shorter latency, as shown in the left panel of figure 3.7B. Indeed, as the expected value of the stimulus increased, the latency decreased linearly (figure 3.7C).

The experiment required the subjects to withhold making a movement until 400 ms had passed after the extinction of the fixation light, then make a movement when the red target appeared in one of two locations. In some trials, at around 200 ms, a green distractor stimulus appeared for 70 ms at various locations. This was to probe whether during the delay period the brain was preparing a movement toward one or both possible targets. If so, the closer the location of the green distractor to the location of the expected red target, the greater the chance that the distractor would trigger a premature movement. An example of a distractor trial is shown in the right panel of figure 3.7B. The distractor appeared at 200 ms, triggering a saccade at around 420 ms.

The probability that the distractor would trigger a saccade depended on whether it was located near one of the two positions where a red target was likely to appear. Importantly, this probability of triggering an erroneous saccade (figure 3.7D) also depended on the expected value of the red target. When the expected value of the left red target was large (top row, figure 3.7D), presentation of a distractor target to the right was ineffective in triggering an erroneous saccade, but presentation of a target to the left was quite effective. When the expected value of the left red target was about the same as the right red target (bottom row, figure 3.7D), presentation of a distractor target at both left and right was effective in triggering an erroneous saccade.

One way to view these results is to imagine that during the delay period, activity increases in neurons that would ultimately need to burst to produce a saccade to the target on the right. Similarly, activity increases in neurons that would need to burst to produce a saccade toward the target on the left. Suppose that the rate of rise of this activity depended on the expected value of each target. When one of the two targets was actually shown, it provided the evidence needed to push the merits of making that movement beyond the threshold. In our next chapter, we will consider the neurobiology of saccadic eye movements and provide evidence for this idea through recordings from neurons in the superior colliculus.

These experiments illustrate that one factor that influenced the vigor of movements was the reward that was expected to be harvested after the completion of the movement. Generally, the higher the value of this expected reward, the greater the movement velocity. The experiments also illustrate that this same factor influenced latency of the movement: the higher the expected value of reward, the shorter the latency.

The results have some real-world implications. If you see a dear friend across the street, the utility for the steps you are about to take toward your friend are higher than if you are about to walk toward someone whom you may not be so fond of. As a result, you will walk faster toward your friend. Reward makes it worthwhile to spend the additional energy so you move faster, enabling you to harvest the pleasure of your friend's company sooner.

Vigor appears to depend mainly on utility, not motivational salience of the stimulus. We may interpret these data in a different way. Perhaps expectation of reward modulates attention or another nonspecific characteristic of the trial. For example, suppose that performing the correct movement results in avoiding an air puff to the face, an example of an aversive stimulus. One would need to pay attention to this aversive stimulus and perform the task correctly in order to avoid it. However, the value of making a movement to not receive an air puff is clearly lower than making that same movement and receiving juice. That is, the utility of avoiding an air puff should be lower than the utility of drinking juice.[1] However, both of the stimuli that signal these consequences require attention. Does vigor reflect utility or attention?

Shunsuke Kobayashi and colleagues (2006) trained monkeys to perform a memory-guided saccade task. The animals viewed a central cue (for example, square, rectangle, or circle) that on each trial determined what would happen if the correct movement was performed: get reward (apple juice), avoid punishment (air puff), or simply hear a sound. After the cue, a target was flashed, and then after a delay, the animal made a saccade to the remembered location of the target. If the saccade was to the correct location, the animal received juice, avoided an air puff, or heard the sound, depending on the cue that had appeared earlier.

Motivational requirements of the trials were highest for reward trials, somewhat lower for punishment trials, and lowest for neutral trials, as evidenced by the fact that correct performance rates were highest in the juice trials and lowest in the sound trials (figure 3.8). This suggests that the monkeys paid a lot of attention during the air puff and juice trials, and less during the neutral trials. In contrast, the utility of the movement was highest for acquiring juice, but lowest for avoiding air puff. (The reason for this is that in the air puff trials, if they performed them correctly, they avoided a mild negative consequence, but for the other two types of trials, they received something, a sound or juice. We are assuming that hearing a sound is in some sense better than nothing.) Kobayashi et al. (2006) found that saccade peak velocity was highest for the reward trials, lowest for air puff trials, and in between for

1. Suppose you are given the option of either reaching and receiving $1 or reaching and avoiding loss of $1. Which option would you prefer? Most would pick the option that involves reward.

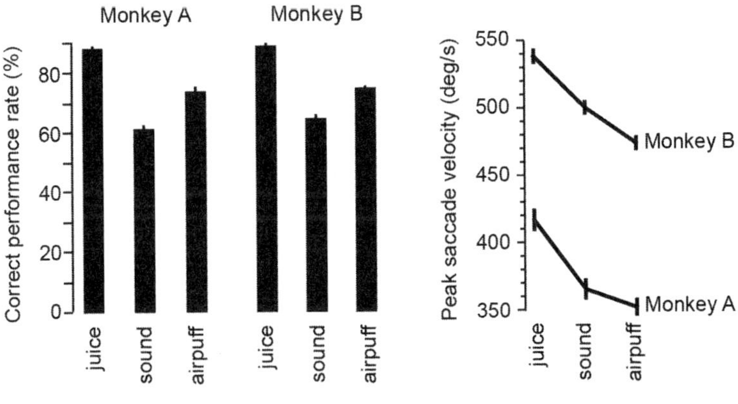

Figure 3.8
Effects of reward and punishment on saccade velocity. Monkeys performed a memory-guided saccade to the remembered location of a target. The cue appeared at a central location and specified whether after the correct completion of the trial, the animal would receive a reward (juice), avoid a punishment (air puff), or hear a sound. The left subfigure shows the correct performance rate in each trial type. This measure serves as a proxy for attention or arousal state of the animal. Animals paid more attention during the juice and air puff trials, less attention during the sound trial. The right subfigure shows saccade velocity in each trial type. The peak velocity was the highest for the trials that provided the most reward (juice). (Data from Kobayashi et al., 2006.)

neutral trials (figure 3.8). This suggests that modulation of vigor is not only a result of motivational salience, but at least partly a reflection of the utility of the movement.

3.5 Effort Modulates Vigor

Suppose a reward of a certain value is associated with completion of a movement. As the target of the movement is placed farther away, the effort associated with the movement increases, thus reducing the utility of the action. For example, in figure 3.4B we have the utility of a reaching movement. As the distance of the reach target increases, the maximum utility decreases. Our framework predicts that reaction time should increase as target distance increases.

The brain needs more time to start an action that requires greater effort.
There is evidence that an increase in effort associated with an action is associated with an increase in reaction time. We measured eye and head movements of people as they reached toward visual stimuli that appeared at various distances (Reppert et al., 2018). Generally, their movements started with a saccade to the target and was then followed by movement of their arm. (Data for a typical subject is shown in the left panel of figure 3.9A.) As the target distance increased, requiring greater effort for the movement, saccade and reach reaction times both increased (right panel of figure 3.9A).

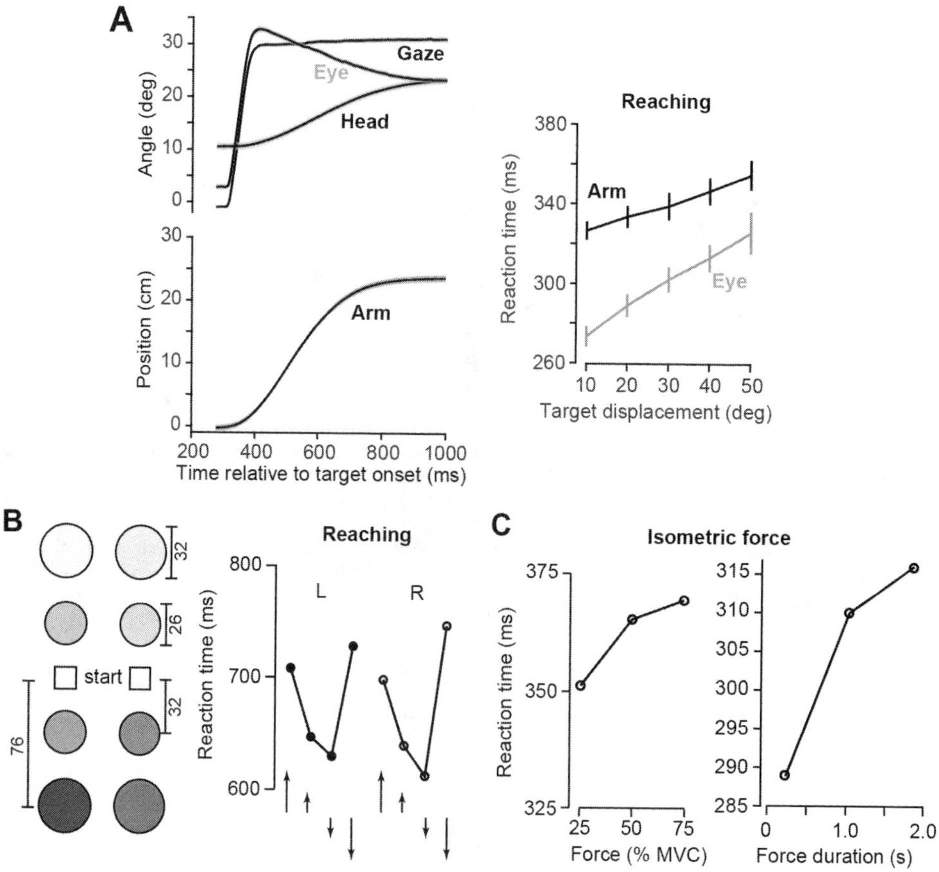

Figure 3.9
Effect of effort on latency of reaching and saccades. **A**. Effects of target distance on reaction time of reaching and saccades. Subjects reached for a stimulus that was positioned at various distances. Left panel shows data from a typical subject. In response to the stimulus, the eyes made a saccade, and then the arm started to reach. Gaze is defined as eye position (in head) plus head position. Right panel shows data across subjects. Reaction times for both the eyes and the arm increased with distance to target. (From Reppert et al., 2018.) **B**. Healthy individuals were asked to put their right and left index fingers on the start location and reach for targets that had unique colors. On each trial, a colored cue at the start location identified the target for that trial. The data show reaction time as a function of target location for the left and right arms. The farther the goal of the movement, the longer the reaction time. (Data from Rosenbaum, 1980.) **C**. Effect of force magnitude and duration in isometric tasks performed with the arm. Left panel: mean reaction time across healthy subjects who were instructed to produce elbow flexion force pulses of various amplitudes. MVC: Maximum voluntary contraction. (From Stelmach and Worringham, 1988.) Right panel: mean reaction time across healthy subjects who were instructed to produce forces with their index finger of constant magnitude but various durations. (From Ivry, 1986.)

David Rosenbaum (1980) asked right-handed volunteers to put their right and left index fingers at a start location (figure 3.9B, left) and then await instructions on where to reach. During each trial, for each hand, there were four possible targets, two near the subject and two farther away. Each target had a unique color. At trial onset, a colored cue appeared at the start location that matched one of the targets, thereby instructing the subject regarding which arm should be used and the direction and distance of movement. Importantly, in this study, the information regarding the movement goal was not presented at the target location, but at the start location. Regardless, reaction time for the reaching movement was shorter when the movement goal was closer to the start position than when the goal was farther away.

George Stelmach and Charles Worringham (1988) instructed people to produce isometric flexion forces at the elbow at various magnitudes (25%, 50%, or 75% of their maximum voluntary force). The subjects viewed a graph that described the required force, then after a random period, heard a tone that served as their cue to begin. Upon hearing this, they generated a contraction of sufficient magnitude so that the peak force reached the required target. They found that as the force requirement became greater, the latency period became longer (figure 3.9C, left subfigure).

Richard Ivry (1986) considered a variant of this task in which subjects were instructed to produce a constant force (7.5 N) with their right index finger for variable periods of time. The required duration was specified before each trial, and a tone served as the cue to begin. The results showed an increase in reaction time as a function of force duration (figure 3.9C, right subfigure).

Hiroshi Nagasaki and colleagues (Nagasaki et al., 1983) asked people to respond as quickly as possible to a tone by extending their elbow against one of three levels of force: small, medium, and large. No instructions were provided regarding movement speed or extent. Reach acceleration increased with the level of force that was used, whereas reach reaction decreased as the level of force increased.

A simple way to alter the utility of reaching is by varying the mass that one has to carry. As we noted in the previous chapter, the primate arm has a mass distribution that resembles a heavy object when it moves in some directions (major axis of the inertia ellipse, figure 3.10A) and a light object when it moves in other directions (minor axis of this ellipse). We can estimate the effective mass of the hand as a function of direction of reach, resulting in quantity $m(\theta)$, where θ is direction of reach (shown via the ellipse centered on the hand in figure 3.10A). (See appendix A, Effective Mass of the Human Arm.) For the arm in the configuration of figure 3.10A, the effective mass is smallest for reaches toward 55°, and largest for reaches toward 145°. Therefore, for a constant amount of reward and a constant distance to the target, the utility of the act is largest when reaching toward 55°, and smallest when reaching toward 145°. In general, the mass ellipse at the hand has its major axis aligned with the forearm, and therefore as one places the arm at various configurations, the mass ellipse rotates with the elbow angle. You can see the

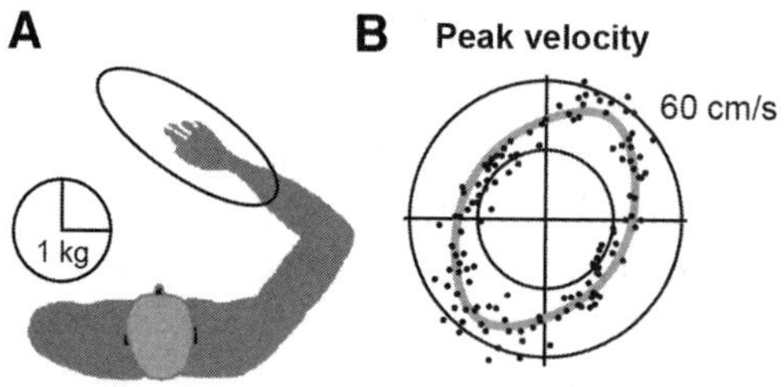

Figure 3.10
Effect of effort on vigor of reaching movements. **A**. Mass properties of the human arm when held in the horizontal plane. The arm has an effective mass that is largest for movements along the direction of the major axis of the ellipse. The effective mass is smallest for movements along the direction of the minor axis. **B**. Peak velocity of reaching movements to targets at 10 cm for various directions. (Data are from Gordon et al., 1994.) Ellipse is model fit from a utility function in which effort grows with the metabolic cost of the reaching movement. Reaching movements are fastest in the direction of lowest effective mass.

consequences of the different effort requirements of reaching in the data presented in figure 3.6A: the subject reaches farther when the target is at the 45° or the 225° quadrant (small effective mass) than when the target is at the 135° or the 315° quadrant (large effective mass).

James Gordon, Felice Ghilardi, and Claude Ghez (Gordon et al., 1994) asked their subjects to place their hand on a horizontal digitizing tablet and make a "single, quick, and uncorrected movement" to a target at 10 cm. The subjects had no time constraints on their movements. They chose to reach with a peak velocity that was around 60 cm/s for some directions, but only 35 cm/s for other directions (dots, figure 3.10B). Hence, there was a large increase in the preferred movement speed because of a change in the direction of the movement. Movements were fastest in directions for which the effective mass of the arm was smallest. (Unfortunately, the reaction times of these reaching movements were not published.)

Taken together, the available data suggest that under a constant reward condition, the brain needs more time to start an action that requires greater effort. Furthermore, as the effort requirements increase, the velocity of the movement is likely to decrease. Increasing effort reduces the utility of the movement. If we imagine that, during the reaction time period, the brain is integrating a decision variable that depends on the movement utility, a lower utility leads to a longer period needed for this decision variable to reach threshold. This in turn results in increased latency to start the movement.

3.6 Rate of Reward Modulates Vigor

When our movement concludes and we arrive at the reward site, we take time to harvest that reward. For example, when you walk to meet a friend, you harvest the reward of being with your friend over the period that you decide to stay with him. When you make a saccade to place an image on your fovea, you harvest the information that that image offers during the period of postsaccadic fixation. Now imagine walking with the goal of meeting your friend, but knowing that he only has a few minutes to spend with you. That is, the harvest period t_h is not controlled by you, but by your friend. Knowing this limitation on the harvest duration, how should you move?

Increased rate of reward produces increases in saccade vigor.

If we are repeatedly provided with opportunities to harvest reward, but the time available for each harvest is brief, then in order to maximize our global capture rate, we should move quickly between reward sites. To illustrate the reason for this, we begin with the assumption that the total amount of reward at a given patch is finite, and we need to spend effort to acquire it (e.g., through metabolic cost of holding still while we gaze at the image). This implies that the harvest function $f(\alpha, t_h)$ is concave downward. That is, with passage of time at the reward site, we acquire a greater fraction of the reward α, but with diminishing returns and increasing effort. For a concave downward harvest function, the presentation of two rewards of equal magnitude in sequence during period $T=2t_h$ will result in a greater harvest per unit time than just one reward during the same period T (figure 3.11A). That is, a concave downward harvest function implies the following inequality:

$$f(\alpha, 2t_h) < 2f(\alpha, t_h) \tag{3.17}$$

By controlling t_h (the harvest duration), we are effectively modulating the rate of reward and, therefore, the capture rate. In this scenario, repeated exposure to reward but with reduced t_h at each harvest location should produce an increase in $\bar{J}$, which in turn should lead to an increase in movement vigor. If during reaction time, one integrates a variable proportional to the utility of the movement, then there should also be a reduction in reaction time.

However, if during image gazing, the harvest function is linear with time, then $f(\alpha, 2t_h)=2f(\alpha, t_h)$. In this case, changes in t_h do not affect the rate of reward. With a linear harvest function, changes in harvest duration should have no effects on movement vigor. Finally, if the harvest function is concave upward with time, then $f(\alpha, 2t_h)>2f(\alpha, t_h)$. In this case, reductions in t_h reduce the rate of reward; thus, movement vigor should decrease.

To test these alternatives, we presented subjects with a small image somewhere in the periphery (Yoon et al., 2018). As soon as their saccade concluded and they had placed that image on their fovea, we controlled the amount of time t_h (gaze duration) that they had for gazing at that image (figure 3.11B). In some blocks of trials, t_h gradually increased from trial to trial, while in other blocks, t_h gradually decreased (or remained constant, figure 3.11C). As t_h decreased, people increased their saccade velocities and reduced their reaction times. That is, when the history of trials indicated that the period available for harvesting of reward was short, the movement started sooner in response to stimulus presentation, and the eyes moved faster to the reward site. These results were consistent with the idea that during image gazing, the harvest function was concave downward.

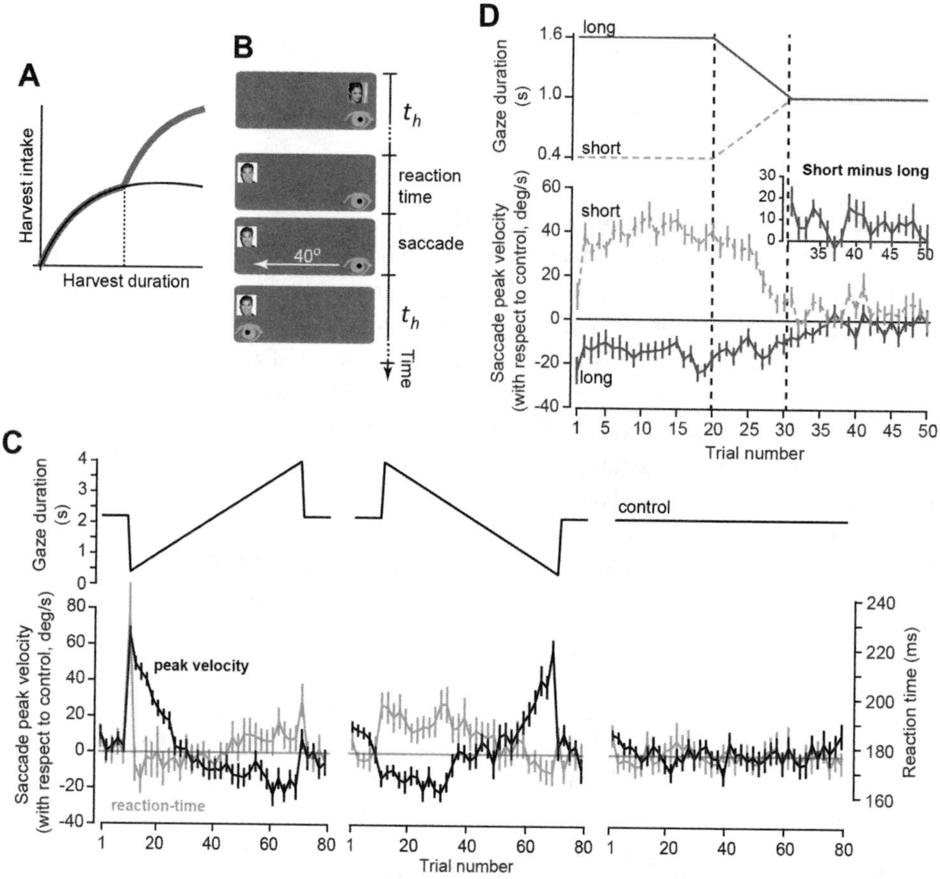

Figure 3.11
Effects of harvest duration on vigor and latency of saccades. **A**. If the harvest function for image gazing is concave downward, then presentation of two images in sequence during period $T=2t_h$ will result in a greater harvest per unit time than that of just one image during the same period T. As a result, reducing t_h should lead to an increase in $\bar{J}$, which in turn should lead to an increase movement vigor. **B**. Experiment design. We presented a small image of a face, located at ±20° with respect to the midline. After the subjects made a saccade to it, we controlled the duration t_h for which they were allowed to gaze at that image before another image was presented. **C**. Harvest duration strongly affected saccade vigor. We measured saccade peak velocity of each subject with respect to that subject's average velocity, which was measured during the control period (constant t_h, right subfigure). An increase in t_h coincided with reduction in saccade peak velocity (left subfigure). A reduction in t_h coincided with increase in saccade peak velocity (middle subfigure). However, increases in reaction time coincided with increases in saccade peak velocity and reductions in the former coincided with reductions in the latter. **D**. History of harvest-modulated saccade vigor. Subjects experienced a history of short or long harvest durations. They were then tested in identical harvest conditions (control trials 31–50). We measured within-subject change in saccade peak velocity with respect to the control trials. Saccade peak velocity was high during the short harvest trials and remained high in the control trials (inset figure, within-subject change in peak velocity during control trials, short harvest versus long harvest). (From Yoon et al. 2018.)

Given this shape for the harvest function, we used the marginal value theorem to predict that after a history of brief harvests t_h, which would result in a high global capture rate ($\bar{J}$), vigor should remain high regardless of the local conditions. That is, vigor should be influenced by history of previous harvests: after a few good harvests, the global capture rate of the subject should be high. As a result, vigor should remain high for the movements that follow, regardless of the local conditions in which those movements are performed.

An experiment was then performed to verify the equation-based results. During some blocks, harvest duration t_h was long for 20 trials (low rate of reward) and then gradually decreased, whereas during other blocks, t_h was brief for 20 trials (high rate of reward) and then gradually increased. All blocks ended with control trials in which $t_h = 1$ s. Long harvests should produce a low rate of reward, which in turn results in smaller global capture rate $\bar{J}$. This should reduce vigor in subsequent control trials. Indeed, the results of the experiment qualitatively confirmed the theoretical predictions: during the control trials, saccade velocity was higher and reaction times were shorter after a history of short harvests (inset of figure 3.11D, short minus long).

A limitation of this work is that the experiment design not only varied the rate of reward by manipulating harvest period t_h, but also varied the rate of saccades that the subjects had to make to harvest those rewards. The increase in the saccade rate implies an increase in the effort expenditure, which reduces the capture rate. This should dampen changes in vigor and reaction time. Despite this, there appeared to be robust increases in these variables. In figure 2.12D we noted the paradoxical result that following expenditure of high effort, movements appear to have greater vigor toward control stimuli. This raises the question of whether the effects of history that we are seeing in figure 3.11D are a reflection of past reward or past effort. The answer remains unknown at this point.

In summary, sometimes we cannot control how long we can stay and harvest reward. Rather, an external agent sets this duration for us. In this case, if the harvest function $f(t_h)$ is concave downward, then two harvests in sequence produce a greater amount of harvest per unit of time than one harvest, a result that leads to a larger capture rate. Yoon et al. (2018) controlled t_h and observed that people responded by changing their saccade vigor: increasing their vigor when gaze duration was short (high rate of reward) and decreasing their vigor when gaze duration was long (low rate of reward). Furthermore, after experiencing a history of short gaze durations (high rate of reward), people continued to move their eyes with greater vigor in the control trials, illustrating that the vigor indeed was affected by the history of capture rates.

3.7 Deliberation and Decision-Making

The tasks that we have considered thus far are of the variety in which there is a single stimulus instructing a movement. The model that we have discussed is called *bounded integrator*, and it is concerned with simple reaction time experiments that involve a single stimulus and its ensuing movement. In our framework, reaction time is due to integration of a rate variable that has a mean proportional to the utility of the movement. With this assumption, we were able to consider the fact that reward and effort characteristics of the movement affect the reaction time.

A critical part of our model was the assumption that whereas this rate (i.e., merit of the movement) could vary between trials, it was constant during the reaction time within a given trial. This allowed us to account for the long-tail distribution of reaction times. However, in our current formulation, this framework cannot account for scenarios in which the reaction time is a period of deliberation, during which information is acquired and compared before coming to a decision. Here, we will expand the framework to include deliberation during decision-making, and then we will consider examples in which the reward and effort associated with the subsequent movement appear to alter the decision itself.

Bounded integrator models of decision-making

A good place to start is an experiment in which subjects are asked to detect not the onset of a stimulus, but rather the difference between two stimuli. Douglas Vickers (1970) showed people two vertical lines and asked them to decide which line was longer. To account for the measured data (reaction time and error rate), he imagined that each line was represented as a sensory stimulus that had a Gaussian distribution with a mean and variance. At each moment of time, the subject would sample the length of line 1, producing an instance of the random variable r_1 from the normal distribution $N(\mu_1,\sigma_1^2)$, and then sample the length of line 2, producing an instance of the random variable r_2 from the normal distribution $N(\mu_2,\sigma_2^2)$. The subject would then compare these two samples. When the comparison at a given moment of time indicated that line 1 was longer, the decision variable for that line increased by one quantum. But, at that moment of time, if the comparison indicated a length difference that was in favor of line 2, then the decision variable for that line increased by one quantum. The two decision variables would race each other, and whichever first reached a set threshold would be declared the longer line on that trial.

Therefore, the essential idea of this *race model* was that at each moment of time, one would sample data from the two stimuli (each a normal distribution with some mean and variance), compare the two samples, and then accumulate the evidence for each stimulus over time in the two independent decision variables until one reached threshold.

To describe these and other models of decision-making, we will begin with the general framework of bounded integrator models, in which the decision-making process involves integration of a random variable that represents instantaneous evidence available for each option. When the integrated evidence for one option reaches threshold, that option is selected. In these models, one assigns a random variable x_i to stimulus i, representing the merit or decision variable corresponding to the choice i. At each moment of time, that stimulus has information in favor of its choice, which we represent with random variable r_i, drawn from normal distribution $N(\mu_i,\sigma_i^2)$. After N time points, each a duration Δt, we have the following:

$$x_i^{(N)} = x_i^{(0)} + \sum_{n=1}^{N} r_i^{(n)} \tag{3.18}$$

The decision variable x_i starts at an initial value $x_i^{(0)}$, which describes the prior information regarding this option, and then grows until it reaches threshold x^*, at which time the subject commits to that option. Over a single sample, representing a small interval of time Δt, the change in the decision variable is the following:

$$\Delta x_i = r_i \tag{3.19}$$

The various integrator models share this formalism, but differ in how they represent the rate and threshold parameters.

In the *independent race model* (Vickers, 1970), there is a decision variable x_i for each option, which depends on the integral of information r_i of that option. The option that reaches the threshold first is selected. In this model, the decision variable for one option is independent of the decision variables for other options.

In the *drift diffusion model* proposed by Roger Ratcliff (1978), there is a single decision variable x that accumulates the information r that in turn is the difference between evidence for options 1 and 2. This model was initially suggested in order to account for data collected in memory-retrieval experiments. In those experiments, subjects were asked to memorize a list of items, then presented with one item termed the "probe," and then asked whether that item was on the original list. To account for the reaction time and accuracy of the response, Ratcliff (1978) proposed a model in which at each moment of time, features of the probe stimulus were compared to each of the memorized stimuli in parallel. For each memorized stimulus, the evidence in favor of matching with the probe is compared to the evidence against it. The decision variable associated with that stimulus changes according to the difference between the evidence for and against. For this model, if we assume that the evidence for and against the match are labeled as r_1 and r_2, we have equation 3.20:

$$\Delta x = r_1 - r_2 \tag{3.20}$$

The decision is made to choose option 1 when $x(t)$ reaches threshold $+x^*$ (probe matches an item in memory) or to reject that option when it reaches the opposite threshold $-x^*$ (probe does not match an item in memory). Therefore, in the drift diffusion model, the decision variable accumulates the difference between evidence supporting hypothesis 1 and that supporting hypothesis 2. As in the simple reaction time experiments, reaction time data in this task also exhibited a skewed distribution, with the median smaller than the mean, a pattern consistent with the predictions based on the model.

In these models, the rate of change in the decision variable does not depend on the value of the decision variable. That is, Δx is independent of x. One can consider a scenario (Busemeyer and Townsend, 1993) in which the rate of change in x depends on its current value, with a magnitude controlled by an additional parameter λ, making x accelerate or decelerate toward a threshold depending on the sign of λ. In that case, we have the following:

$$\Delta x = \lambda x + r \tag{3.21}$$

In the *mutual inhibition model* (Usher and McClelland, 2001), the decision variables associated with stimuli 1 and 2 inhibit each other:

$$\begin{aligned} \Delta x_1 &= -\lambda_{1,1} x_1 - \lambda_{2,1} x_2 + r_1 \\ \Delta x_2 &= -\lambda_{1,2} x_1 - \lambda_{2,2} x_2 + r_2 \end{aligned} \tag{3.22}$$

In another variant called *feedforward inhibition model* (Shadlen and Newsome, 2001), the decision variable for option 1 receives inhibition from the rate signal that drives the decision variable for the other option:

$$\begin{aligned} \Delta x_1 &= r_1 - k r_2 \\ \Delta x_2 &= r_2 - k r_1 \end{aligned} \tag{3.23}$$

Bogacz et al. (2006) have formally analyzed these models in the context of a two-alternative forced-choice paradigm and concluded that all except the race model can be reduced to the drift diffusion model.

Finally, in the *urgency model* (Cisek et al., 2009), there is a buildup of activity toward a threshold, but the buildup is not due to integration of momentary evidence. Rather, the decision variable is due to the multiplication of the current evidence (averaged over some recent period of time) with a rising urgency signal $u(t)$:

$$x(t) = (r_1(t) - r_2(t))\, u(t) \tag{3.24}$$

If the evidence during the trial is constant, and the urgency signal is proportional to time $u(t) = t$, then the urgency model produces results that are similar to the drift diffusion model. The urgency model's primary difference with earlier models is that a decision is reached because as time passes, an urgency signal proportionately increases the weight of the most recent evidence. Hence, when there is a limited window of time to make a decision, the model does not need to wait for the decision variable to reach a bound through linear accumulation. Furthermore, the urgency signal is not specific to each potential action, but is shared by all possible actions. In our next chapter, we will compare the activity of cells in the superior colliculus in the contexts of the urgency model and the drift diffusion model, and find that in this example, the data appear more consistent with the urgency model.

Utility of the movement affects the decision-making process.

Importantly, in all of these models, the decision-making process is independent of the utility of the act that will signal the decision. To explain this idea, suppose that we are asked to compare lengths of two lines, and report which line is longer. We are instructed to use our left arm to report that line 1 is longer but our right arm to report that line 2 is longer. The models of decision-making that we considered would predict that even if moving the left arm requires more effort than moving the right arm, the reduced utility of the left arm should make no difference in percentage of times that we pick line 1. That is, in these models, the information associated with each option is accumulated, and then we report the winning option regardless of how much it will cost us in terms of reward and effort to make this report. Does the utility of the act that is linked to selecting an option play a role in selecting that option?

As an example, consider a classic perceptual discrimination task (Britten et al., 1992) performed by Nobuhiro Hagura, Patrick Haggard, and Jörn Diedrichsen (Hagura et al., 2017). In this experiment, this stimulus was random dot motion presented in a small diameter circle aperture (top part of figure 3.12B). The dots moved at a nominal speed (say 10 degrees/s). During each trial, a fraction of the dots moved leftward or rightward (for each trial, 0%, 3.2%, 6.4%, 12.8%, 25.6%, or 51.2% move rightward or leftward); this fraction defined the coherence of the motion. (All other dots moved in a random direction, picked for each dot separately from 0–360 degrees.) The subject had a limited period of time (750 ms) to report the motion of the dots. They reported their perceived direction of motion (left or right) by reaching upward with their left or right arm (figure 3.12A). In the baseline trials, the resistance against the reach was the same for the two arms. Now the authors varied the effort it took to report the decision with each arm. In the induction trials, the resistance for the left arm gradually increased until it was about twice as strong as the resistance for the right arm. (They continued to make perceptual decisions during this period.) During the test trials that followed, the resistance was held high for the left arm. They plotted the proportion of rightward judgments as a function of stimulus intensity and estimated the point of subjective equivalence (the point at which each subject judged 50% of the trials to go rightward). The increase in the effort required to express leftward judgments resulted in an increased avoidance in making leftward motion judgments (figure 3.12B). That is, the subjects became biased, somewhat avoiding perceptual choices that required a high-effort motor response. The decision regarding perception of visual motion direction was affected by the effort of the action that was used for reporting that decision.

To model their results, the authors considered a drift diffusion process in which at each moment of time, evidence r_1 for rightward motion was a random variable drawn from a normal distribution $N(\mu_1, \sigma^2)$, where μ_1 was proportional to the coherence of the rightward motion. This random variable is integrated (as in equation 3.18) until either (1) it reaches the upper bound $x^* = A$, at which point the decision is that the random dots are moving rightward or (2) the variable reaches the lower bound $x^* = -B$, at which point the decision

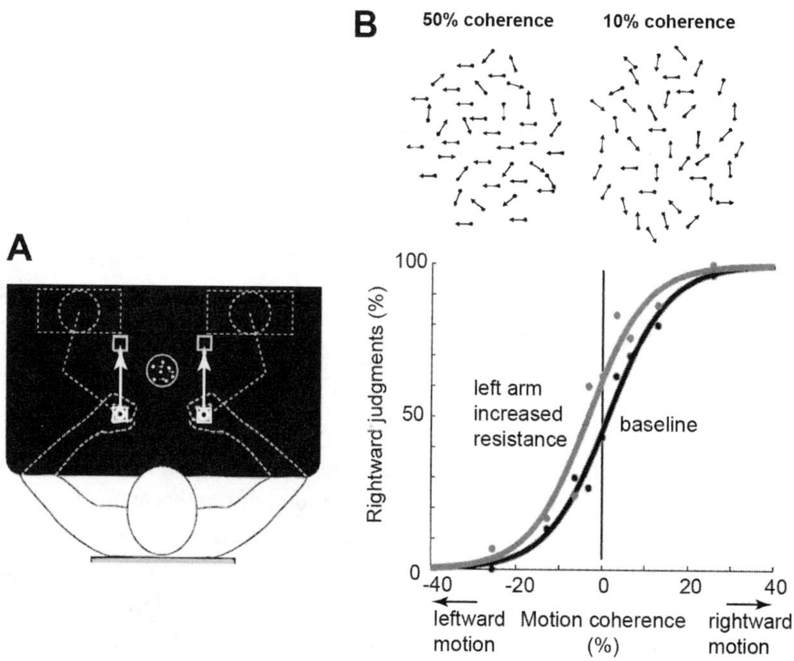

Figure 3.12
Effort of the movement that signals the decision biases the decision. **A**. In a perceptual decision-making task, the subjects observed moving dots and then with the appropriate arm, signaled their decision as to whether the motion was rightward or leftward. If the judgment was that the motion was leftward, the subject used the left arm to reach upward to the target. The motion of the left arm required more effort because a force field resisted the reach. **B**. The coherence of the motion was modulated, affecting the judgment. However, the resistance imposed on the left arm biased the judgment toward greater probability of rightward choices. (From Hagura et al., 2017.)

is leftward. For this process, the probability of the integrated variable reaching the upper bound has been described elsewhere (Palmer et al., 2005) and is the following:

$$\Pr(A) = \frac{\exp(2\mu_1 B / \sigma^2) - 1}{\exp(2\mu_1 B / \sigma^2) - \exp(-2\mu_1 A / \sigma^2)} \tag{3.25}$$

A larger coherence μ_1 for rightward motion makes it more probable that one can detect a rightward motion. The average time it takes for the integration process to stop at the upper bound has also been described (Palmer et al., 2005):

$$T_A = \frac{A+B}{\mu_r} \coth\left(\frac{(A+B)\mu_1}{\sigma^2}\right) - \frac{B}{\mu_r} \coth\left(\frac{B\mu_1}{\sigma^2}\right) \tag{3.26}$$

Nobuhiro Hagura et al. (2017) measured the probability of detecting rightward motion as well as the reaction time for the decision as a function of motion coherence. They fit these equations to the measured data gathered under baseline conditions as well as the data gathered under test conditions (when the cost of reporting with the left arm had been elevated). They concluded that the increased cost of reporting a leftward motion was akin to an increase in parameter B. That is, if choosing leftward motion required production of a costlier movement, one required more evidence for the decision to conclude in favor of leftward motion. The result was a bias that favored the detection of rightward motion.

These results suggest that during deliberation, the utility of the act that will report the result of deliberation plays a role in the decision-making process. In the framework of drift diffusion models of decision-making, the utility of the movement may bias the evidence for that stimulus:

$$r \sim N\left(\mu + k\, J^{(n)}\Big|_{t_r^{(n)*},\, t_h^{(n)*}}, \sigma^2\right) \tag{3.27}$$

As a result, the propensity of choosing an option will depend on the momentary evidence for that option as well as the utility of the movement that is required to acquire that option. Alternatively, the utility of the movement may bias the baseline $x(0)$, or equivalently, the threshold, for choosing that option. In all of these cases, the probability of choosing an option will be biased by the utility of the movement that signals that option.

In summary, it is thought that during deliberation, the brain integrates the momentary evidence for each option and reaches a decision when the decision variable associated with one of the options reaches a threshold. In these models, the momentary evidence is a random variable that is drawn from a distribution associated with the evidence for that option. These models have been applied extensively to perceptual decision-making, in which the objective is to compare the evidence for two potential options and then signal which option is more likely. However, recent evidence suggests that the utility of the movement that signals the decision affects the choice probability: people avoid selecting the option that requires a movement of greater effort. This implies that during the perceptual decision-making process, one integrates both the evidence for the option and the utility of the movement that will signal that option. In this framework, decision-making involves integration of momentary evidence for each option, but that evidence is biased by the utility of the movement that is required to acquire that option.

3.8 Modulation of Saccade Vigor during Decision-Making

The idea that utility of the movement (reward and effort) affects movement vigor raises an interesting possibility: perhaps one can use vigor as a way to understand how much the brain implicitly values a particular option. For example, suppose you are at the supermarket and are trying to decide between two different brands of jam. As you stand there looking at the jars, you make saccades that bring your eyes from one jar to the other. After a few saccades, you decide which one you prefer, then reach for and pick up the jam you have selected. As you deliberated, you made saccades to accumulate information about your options. Did the velocity with which you moved your eyes toward each stimulus reflect the real-time value that your brain assigned to that option? That is, did you make faster saccades to the jam that you decided to buy?

During deliberation, saccade velocity is modulated by the subjective value of each option.

In a collaboration with Paul Glimcher, Thomas Reppert and Karolina Lempert (Reppert et al., 2015), we performed a decision-making experiment in which participants were offered monetary rewards. The goal of the experiment was to ask whether the subjective value that a person assigned to a potential option affected the vigor with which they moved their eyes toward that option. A critical part of the experiment was that the movement (the saccade) had no bearing on the reward itself: that is, people were not rewarded for making saccades. Rather, the saccades were a natural mechanism with which participants acquired information for the purpose of decision-making (the usual reason why we make saccades).

The subjects were engaged in a decision-making task in which they chose between small amounts of money that they could have immediately and larger amounts that they could have in 30 days (figure 3.13A). The two options were displayed as text on a computer screen, and the subjects indicated their choice by pressing a button. The options were fictitious in that the subjects did not actually receive their chosen rewards after they indicated their choice. Rather, they were instructed that after the experiment concluded, one trial would be selected at random and they would receive the amount that they had chosen in that trial. If they had chosen the immediate reward, they would receive the cash immediately. If they had chosen the larger, delayed reward, they would receive a debit card with the amount of the larger reward, but the card would not be activated until 30 days had passed.

As the subjects considered their two choices, they moved their eyes from one option to another and then selected their preferred option by pressing a key. After they indicated their choice, they continued to make a few more saccades for the duration of the decision period, which was always 6 s (figure 3.13B). Their saccades had various magnitudes and velocities, as shown by the data for a single subject in figure 3.13C. To represent this velocity-amplitude data, we fitted the peak saccade velocity v for each subject n (separately for temporal and nasal saccades of each eye) across all trials to a hyperbolic function of amplitude x:

$$v_n = a_n\left(1 - \frac{1}{1 + b_n x}\right) \tag{3.28}$$

For saccade amplitude x, the expected saccade velocity is $\hat{v}_n(x)$. For each saccade, we computed the ratio between the measured peak velocity and the expected velocity: $v_n/\hat{v}_n$. This ratio defined a *within-subject* measure of saccade vigor. When this ratio was greater than one, the saccade had a velocity that was larger than expected for that amplitude,

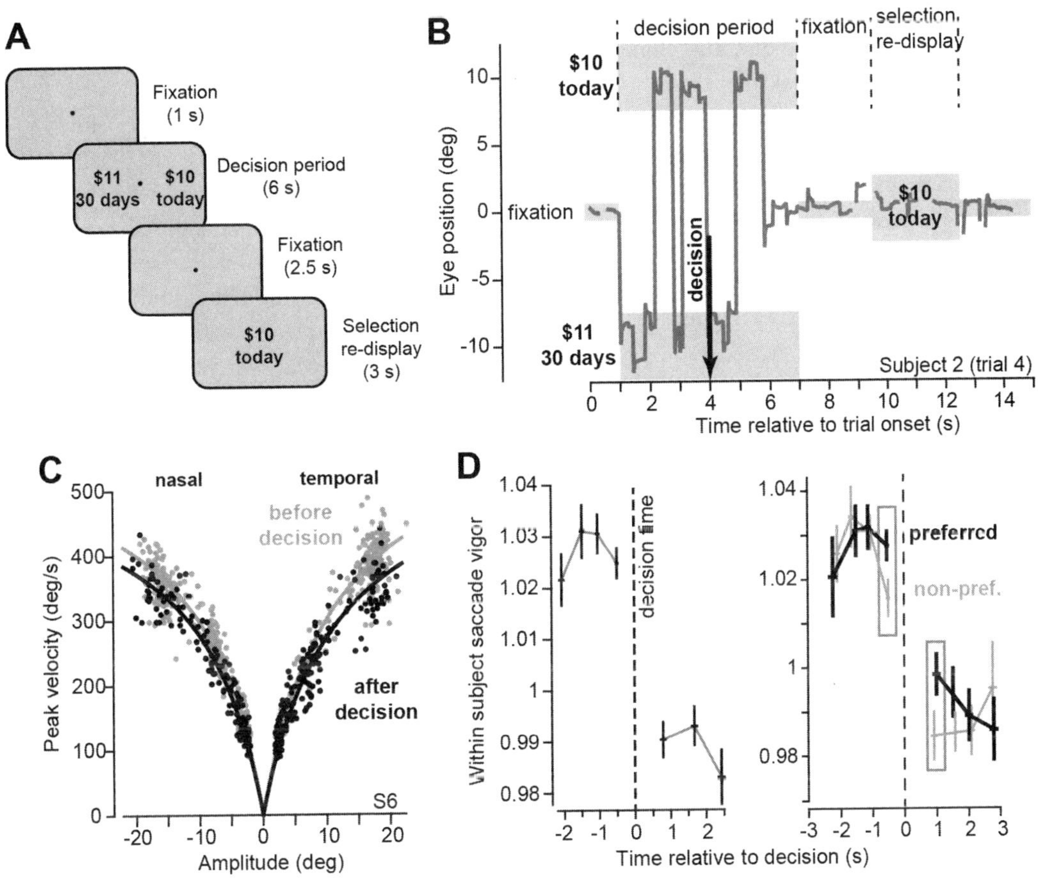

Figure 3.13
Saccade vigor is modulated during a decision-making task. **A**. Participants completed a decision-making task in which they made a choice between a small, immediate monetary reward and a larger, delayed monetary reward to be attained at 30 days. They indicated their choice by pressing a key. Regardless of when the participants indicated their choice, the options remained on the screen for 6 s. **B**. Trace of eye position during a single trial. **C**. Saccade velocity-amplitude relationship during the decision-period for a single subject. Data were fit to a hyperbolic function, separately for nasal and temporal saccades. Saccades made before decision time (gray dots) appeared to have a greater velocity than those made after decision time (black dots). **D**. Within-subject saccade vigor as a function of timing of saccade with respect to the press of the key. The number on each data point represents saccade number with respect to the press of the key. Around the time of decision, vigor of saccade made to the preferred option was higher than vigor of saccade made to the non-preferred target. (From Reppert et al., 2015.)

reflecting a greater than average vigor for that movement. We used this within-subject measure of vigor to quantify changes in saccade peak velocity as a function of time during the decision-making period.

While the subjects sampled their two options and deliberated, vigor was high. However, once they made their decision, vigor dropped (figure 3.13D, left subfigure). That is, saccade vigor was higher if the goal location contained information that was needed for the purpose of decision-making. Once that information was no longer valuable (the decision had been made), vigor dropped.

After the subjects indicated their choices, we had data regarding which option they valued more. We separated the saccades on the basis of whether they were directed toward the more valued or the less valued option, and we found that saccades made before decision time did not differ between the two options, except for the last saccade just before the key was pressed (figure 3.13D). This final saccade took place at around 0.5 s before the key was pressed and had a higher vigor if it was directed toward the more valued stimulus. The saccade immediately after the decision was also made with greater vigor when it was directed to the more valued stimulus. Outside of this window around decision time, there were no differences in the vigor as the eyes moved toward each option.

If we imagine that during the deliberation period, a decision variable (reflecting the merit of each option) is rising, one option with a small rate and the other with a larger rate, and that the decision is made when one of these variables reaches threshold, the results here suggest that vigor of the saccades during the deliberation period may be partly a reflection of the integral of these rates. This raises the exciting possibility of using saccade vigor as a real-time behavioral proxy for tracking the decision variable. Much more work is needed to test this hypothesis.

If one moves the eyes with greater vigor toward the option that is more valuable, then perhaps by considering saccade vigor around decision time, we may be able to infer the subjective value of each option. To compute subjective value of the delayed reward, we analyzed the choices that the participants made. Figure 3.14A illustrates the choices made by two participants. Participant S21 (left subplot) often picked the delayed reward when the dollar amount of that option exceeded that of the immediate option by more than \$5. In contrast, participant S41 picked the delayed reward less frequently, generally preferring this option when the dollar amount of that option exceeded the amount of the immediate option by more than \$20. For each subject, we fitted these data to the following equation:

$$\Pr(\text{choice} = \alpha_d) = \frac{1}{1 + \exp(-b(\alpha_d - \alpha_i - a))} \tag{3.29}$$

In equation 3.29, α_d is the delayed reward amount, α_i is the immediate reward amount, and a represents the point of subjective equivalence between the delayed and immediate options (the term b affects the rate with which the probability changes with reward

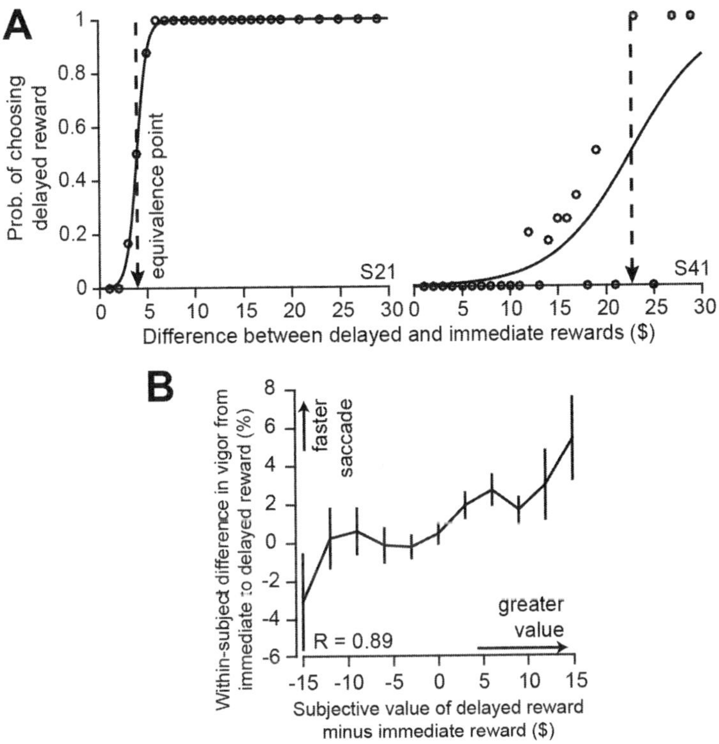

Figure 3.14
Saccade vigor encodes choice preference just before time of decision. **A.** Quantifying point of subjective equivalence. The left and right subfigures show the probability of choosing the delayed reward as a function of the difference in magnitude of the rewards offered for two subjects. Subject 41 required a larger amount of delayed reward in order to switch preference from the immediate to delayed reward, and thus tended to favor the immediate reward more than Subject 21. A two-parameter sigmoid function was fitted to the subjects' choice data, resulting in an estimate of the point at which subjects switched preference from the immediate to delayed reward, labeled as the point of subjective equivalence ($3.97 for S21 and $22.71 for S41). **B.** For saccades made immediately before and after the decision, within-subject difference in saccade vigor was related to within-subject difference in subjective value of the delayed and immediate rewards. (From Reppert et al., 2015.)

values). Therefore, in a trial in which $\alpha_d - \alpha_i = a$, the participant was equally likely to pick the delayed or the immediate option (this is indicated by the dashed arrow in figure 3.14A). For participant S21, a difference of $4 made the delayed reward equivalent to the immediate reward. For participant S41, a difference of $23 was needed. Participants who preferred the immediate reward more often, and thus were in a sense more impulsive (less willing to wait) in their decision-making, had larger values of a.

To estimate each participant's subjective value of the delayed reward, we picked all of that participant's trials in which the immediate and delayed rewards were near a. Using the values of the immediate $\alpha(0)$ and delayed rewards $\alpha(T)$ in these selected trials and the time to delayed reward ($T = 30$ days), we estimated the temporal discount rate γ_n for each subject by using the following relationship:

$$\alpha(0) = \frac{\alpha(T)}{1 + \gamma_n T} \tag{3.30}$$

For the temporal discount rate for each participant, we computed the subjective value of the delayed reward $\alpha_d/(1 + \gamma_n T)$ for each trial and compared it with the immediate reward α_i that was available during the same trial. This is plotted along the x-axis of figure 3.14B. We then focused on the two saccades made immediately before and after decision time and measured vigor when the participants directed their gaze toward the immediate reward and compared it to vigor when they directed their gaze at the delayed reward. The differences in vigor are plotted on the y-axis of figure 3.14B. Vigor increased from the immediate to the delayed reward as a function of the difference in the subjective value of the delayed reward over that of the immediate reward. That is, around the time of the decision, the vigor of the saccade that directed the gaze toward an option was positively correlated with the subjective value that the brain assigned to that option.

In summary, results of this experiment suggested that as the deliberation process started, the eyes moved with the same vigor toward the two options. However, at about half a second before the decision was made, saccade vigor became greater for the option that the subject valued more. Therefore, during decision-making, the vigor with which the brain moved the eyes toward an option appeared to be roughly a reflection of the value that it assigned to that option at that time. This would suggest that there may be a shared element between the neural system that assigns value to a stimulus during the process of decision-making and the neural system that controls movements toward that stimulus. If these results are confirmed, vigor might act as a proxy for a real-time measure of value that is being computed as the brain deliberates among various options.

3.9 Decision Uncertainty and Vigor of Movements

During deliberation, we weigh the information regarding our options and then come to a decision. This process is akin to a race to threshold between two rising variables, each representing the merits of one of the options, with the utility of each option reflected in the rate of rise. A central question is whether there is a relationship between the confidence that we have in our decision and the vigor with which we express that decision. The prediction would be that if we have more information about option 1 than about option 2, then we are more confident about option 1. If that information indicates that option 1 is likely to be rewarding, then the rate of rise in the variable reflecting option 1 may be faster, reaching threshold sooner, making us more likely to choose option 1. However, because the rate of rise is faster for option 1, we should also move more vigorously toward that option. To test for this, one needs to carefully control the amount of information available for each option, then measure the relationship between the confidence that the subject had at the time of the decision and the vigor of the ensuing movement.

Vigor is modulated by uncertainty of reward.

Joshua Seideman, Terrence Stanford, and Emilio Salinas (Seideman et al., 2018) trained monkeys to watch a monitor in which two yellow targets appeared to the left and right of a red or green fixation point (figure 3.15A). The color of the fixation point indicated the color of the rewarding target. After a short fixation period, the fixation point disappeared, cuing the animal to go. At this time, the identities of the rewarding target and the distractor were unknown. After a gap period of 25–250 ms, one of the targets turned green and the other red, identifying the rewarding target (the one whose color matched the previously viewed fixation point). Monkeys had a maximum of 450 ms after the cue to start their saccade.

In this task, reaction time is the period from the go cue to saccade onset, and cue-processing time refers to the period from the cue onset to saccade onset. Therefore, cue-processing time indicates the amount of time that the monkeys had to acquire information about identity of the rewarding target: the longer this time period, the more certain that

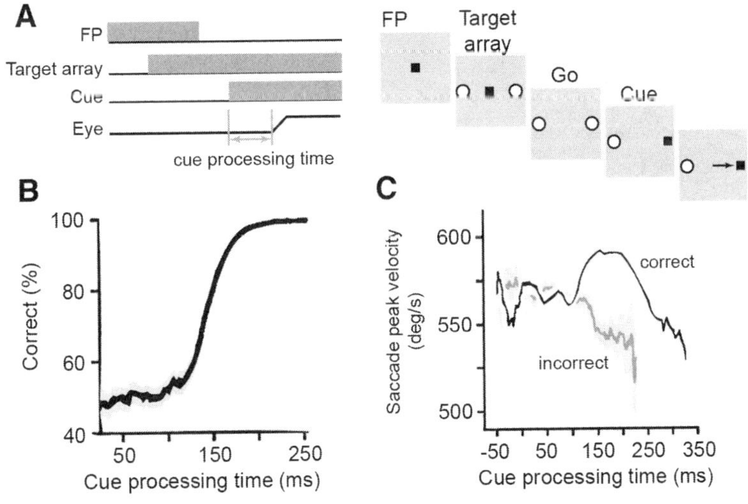

Figure 3.15
Effect of uncertainty on saccade vigor. **A**. The subjects were trained to fixate on a spot. The color of the spot indicated the identity of a future rewarding option. The target array consisted of two dots, one a distractor and the other a rewarding target. However, at this time, the identities of the distractor and target were not known. The go cue was removal of fixation spot. At some point later, one of the target array members turned to a color that matched the previously viewed fixation spot. Cue-processing time refers to the time from onset of the cue to the onset of saccade. **B**. Probability of correct choice increased with cue-processing time. **C**. Saccade vigor toward the correct target tended to increase with cue-processing time. (From Seideman et al., 2018.)

the animal would be regarding its decision. Indeed, the probability of correctly choosing the target rose as cue-processing time increased (figure 3.15B). The monkeys required a minimum of 110 ms of cue processing so that their decisions became better than chance. Therefore, if the cue-processing time was less than 110 ms, it was not useful to the subject. However, if that period was longer, then the animal could use it and the decision could be affected by it. As the cue-processing time increased, the animal had more time to deliberate on the information and becoming more confident about the decision (as reflected in the increased probability of success). But if the cue-processing time became too long, then that information was not useful because the reaction time would exceed the maximum allowed 450 ms, removing the possibility of reward.

When cue-processing time was less than 110 ms, the animal had maximum uncertainty regarding identity of the rewarding target. Salinas and colleagues found that saccade velocities for these decisions did not distinguish between correct and incorrect choices (figure 3.15C). As cue-processing time increased, the probability of success increased, suggesting that the animals had greater confidence in their decision. Interestingly, saccade vigor also increased as cue-processing time increased, suggesting that reduction in the uncertainty resulted in increased saccade velocity. However, as cue-processing time became too long, vigor decreased. It became less likely that the animal would be rewarded, regardless of which target the animal chose, because the animal had waited too long. (The choice had to be made within 450 ms following cue onset.) At that point, vigor returned to baseline.

The authors proposed a model in which during the gap period, two motor plans raced toward a threshold with constant buildup rates. If the cue did not arrive to influence these rates, the saccades that took place would have similar velocities for correct and incorrect choices. However, if a cue did arrive, then the rate for the saccades toward the correct choice got a boost and thus reached threshold sooner. That bumped-up rate resulted in the saccades for the correct choice winning the race. More interestingly, the greater rate during deliberation was followed by greater saccade vigor.

These results suggest that the vigor of the movement that indicates the decision is biased by the amount of information that the subject had during the deliberation period: the more information that was available, increasing the certainty of reward, the greater the saccade velocity that indicated the choice. If we view this result in the framework of a rising decision variable to threshold, then we can infer that within a single trial, the rate of rise in the decision variable affects not just the choice that is made, but also the vigor of the movement that brings us to the chosen stimulus.

3.10 Between-Subject Differences in Vigor

Among healthy people, there is surprising diversity in how we move: some people tend to consistently move fast, whereas others tend to move slower. Is vigor a trait of individuality?

Some individuals move with high vigor, whereas others move with lower vigor.
To explore this question, we began with the simplest voluntary movement: saccades. Jenny Choi and Pavan Vaswani (Choi et al., 2014), along with Thomas Reppert in collaboration with Oleg Komogortsev (Reppert et al., 2018) in our lab, asked a large number of healthy young subjects (about 300) to make saccadic eye movements from one visual target to another, covering various distances. We found that some individuals persistently moved their eyes with high speed, while others made the same amplitude saccade but at a lower speed (figure 3.16A). These velocity differences were present across various amplitudes, and reproducible across days. That is, if a subject had higher than average peak speed for one movement amplitude, that subject also had a higher peak speed for other amplitudes. Similar observations have been made in a number of other studies (Rigas et al., 2016; Bargary et al., 2017). We will examine the difference in vigor between subjects, with the aim of first quantifying the differences across various modalities of movement, then asking why might there be differences between people.

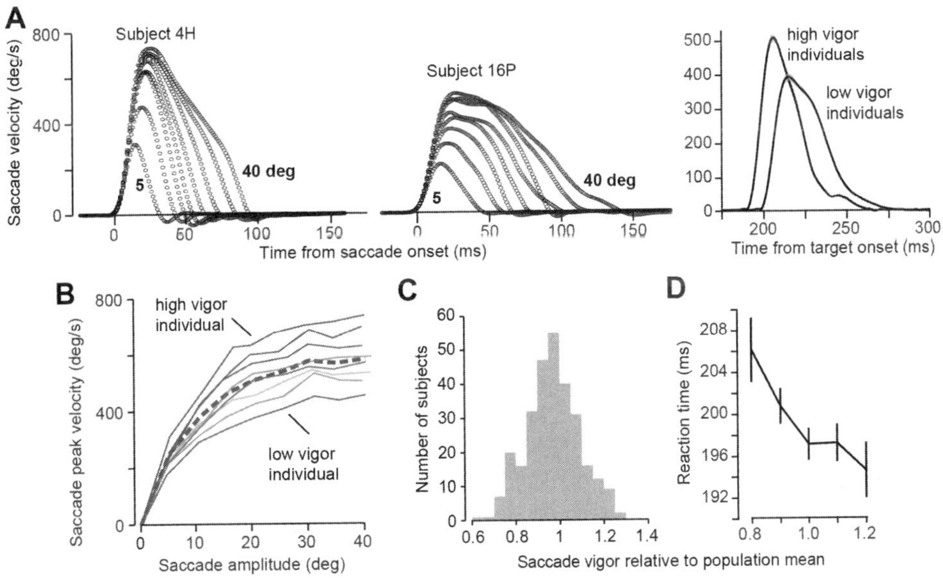

Figure 3.16
Between-subject differences in vigor of saccades. **A**. Saccadic eye movements of two healthy subjects for various amplitudes. Subject 4H's saccades exhibited higher peak velocity than did subject 16P's saccades. Right subfigure shows population data aligned to stimulus onset. People with higher saccade vigor have shorter reaction times. **B**. Peak saccade velocity as a function of amplitude for a sample of healthy human subjects. For all individuals, peak speed increases with amplitude, but the speed is consistently higher for some than for others. The thick dashed line is the mean speed-amplitude relationship for the population of 23 subjects. **C**. Distribution of saccade vigor across approximately 300 healthy people. **D**. Relationship between reaction time and saccade vigor. (Data from Choi et al., 2014 and Reppert et al., 2018.)

We can define *between-subject vigor* as the relationship between the velocity-amplitude function for movements of one individual with respect to those of the mean of the population. To explain this, consider the saccade data shown in figure 3.16B. It appears that the between-subject differences in saccade peak velocity may be accounted for by a scaling factor, multiplicatively affecting behavior. We label the across-subject mean of the velocity-amplitude function as $\overline{v}(x)$, where x is saccade amplitude and $\overline{v}(x)$ describes the relationship between amplitude and average peak saccade velocity across the population, reflecting the between-subject mean of the velocity-amplitude relationship (thick dashed line, figure 3.16B). Choi et al. (2014) found that each subject's velocity-amplitude relationship was a scaled version of this function. That is, for subject n, peak velocity v for displacement x was well described by this equation:

$$v_n(x) = k_n \overline{v}(x) \tag{3.31}$$

The scale factor k_n is our proxy for vigor of saccades for subject n. A subject with average vigor has $k=1$. We say that a given subject has high vigor if their velocity-amplitude curve lies above that of the mean of the population ($k>1$). The distribution of saccade vigor, as measured across about 300 subjects, is illustrated in figure 3.16C.

People who move with high velocity also tend to have shorter reaction times.
The right panel of figure 3.16A illustrates saccade velocities for 12°–14° displacements in a group of individuals who had low saccade vigor ($k_n<0.85$) and in a group of individuals who had high saccade vigor ($k_n>1.15$). Reppert et al. (2018) found that those who had faster saccades also responded earlier to the visual stimulus. The relationship between reaction time and vigor is quantified in figure 3.16D: as saccade vigor increased, reaction time tended to decrease.

Similar to the between-subject differences in saccade vigor, Reppert et al. (2018) found that there were consistent differences in the way people moved their heads and arms. Hand velocity during a 20 cm reach is shown for two subjects in left panel of figure 3.17A, and the peak velocity of their movements is plotted as a function of displacement (target distance) in figure 3.17B. Like saccades, hand peak velocity during a reach was a function of target distance, and between-subject differences in peak velocity could be well described as a scaling factor with respect to the mean velocity of the population. The distribution of this scaling factor, that is, distribution of reach vigor, is shown in figure 3.17C. Interestingly, individuals who moved their arms faster than average also tended to start those movements with a shorter reaction time (right panel of figure 3.17A; data are for 25 cm reaching movements). Therefore, people who moved their arms with greater velocity also tended to react sooner to the stimulus, starting their arm movements after a shorter latency (figure 3.17D).

Reppert et al. (2018) compared vigor across various modalities of movements within each individual and found a strong positive relationship between the vigor of arm move-

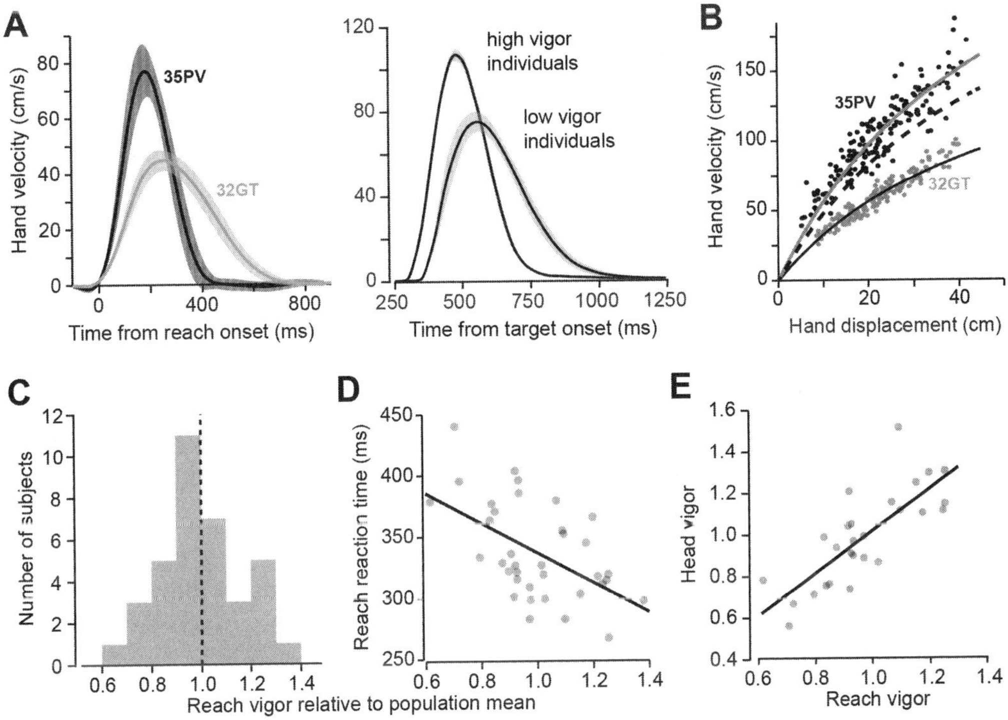

Figure 3.17
Between-subject differences in vigor of reaching movements. **A**. Reach velocities of two subjects and of the population. High-vigor individuals not only reach faster, but they also tend to have shorter reaction times. **B**. Velocity-displacement properties of the two subjects. Dashed line is the mean relationship in the population. **C**. Distribution of the reach vigor in the sample population. **D**. Reaction time tends to decrease with vigor. **E**. Head vigor tends to be highly correlated with reach vigor. (From Reppert et al., 2018.)

ments and the vigor of head movements (figure 3.17E), but a weaker positive relationship between the vigor of arm movements and the vigor of eye movements. That is, individuals who moved their arms faster than average also moved their heads faster than average. Saccade vigor was not a strong predictor of head or arm vigor, but arm vigor was a strong predictor of head vigor.

We next considered the hypothesis that the various variables that we had measured in each subject (eye, head, and arm velocity; eye, head, and arm reaction times) were generated by a single latent variable, a trait vigor of that individual. We estimated that with 65% (±1.2%) probability, the velocity and reaction times of eye, head, and arm movements in each subject were conditioned on a latent variable in that subject, their trait vigor. Notably, vigor conservation was much stronger within skeletal movements (arm and head) than between skeletal and non-skeletal movements (arm and eye).

Vigor may be a trait of individuality, reflecting a propensity to expend effort. Why does one individual consistently move her eyes or arms with higher vigor than another person? An obvious factor is differences in biomechanics, such as limb size and inertia, which may then translate into differences in cost of motion. For example, if the effort it takes to move a body part is greater for one person (their arm is heavier than another person's arm), then that person may indeed opt to move more slowly. Bastien Berret and colleagues (Berret et al., 2018) explored this question in the context of reaching movements. They asked 37 young subjects to point with their fully extended arm to visual targets that appeared on a vertical screen at various amplitudes. Many of the subjects were tested repeatedly on different days. They fitted a line to each individual's reach duration-amplitude data and defined vigor as the inverse of the slope of that line. They found that reach vigor differed between subjects, but tended to be highly reproducible within each subject: 72% of the variance in the data was due to between-subject differences, whereas 11% was due to within-subject variability (as measured in repeated sessions). Therefore, the between-subject differences in vigor was the dominant source of variability in the population data.

In order to produce a biomechanics-based measure of effort, Berret et al. (2018) calculated the joint torque that a given subject might produce in order to make a given movement: $\tau = I_n\ddot{\theta} + b\dot{\theta}$, where I_n is the inertia of the arm about the shoulder (estimated for each subject based on mass of their arm), and b is viscous friction, fixed across individuals. They then assumed that the effort expended during a movement was the integral of the squared torques plus another term that depended on the squared jerk of the motion, which was also fixed:

$$g_n(T) = \int_0^T (I_n\ddot{\theta} + b\dot{\theta})^2 + c\dddot{\theta}^2 dt \tag{3.32}$$

(The rationale for the squared jerk is not from biomechanics; rather, it is a mathematical method used to enforce movement smoothness.) The idea is that given this cost of movement, if two people had the same arm inertia, then the effort expended for a particular reach trajectory would be the same across the subjects. The researchers then assumed that the actual movement that the subject made was the one trajectory that minimized a total cost, the sum of the cost of biomechanics and a cost-of-time. The cost-of-time was a subject-specific function that increased monotonically with movement duration (a sigmoid) and had some parameters that described its shape. Given the subject-specific biomechanics cost, the researchers then found the parameters for the cost-of-time function that could reproduce the reach trajectories that the subject had actually generated. That is, having accounted for the effects of between-subject differences in biomechanics, they looked for another cost (cost-of-time) that could explain the between-subject differences in vigor.

They found that if the person reached with high vigor, then their cost-of-time tended to be higher (a parameter that described the saturating value of cost-of-time accounted for

89% of vigor variability). Therefore, this study showed that for reaching, people exhibited large between-subject differences in vigor. Once biomechanics-related costs were accounted for, the differences in vigor appeared to be related to a cost-of-time function: the cost-of-time rose faster in individuals who exhibited greater vigor.

An alternate view is that vigor differences reflect a speed-accuracy tradeoff: people who move vigorously are sacrificing accuracy for the purpose of arriving sooner. Reppert et al. (2018) considered this hypothesis and found that in both saccades and reaching, there was variability near the midpoint of the movement, and this variability was greater in people who moved more vigorously. However, the variability in peak velocity did not translate into variability in movement endpoint. That is, the variability at the beginning of the movement was partially corrected as the movement unfolded. (This within-movement process of correction likely depends on the cerebellum [Xu-Wilson et al., 2009a].) Therefore, while some individuals moved twice as fast as others and produced greater variability near the midpoint of their movements, their elevated vigor did not translate into increased endpoint variability. As a result, it appeared that speed-accuracy tradeoff was not the main factor that contributed to the between-subject differences in vigor: people who moved fast did not appear to be sacrificing accuracy for speed. Rather, they appeared more willing to exert effort to arrive at the goal sooner.

In summary, some people move with low vigor, while others move with high vigor. Those with higher vigor tend to react sooner to a visual stimulus; they move both their eyes and arms after a shorter reaction time. Arm and head vigor appear to be tightly linked: individuals who move their heads with high vigor also move their arms with high vigor. However, eye vigor does not correspond strongly with arm or head vigor. Once one accounts for differences in the biomechanics of the arm, persistent differences remain in the vigor of movements. This finding suggests that people who move fast may have a cost-of-time that rises more rapidly than those who move slowly. Importantly, in all modalities, vigor appears to have no impact on endpoint accuracy, demonstrating that differences in vigor are not due to a speed-accuracy tradeoff. These results hint that movement vigor may be a trait of individuality. They reflect a propensity to expend effort, not a willingness to accept inaccuracy.

3.11 Decision-Making and Movement Traits of Individuals

The willingness to exert effort is an intriguing aspect of individuality. For example, in a task in which subjects were asked to press a key a number of times for a given amount of money, some preferred the low-reward/low-effort option, while others chose the high-reward/high-effort option (Treadway et al., 2009). If we assume that the individuals valued a certain amount of money similarly, the results imply that the degree to which people are willing to exert effort varies between healthy individuals. Importantly, these differences were correlated with between-subject differences in the neural circuits that evaluate reward and effort, particularly circuits that regulate dopamine transmission. Michael Treadway et al. (2012) used position emission tomography (PET) to measure dopamine binding potential (D2 receptor availability) in the brain and found that individual differences in this binding potential in the left striatum and the prefrontal cortex were positively related to the willingness to expend greater effort for larger monetary rewards, particularly when the probability of success was low. However, a more recent report could not replicate this finding (Castrellon et al., 2019). At the time of this writing, we are still unclear on the neural basis for differences in the willingness to exert effort.

As we will see in chapter 6, many of the circuits that encode reward, particularly in the striatum, are also involved in control of movement, and hence modulate the vigor with which a movement is performed (Kravitz et al., 2010; Pasquereau and Turner, 2013; da Silva et al., 2018). Furthermore, changes in these circuits due to disease or drug use affect not only patterns of decision-making, but also the vigor of elementary movements. For example, people who suffer from depression are less willing to exert effort, and people who have ingested amphetamine are more willing to exert effort. Saccades are also slower among people with depression and faster among those under the influence of amphetamines (Shadmehr et al., 2010). These limited data raise the possibility that there may be a relationship between how individuals evaluate reward, effort, and time for the purpose of decision-making and how they evaluate these same variables for the purpose of making a movement.

During the teenage years, movements exhibit their greatest vigor.

To quantify decision-making traits of individuals, one can ask them to consider a rewarding state that can be attained soon and a more rewarding state that can be attained later (Millar and Navarick, 1984; Myerson and Green, 1995). For example, suppose that you purchase a cell phone from a store. At the check-out counter, the salesperson makes you an offer: you can have the phone now, or you can wait for a day and have the phone engraved with your name on it for free. Assuming that two people value the immediate acquisition of the un-engraved phone the same, the person with a steep temporal discount function would be unwilling to wait and choose to take the phone immediately. On the other hand, the person with a shallow temporal discount function would choose to wait for a day to receive the engraved phone.

To consider these two options formally, suppose that you are offered a small reward α_s, but must wait for an amount of time T_s. (You do not need to do any work to receive this reward.) As you wait, there is some cost associated with the passage of time kT_s. If you were to receive this reward immediately (i.e., $T_s=0$), its utility should be α_s. A common way to achieve this is via the following formulation of utility:

$$J(\alpha_s,T_s)=\frac{\alpha_s-kT_s}{1+\gamma_n T_s} \tag{3.33}$$

You are also offered a large reward, α_L, but must wait for a longer amount of time, $T_s + \Delta$. The resulting capture rate for this option is $J(\alpha_L, T_s + \Delta)$. For an impulsive person, the temporal discount term γ_n is large, implying that the act of waiting substantially degrades the value of the reward. As we saw in figure 3.14, in order to estimate γ_n from the decision-making patterns of the individual, a common approach is to fix α_s, T_s, and Δ, and then vary α_L until the subject is equally likely to select α_s or α_L. The term α_L^* represents the *indifference point*, the point at which the capture rates of the large and small rewards are equivalent: $J(\alpha_L^*, T_s + \Delta) = J(\alpha_s, T_s)$. This leads us to the following equation:

$$\alpha_L^* = \frac{\alpha_s(1 + \gamma_n \Delta + \gamma_n T_s) + k\Delta}{1 + \gamma_n T_s} \tag{3.34}$$

As figure 3.18 illustrates, a person who has a steep temporal discount function (i.e., γ_n is large, person is impulsive) will require a larger delayed reward.

Leonard Green and colleagues measured the temporal discount rate in various populations (Green et al., 1999). A typical experiment involved presentation of pairs of hypothetical immediate ($600 now) and delayed rewards ($1000 in 3 years). The delays for the $1000 reward ranged from 1 week to 25 years. The amounts for the immediate reward ranged from $1 to $1000. For each of the delays used for the $1000 reward, the immediate amount was varied until an indifference point was measured. The results produced an estimate of the subjective value of the delayed reward at present. The authors found that in general, the subjective value declined hyperbolically with time. They fitted a hyperbolic function with a temporal discount parameter to the data from each individual and found that the temporal discount rate was highest for children (mean age of 12 years), medium for young adults (mean age of 20 years), and lowest for old adults (mean age of 69 years).

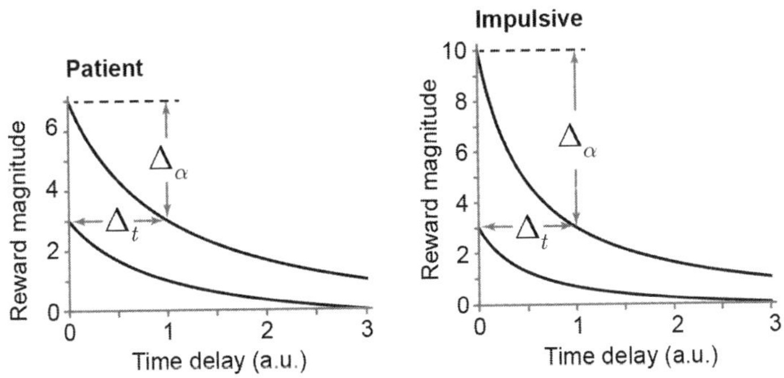

Figure 3.18
Decision-making characteristics as measured in a task in which people make choices between immediate and delayed rewards. Two hypothetical individuals, one labeled as impulsive and the other labeled as patient. For the impulsive person, the utility of reward falls rapidly as a function of time. The term Δ_α specifies the difference in reward that is required to produce indifference between a small reward now and a large reward after time delay of Δ_t. For the impulsive individual, Δ_α is large.

Thus, impulsivity, as measured via temporal discounting, appears to decline with age. (This is a paradoxical result, because the elderly have less time left to live, and thus they should prefer to take smaller rewards rather than wait for the larger, later reward).

Does movement vigor change with age? To answer this question, Irving et al. (2006) asked healthy children and adults to make saccadic eye movements in response to the presentation of a small dot that appeared randomly at various locations on the screen. The researchers measured peak saccade velocity as a function of amplitude and then fitted the data with an exponential function. As a result, they found a parameter that estimated the asymptotic peak velocity for each subject: peak velocity increased from early childhood, reached a maximum in the teen years (figure 3.19), and then declined with aging. Saccade latency showed the opposite trend: latency was highest in very young children, reached a

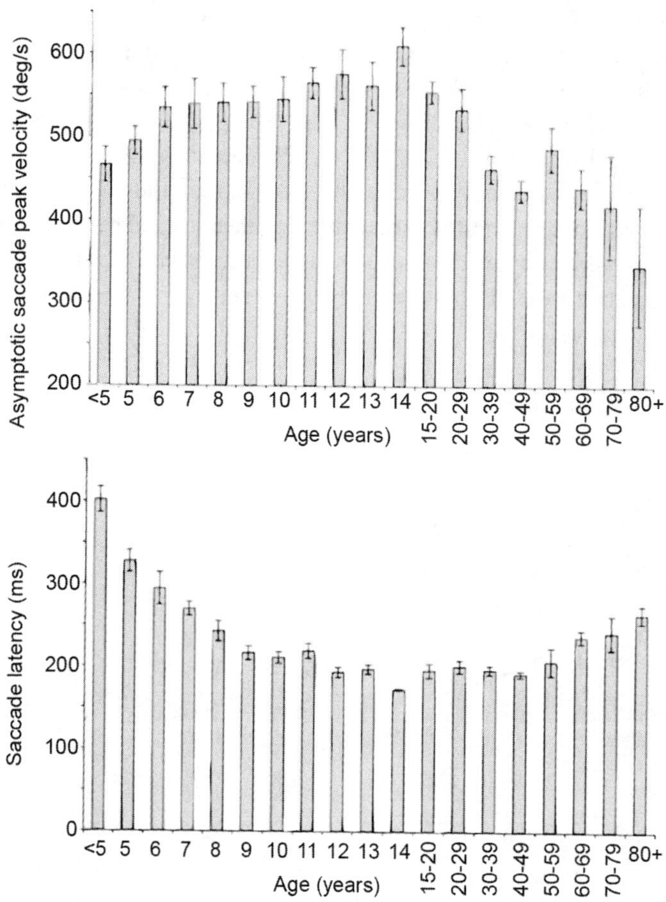

Figure 3.19
Effects of development and aging on saccade velocity and reaction time. (From Irving et al., 2006.)

minimum in the teen years, and then increased again with aging. If we consider the data on effects of development on impulsivity and vigor together, we arrive at the following conjecture: during the teen years, when people in general tend to show their greatest impulsivity, their movements also exhibit the greatest vigor (highest saccade velocity, lowest latency).

Disease can also alter the temporal discount rate γ. For example, in people with schizophrenia, there is an increase in the temporal discount rate (Kloppel et al., 2008), meaning that these patients exhibit greater-than-normal impulsivity during decision-making: given an option between a small amount of money now and a larger amount later, they tend to take the smaller, sooner reward. This population also exhibits higher-than-normal peak saccade velocities (Mahlberg et al., 2001).

An important limitation is that these observations regarding patterns of decision-making and movement vigor were made across populations, not in the same individual. Therefore, the critical question is whether between-subject differences in vigor are partially reflected in between-subject differences in the temporal discounting of reward.

High vigor individuals appear to be less willing to wait to improve their odds of success. A commonly used method to assess decision-making traits of individuals is via questionnaires. A typical questionnaire (Barratt Impulsiveness Scale [BIS]) determines the response to queries such as these: "Do you often buy things on impulse?" and "Do you mostly speak before thinking things out?" Another questionnaire (Boredom Proneness Scale [BPS]) asks subjects to agree or disagree with statements such as these: "Time always seems to pass slowly" and "I am good at waiting patiently." In Choi et al. (2014), we measured saccade vigor and then asked the subjects to fill out the BIS questionnaire. A high score in the questionnaire suggested a psychological profile for impulsivity. In our subjects, impulsivity was not a good predictor of movement vigor (BIS versus saccade vigor, $r=+0.17$, $p=0.45$). Berret et al. (2018) measured reach vigor and then asked their subjects to fill out the BIS and the BPS. They also found that impulsivity as measured by the BIS was not a good predictor of vigor (BIS versus reach vigor, $r=+.25$, $p=0.15$), but proneness to boredom as measured by the BPS, roughly representing a cost-of-time, was weakly related to vigor (BPS versus reach vigor, $r=+.33$, $p=0.04$). The positive correlation values indicate that in general people who score as more impulsive on the questionnaires and are more prone to boredom tend to have higher vigor. However, these relationships between movement vigor and survey-based estimates of impulsivity appear to be quite weak.

A different way to approach the problem is to actually measure decision-making behavior of subjects. In general, there are two broad classes of experiments that one can do to measure patterns of decision-making (Navarick, 2004). In one class, subjects are presented with potentially rewarding outcomes, the resulting choice is measured, and then the consequences are immediately applied. The key element of this operant class of experiments is that the choices have real and immediate consequences that are experienced before any other choices are made. These consequences act as reinforcements or punishments, which then affect the next choice. For example, a subject is given the option of receiving a small reward now or receiving a larger reward a few seconds later. Regardless of the choice, the consequences are experienced before the next trial starts. If a subject chooses the larger reward, that subject waits for it and then consumes it before being asked to make another choice. Most experiments in animals and some experiments in humans (Jimura et al., 2009) fall under this class.

In the non-operant class of experiments, rewarding states are presented (e,g, a small amount of money soon versus a larger amount later) and a choice is measured, but the consequences of that choice are not experienced before the next choice is made. This is because the delay associated with the two rewarding states is typically days or weeks rather than seconds, as in the operant experiments. Furthermore, nearly all rewards are hypothetical. In that case, participants are instructed that a couple of their choices will be selected for real payment after the session. Importantly, because all decisions are made before the money is received, the reward or punishment is not a reinforcement that affects the subsequent choices that are made in the experiment.

In Reppert et al. (2015), we performed a non-operant version of the decision-making task (figure 3.13A). The rewards were largely fictitious (except in one trial), and the consequences of the decisions were not experienced before the next decision was made. In that experiment, we found that although there were large differences in willingness to wait for the larger reward, there was no relationship between the temporal discounting rate and saccade vigor.

In contrast, in Choi et al. (Choi et al., 2014), we performed an operant version of the decision-making task, one in which each decision had an immediate and real consequence. In the operant experiment, we found that people who were willing to wait to improve their chances of reward (i.e., exhibiting a propensity for patience) also tended to move with lower vigor.

In Choi et al. (2014), we considered a task in which subjects received instructions to perform an action but would improve their odds of success if they waited for a second instruction (figure 3.20A). The task was of the operant type, in which each decision was associated with an immediate and real consequence that affected behavior during the next trial. In the experiment, subjects looked at a central fixation point and saw two targets, one to the left and one to the right (figure 3.20A). They were told that if the central fixation point showed an "X", they should look at the target on the right. A "Y" instructed them to look at the target on the left. After they learned this, they were provided with new instructions: "For some of the trials, the first instruction may be followed, after a delay, by a tone, alerting you of a second instruction. Occurrence of this tone means that the first instruction has been canceled and replaced with new instructions. In that case, the new instruction is that you should not move [your eyes]. Keep fixating on the center target."

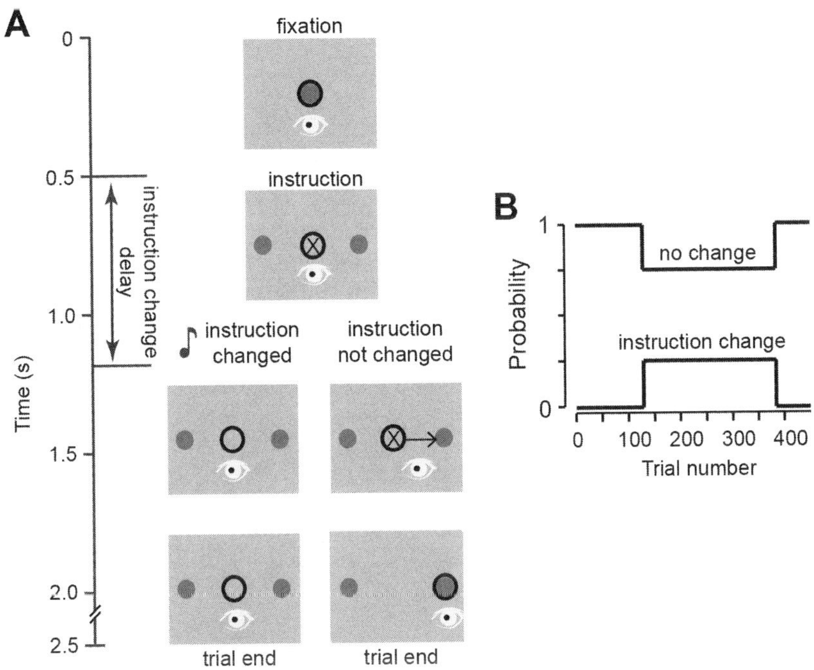

Figure 3.20
Measuring the willingness to wait in a decision-making task. **A**. The trial began with a central fixation spot. Two targets were presented at 20° from fixation along with an instruction at the fixation spot indicating which target was the direction of the correct saccade. In some blocks, there was a 25% probability that after a variable delay period, a second instruction would be given indicating the previously instructed saccade should be canceled. The delay period was adaptively adjusted to the success and failure of the subject on previous trials: success made the delay period 30 ms longer. The experiment attempted to measure the length of time the subject was willing to wait to improve the probability of success. **B**. Schedule of instruction probabilities. (From Choi et al., 2014.)

Therefore, if the subject followed the first instruction and made a saccade and the instruction did not change, then that trial was a success. If the subject followed the first instruction even though the instruction changed, then the trial was a failure.

The task was designed so that if subjects reacted only to the first instruction, they were successful with 75% probability. Waiting for the second instruction improved the probability of success by 25%. We wanted to quantify how long each individual would be willing to wait to improve his or her odds. The idea was that the person who has a steep temporal discount function will be willing to wait a little to improve their odds of success, but not as long as the person who has a shallow temporal discount function.

For each trial, the instruction-change delay period started at 200 ms for all subjects. We then measured how long the subjects waited on that trial. If during the instruction-change trial the subject was successful (i.e., the subject waited), the instruction-change delay increased by 30 ms, requiring that subject to wait longer in the future. If during the instruction-change trial the subject failed, then the instruction-change delay decreased by 30 ms. Therefore, with this adaptive algorithm, we could find the maximum amount of time the subject was willing to wait to increase the odds of success (asymptotic wait time).

In addition, we measured how the subjects reacted to a failed trial. We recorded the trial-to-trial change in the length of time that the subject waited after a failed trial. Importantly, each trial was 2.5 s in duration, and this duration was fixed regardless of the events that occurred in that trial. In this way, the subject who waited a brief period of time and the subject who waited a long period both experienced the same total experiment time. Do people who have high movement vigor, as measured via the speed of saccades, tend to be less willing to wait?

The movement latencies of two subjects during the decision-making task are shown in figure 3.21A. These subjects are the same ones for whom we displayed saccade velocities in figure 3.16A. During the first two blocks (130 trials), the probability of a second instruction was 0. In the subsequent four blocks (additional 260 trials) this probability increased to 0.25 (figure 3.20B). At the start of the third block, Δ, representing the delay to the second instruction, was 200 ms. By the end of the sixth block, Δ had increased to around 250 ms for subject 4H (gray dots), whereas it had increased to around 900 ms for subject 16P. Therefore, the subject who had low saccade vigor (16P) was willing to wait longer, whereas the subject who had high saccade vigor (4H) was unwilling to wait. For this high-vigor subject, a 25% improvement in success was not worth the added time of waiting.

For every failed trial in the first block in which the second instruction occurred, we computed the change in latency from the single-instruction trial before to the single-instruction trial after. We found rather modest correlations between decision-making patterns and vigor: subjects who had more vigorous saccades tended to have a smaller increase in latency after the failed trial (figure 3.21B). That is, people with high vigor were less willing to wait longer to improve their odds of success. A second variable of interest was the asymptotic value of the instruction delay. People who had higher saccade vigor tended to be less willing to wait (figure 3.21C).

In summary, decision-making and vigor both rely on evaluation of the option based on subjective measures of reward, effort, and time. Disease, drug-abuse, and development alter patterns of decision-making. These same factors may also lead to changes in movement vigor. For example, teenagers tend to be more impulsive, exhibiting a steep temporal discount rate. Aging reduces this impulsivity. Saccade vigor is highest during the teen years, then decays with aging. Recent experiments have begun to quantify decision-making traits and vigor patterns. The results of these experiments currently present a weak but positive correlation between temporal discounting parameters that are reflected in patterns of decision-making and the vigor of movements.

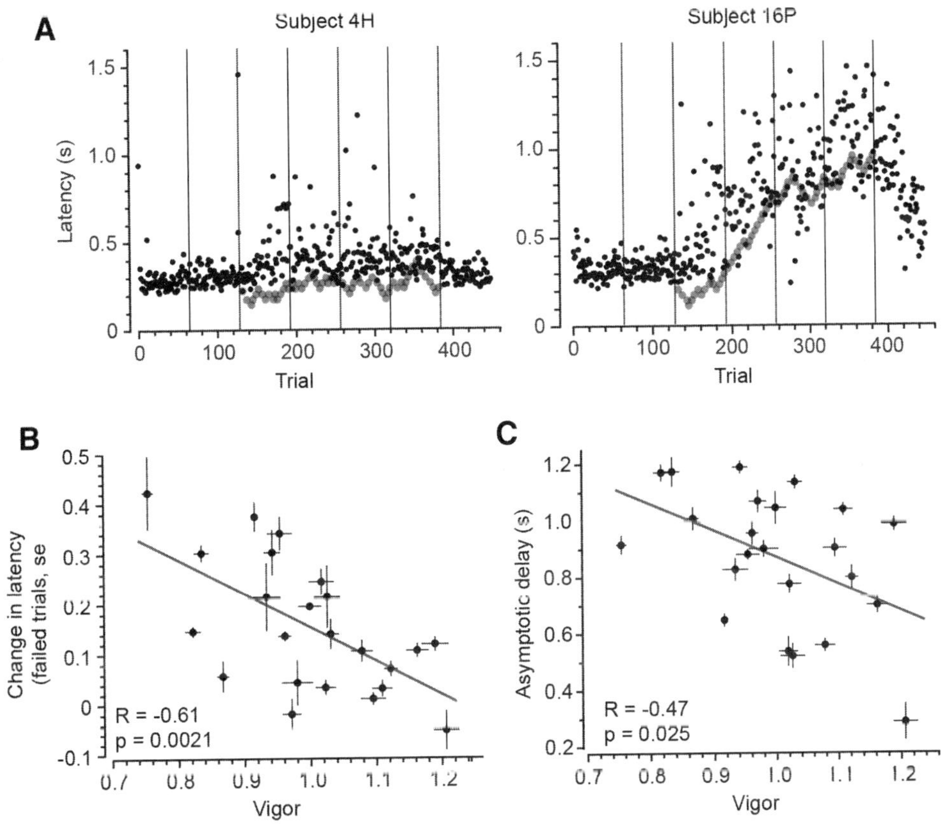

Figure 3.21
Relationship between the willingness to wait and the vigor of movement. **A**. Latency of saccades and delay of the second instruction for two subjects, one with high saccade vigor (subject 4H) and one with low saccade vigor (subject 16P). The vertical lines denote breaks between blocks. Saccade latencies are shown with the black dots and latencies of the instruction-change cue are shown by the gray dots. Subject 4H is unwilling to wait, whereas subject 16P is more willing to wait. **B**. Trial-to-trial change in saccade latency in response to an error-trial (i.e., a trial in which a second instruction occurred but the subject chose not to wait). The latencies were measured in the third block of the experiment (first block in which the second instruction occurred). After an error trial, people who exhibited lower vigor tended to increase their wait time by a larger amount. **C**. Relationship between asymptotic delay of the second instruction and saccade vigor. The delays were measured during the last half of the final block of trials in which there was a second instruction. (From Choi et al., 2014.)

Limitations

After movement to the first harvest site, every movement toward a new harvest site is an act of abandoning the current harvest. When a stimulus appears that elicits a movement, the latency of that movement will depend on the reward expected at the new harvest site and the effort needed for travel to and acquisition of that harvest. However, it is likely that this latency also depends on the harvest at the current site. That is, one reacts to a stimulus that instructs one to move in a way that depends on both the utility of moving as well as the utility of not moving (holding still at the current site). For example, if you are gazing at an interesting image, you are likely to react more slowly when another image appears elsewhere (as compared to the case where you are gazing at a boring image). In the next chapter, we will see examples of this in experiments that manipulate the stimulus at fixation and measure latency of the saccade in reaction to presentation of another stimulus. The idea that emerges is that the decision to move depends partly on the value associated with moving and partly on the value associated with staying. In our discussion here, we did not focus on the value associated with staying.

A related issue is that in our formulation of utility, we explicitly included periods of time devoted to moving and harvesting, but we did not include the reaction time period. In the simple tasks that we have considered, reaction time is implicitly part of the harvest period, during which time the subject is generally holding still. A more accurate representation of utility would include the reaction time period as a factor that influences the global capture rate.

Finally, we have imagined that vigor is modulated by two primary factors: utility (as defined with respect to reward and effort) and salience (as defined with respect to qualities such as stimulus luminance, contrast, and motion). However, there are other factors that affect vigor. For example, in a task in which people reach for random targets, having prior knowledge about target positions from previous trials reduces the reaction time and reach duration of trials in which target position is random (Wong et al., 2015). That is, having information from prior trials appears to be a motivating factor, enhancing vigor of movements for subsequent trials. This result may be due to changes in the expected value of the stimulus, as in figure 3.7. The number of variables that affect vigor is likely to be as large as the number of variables that affect utility.

Summary

It takes time to start a movement. This latency is due to a decision-making process in which a neural signal representing the merit of a movement rises toward a threshold. The rate of rise of this signal appears to be proportional to the utility of the movement. As a result, an increase in reward reduces reaction time, whereas an increase in effort increases reaction time. During deliberation, we move our eyes between the various options, and the vigor of each saccade reflects the current subjective value of that option. Thus, vigor is an indirect measure of how much we currently value the item we are moving toward. Results from a small number of experiments currently suggest a weak but positive correlation between impulsivity and movement vigor.

We will next turn our attention to the neural basis of vigor and consider how the brain moves our eyes toward stimuli of value, how utility of these movements is computed by cortical structures, and how vigor is modulated by the basal ganglia and the availability of dopamine and serotonin.

4

Neural Prelude to a Movement

Desire itself is movement.
Not in itself desirable,
Only the cause and end of movement.
—T. S. Elliot

The concept of utility and the idea of a decision variable that rises until it reaches a threshold may be useful to describe certain behavioral aspects of decision-making and motor control. However, these ideas are abstract descriptions of action. Do these concepts provide us with a framework to view the neural data that have been gathered as the brain makes decisions and performs movements? For example, is there somewhere in the brain where we might find a signal that indicates that we are satisfied with our harvest at the current location and are contemplating leaving for another location? Is there a neural correlate of a signal that resembles a decision variable, rising until it reaches a threshold and commanding the generation of a movement?

The superior colliculus may be a good place to begin our search for the neural correlates of utility and vigor. This structure plays a central role in generating saccadic eye movements. These movements are worth studying because saccades are plentiful, demonstrating two or three times per second an expression of what our brain currently values. In addition, saccades are generated entirely via structures in the brainstem, the cerebellum, and the cerebral cortex, structures that are far more accessible for neurophysiological recordings than the spinal cord (which is required for generation of reaching movements or walking).

In this chapter, we will consider the neural events that take place in the superior colliculus around the time that an animal makes a decision to move its eyes away from its current location and gaze at another location. In the superior colliculus, there are cells that encode the utility of staying (continue fixating, gathering information from the stimuli on the fovea), and there are cells that encode the utility of moving (move the eyes so that one can gather information from another location). The cells that encode the utility of staying discharge at a high level when the eyes are holding still, gathering information from the visual stimuli around the fovea. These cells reduce their activity as the brain contemplates making a movement, and then pause during the eyes' saccade to another location. The rate of decline in the activity of these cells corresponds to the reaction time of the saccade. Once the eyes arrive at their destination, the cells once again resume their high levels of discharge. In contrast, the cells that encode a particular movement have a discharge that builds up during the period of reaction time. The buildup rate is faster when the utility of the movement is higher. The faster rise in the neural activity of these movement related cells coincides with an earlier reaction time. Therefore, holding and moving are active processes, each with their own utility, and each reflected in the discharge of specialized cells in the colliculus.

4.1 Superior Colliculus

The midbrain contains a structure called the superior colliculus. In its superficial layers, the superior colliculus contains neurons that receive direct inputs from the retina and the visual cortex. In its deeper layers, there are neurons that project to brainstem premotor neurons (called burst generators), which in turn convey commands to the extra-ocular muscles (figure 4.1). The inputs from the retina and the visual cortex suggest that the organization of the superior colliculus is tightly linked to the organization of the visual system. Indeed, there is a map of the visual space with respect to the fovea on each colliculus, with the left visual field mapped onto the right colliculus, and the right visual field

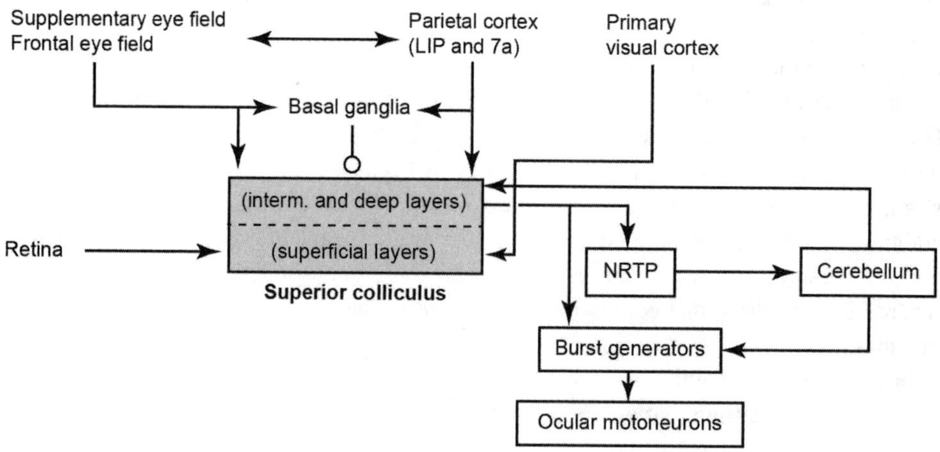

Figure 4.1
Schematic anatomy of the inputs and outputs from the superior colliculus. Filled arrows indicate excitatory connections, and open circle indicates inhibitory connection. (LIP, lateral intraparietal; cortex; NRTP, nucleus reticularis tegmenti pontis)

mapped onto the left colliculus. We will describe the events that take place in the superior colliculus as the eyes transition from looking at one point to another.

During fixation, fovea-related neurons in the rostral pole are active.
Consider a task in which an animal is holding the eyes still, maintaining gaze at a point of fixation (figure 4.2A). The point of fixation has nothing particularly interesting in it, and so the animal would normally not be motivated to look at it. However, the animal has learned that if it maintains fixation, it will be rewarded with a few drops of juice at the end of the trial. Therefore, fixating on this point has utility.

During fixation on a valuable stimulus, the activity of the cells is sustained in the rostral pole of the superior colliculus (Munoz and Wurtz, 1993a). In the rostral pole, the cells have a visual response field close to the fovea: when the eyes are held still fixating on the object of interest, these cells are very active. As we will see, the amount of activity in this region of the colliculus has something to do with the value of the stimulus on the fovea: if the eyes are held still but there is no reward associated with the act of fixation, the fovea-related cells of the colliculus tend to show much less activity. Artificial deactivation of these fovea-related cells impairs the ability to hold fixation and suppress unwanted saccades (Munoz and Wurtz, 1993b). (To deactivate these cells, researchers injected chemicals that increase efficacy of inhibition.) Indeed, for the eyes to perform a macrosaccade (amplitude of 3° or more), the cells on the rostral pole of the colliculus, the fovea-related cells, must pause their activity.

Suppose that while the subject is viewing the rewarding fixation point, two stimuli appear (A and B), one to the left of fixation and the other to the right (figure 4.2B). The appearance of the two stimuli activates cells on the retina, which subsequently excite two groups of cells in the superficial layers of the superior colliculus at a latency of 70 ms (Mays and Sparks, 1980). One of these groups of cells is located in the left colliculus, corresponding to the stimulus on the right of the fixation point, while the other is located in the right colliculus, corresponding to the stimulus on the left of the fixation point. The location of the activated cells on the superficial layer of the colliculus corresponds to the distance of each visual stimulus from the fixation point. In this way, each colliculus (left and right) has a map of the contralateral visual field, with the activated cells describing the position of each stimulus with respect to the fixation point on the fovea.

However, despite the fact that the colliculus is activated at 70 ms after the presentation of the two stimuli, usually it takes another 80–100 ms or longer before onset of the saccade. This long delay is not due to the response of the eye muscles, but instead it is because it takes time for the brain to decide which action to perform: whether to move the eyes toward one stimulus, move it toward the other stimulus, or ignore both and continue fixating.

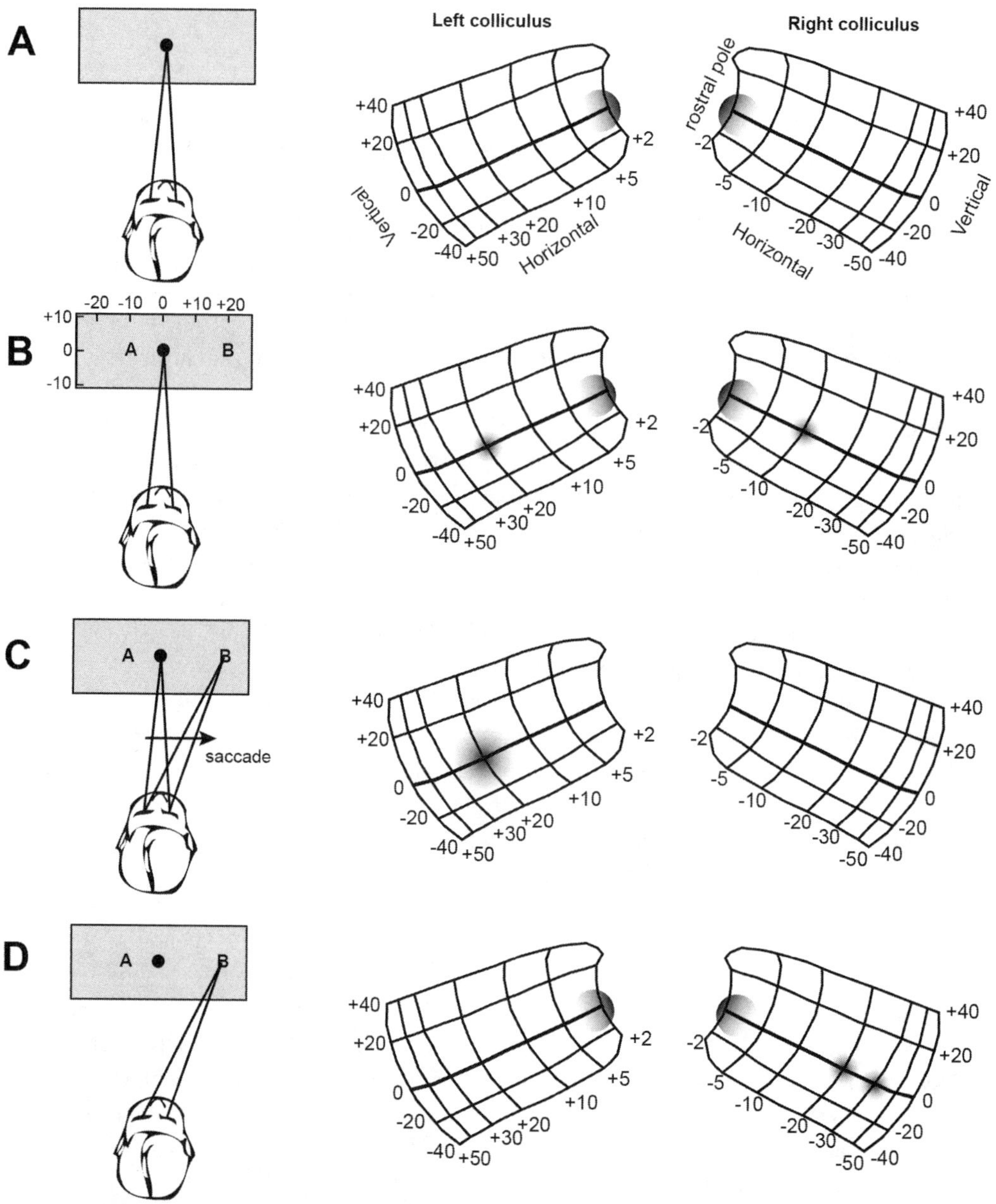

Figure 4.2
Schematic description of activity among cells of the superior colliculus before and after a visually guided saccade. In the task, the animal had been trained to associate fixation on the center point with the acquisition of a reward. The animal had also been trained to expect a reward if it made a saccade to the letter *B* and not *A*. **A**. During fixation on the center target, cells in the rostral pole of each colliculus, responding to valuable visual stimuli around the fovea, showed sustained discharge at high rates. **B**. Presentation of two visual stimuli, noted by letters *A* and *B*, at a delay of around 100 ms produced activity in cells in the caudal regions of each colliculus. **C**. In the period before the saccade, activity in the cells in the rostral pole (fovea-related neurons) declined to near 0 spikes/s, while the activity in the caudal region of the left colliculus, associated with the rewarding movement, rose and reached a threshold, producing a saccade to place the letter *B* on the fovea. **D**. After completion of the saccade, activity rose in the cells in the rostral pole. At that point, the two visual stimuli, the center target and letter *A*, were to the left of fixation. The presence of these visual stimuli resulted in activity in the right colliculus.

During a saccade, fovea-related neurons pause, and saccade-related neurons burst. During the reaction time, as the brain is considering moving the eyes toward one stimulus or the other, two events take place simultaneously: activity in the fovea-related neurons (cells in the rostral pole of the colliculus) starts to decline, and activity in neurons that encode one of the movement vectors (in this case, stimulus B) starts to buildup (cells in the caudal colliculus, intermediate and deep layers) (figure 4.2C). The buildup continues until it reaches a threshold, at which point the cells that had exhibited the buildup burst (Mohler and Wurtz, 1976). The burst leads to a saccade toward the stimulus that was selected. The time from the burst onset to the start of the saccade is around 20 ms (Munoz and Wurtz, 1995), but the buildup may take 80 ms or more.

This burst is made possible because cells in an output nucleus of the basal ganglia (SNr, substantia nigra reticulata) stop inhibiting the caudal regions of the superior colliculus (Hikosaka and Wurtz, 1983). Effectively, during fixation, the output nucleus of the basal ganglia is suppressing all movements. It must disengage this suppression before the colliculus is able to transition from maintaining fixation to generating a saccade. It is this suppression that is partly responsible for the fact that we do not usually make saccades at latency of 80 ms, even though the visual information has arrived at the colliculus. Rather, we need time to compute utilities for the various available actions, and select the action that is subjectively best for us. Once the basal ganglia release their suppression, a specific group of cells in the intermediate and deep layers of the superior colliculus shows a buildup that culminates in a burst.

In a sense, the superior colliculus is quite capable of producing saccades, but it is not trusted with that responsibility: it requires supervision from the basal ganglia and the cortex. The reason for this is that the colliculus is ill-informed; it knows about the locations of the various stimuli (via the activity that the stimuli produce in the cells of its superficial layer), but it does not know what these stimuli are or their utility. The utility is computed by the cerebral cortex, which directly through its projections to the superior colliculus and indirectly via the basal ganglia imposes control over which movement it allows the colliculus to execute and the vigor of that movement. Otherwise, the colliculus would saccade to whichever stimulus activated it more strongly (perhaps the brightest or most salient stimulus).

An example of this cortical control is a test that neurologists employ in examining patients with frontal lobe damage: the patient is asked to place both hands in front of them and close their eyes. The examiner will lightly touch one hand, but the patient is instructed to raise the hand that is not touched. Patients with frontal lobe damage are more likely to raise the hand that received the sensory stimulation. In an analogous test of the saccadic system, the patient is asked to look at a fixation point. A target appears at one side of fixation, but the patient is instructed to direct their gaze toward the opposite direction (an "antisaccade" task). Patients with damage to the frontal lobe, particularly the frontal eye field and/or the dorsolateral prefrontal cortex, have trouble inhibiting the reflexive saccade to the visual stimulus. They make errors, directing their gaze toward the stimulus rather than away from it. A similar behavior is exhibited when the output nucleus of the basal ganglia is deactivated: animals make reflexive saccades toward any stimulus that appeared in the contralateral visual field, despite the fact that they are being rewarded to maintain fixation.

What appears to be happening is that with the damage to the frontal lobe or the deactivation of the output nucleus of the basal ganglia, subjects have a hard time inhibiting the reflexive saccades that the superior colliculus is eager to produce.

4.2 Bursting to Move

The decision variable that we described in the previous chapter had a rate with which it rose to reach a threshold. This rate, we imagined, was proportional to the expected utility of the action, so that if the action had a large utility (high reward), the decision variable associated with it increased faster, reached threshold sooner, and produced an earlier movement (a shorter reaction time). Here, let us consider the activity of neurons in the superior colliculus and ask whether utility of a saccade is reflected in the rate of increase in the discharge of cells that encode that movement.

Superior colliculus activity controls saccade onset and velocity.

David Sparks (1978) trained a monkey to fixate on a point, labeled O, and then over the duration of 100 ms turned on a light somewhere away from fixation (figure 4.3). Because the light at point A flashed for such a brief period of time, in some trials the monkey made a saccade to A (trials with a saccade, top row, figure 4.3), but in other trials the monkey simply kept looking at the fixation point O and did not make a saccade (trial without a saccade, bottom row, figure 4.3). In trials in which the animal decided to make a saccade, the saccade took place at around 180 ms after the onset of target A. Sparks placed an electrode in the superior colliculus and recorded from neurons that responded to the flash of the light. Did the activity in the superior colliculus differ between the trials in which the animal decided to make a saccade and the trials in which the animal did not?

In the superficial layers of the colliculus, Sparks found cells that responded to the light (if the light fell in their receptive field). This response occurred at about 80–90 ms after light onset. However, this early response did not differentiate whether the animal would make a saccade to the target or not (figure 4.3, left column).

In slightly deeper layers of the superior colliculus, he found cells that not only responded at around 80 ms to target onset, but also had a burst of activity around 20 ms before the saccade (figure 4.3, right column). Importantly, these cells had a burst only if the saccade took place. Sparks named these visuomotor cells. These cells were located in the intermediate and deep layers of the superior colliculus.

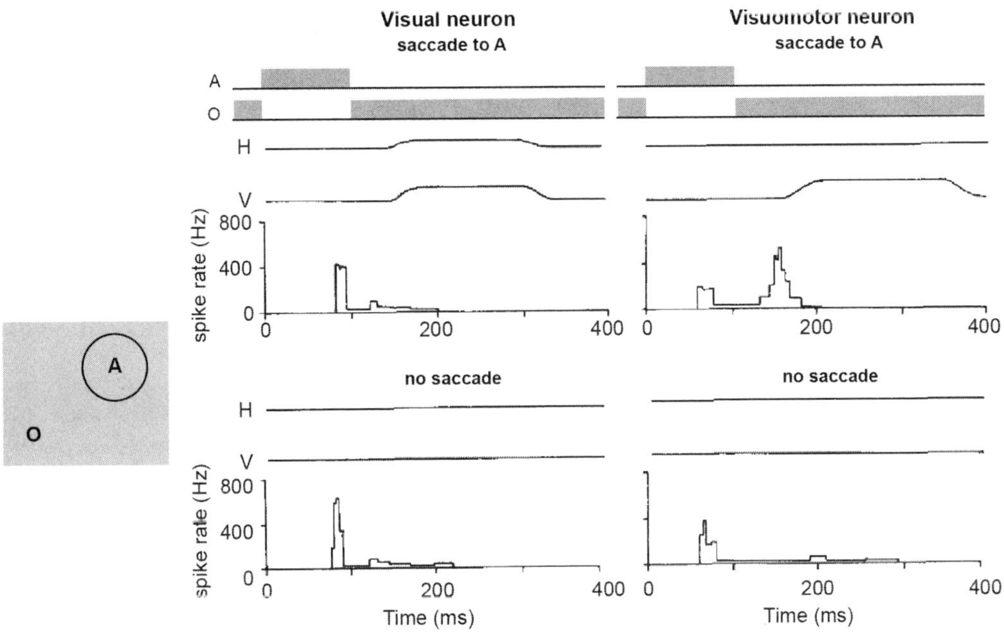

Figure 4.3
Discharge of cells in the superficial and deeper layers of the superior colliculus. **A**. Discharge patterns recorded from a neuron in the superficial layers. A burst was observed after the presentation of the visual stimulus in the receptive field of the neuron, whether or not a saccade was made to acquire the target. H and V indicate the horizontal and vertical positions of the eye. **B**. Discharge patterns recorded from a neuron in a deeper layer. Both a target-related and a saccade-related discharge were observed in trials in which target appearance in the receptive field of the neuron was followed by a saccade. When the saccade was not made, the cell discharged after only the presentation of the visual target. (From Sparks, 1978.)

The visuomotor cells had a burst of activity right before a saccade to a target in their movement field. However, these cells were also active in the trials when the animal decided not to make a saccade, but this activity was not as strong as the burst that they produced right before a saccade. An example of this is shown in figure 4.4A. In this example, the animal was fixating on point O when target A was presented for varying periods of time in the movement field of the neuron. When the presentation duration was long, the animal made a saccade (bottom panel, figure 4.4A), but when the presentation duration was short, the animal decided not to make a saccade (top panel, figure 4.4A). In the trial in which the saccade was made, there was a gradual increase in the number of spikes in this cell, until there was a burst right before the saccade. In the trial in which the saccade was not made, there were a few spikes akin to a buildup of activity, but the buildup was weak and the cell did not exhibit a burst. Importantly, note how in both cases there was a response to the onset of stimulus A at latency of about 80 ms, and then a

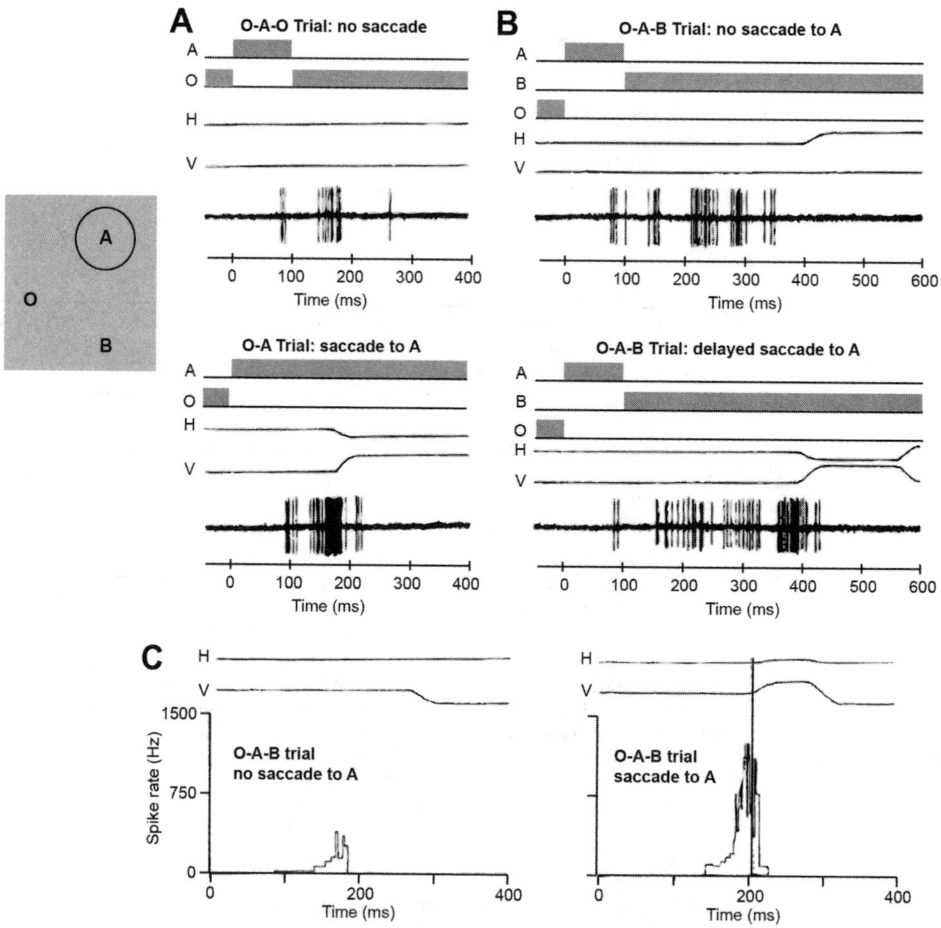

Figure 4.4
Discharge of saccade-related superior colliculus cells. The animal fixated on the O target while stimulus A was displayed in the receptive field of the cell. **A**. In the top panel, the animal made a saccade to stimulus A. In the bottom panel, stimulus A was presented only briefly, and so the animal did not make a saccade to A. In the case in which a saccade was made, the cell showed a burst just before saccade onset. In both conditions, there was a buildup of activity in the cell, but only in the case of a saccade did the buildup culminate in a burst. **B**. In the O-A-B trials, stimulus A was briefly presented (inside the receptive field of the cell), followed by stimulus B (outside the receptive field). The top panel shows a trial in which the animal did not make a saccade to A but made a saccade to B. The bottom panel shows a trial in which after a delay, the animal made a saccade to A. In both cases, there was a buildup of activity in the cell, but only in the case of a saccade did the buildup culminates in a burst. **C**. Spike density of a different cell in an O-A-B trial in which there was no saccade to A, and in an O-A-B trial in which the animal decided to make a saccade to A. The burst is present when there is a saccade to A. (From Sparks, 1978.)

second response about 80 ms later. When the second response was weak, there was no saccade. When the second response was strong, a saccade followed. That is, the cell displayed a response to the visual stimulus, then paused; after the pause, there was a buildup of activity that in some occasions resulted in a burst, which was then followed by a saccade.

To check how tightly the timing of the burst was related to the onset of the movement, Sparks cleverly varied the duration that the animal needed to make a saccade. He had the animal fixate on a central location, and then he presented target A at the center of the movement field of the neuron, and then target B at a different location, as illustrated in figure 4.4B. In these O-A-B trials, sometimes after a delay the animal made a saccade to A (lower panel, figure 4.4B), but sometimes the animal made a saccade only to B (upper panel, figure 4.4B). When the monkey decided to make a saccade to A, there was a buildup of activity in the cell, and then a burst (lower panel, figure 4.4B). When it decided not to make a saccade to A, there was again a buildup of activity, but now this buildup was weaker, and did not culminate in a burst (top panel, figure 4.4B).

The density of the spikes that was generated after presentation of target A is shown for a different cell in figure 4.4C. This cell did not exhibit a visual response to target onset, only a buildup in prelude to a saccade. The buildup was followed by a burst when there was a saccade (bottom panel, figure 4.4C), but the buildup was weaker when the animal decided not to make a saccade (top panel, figure 4.4C). If there was a burst, then a saccade followed.

By taking advantage of the natural variability in the time the animal took to make a saccade, David Sparks and Xintian Hu (Sparks and Hu, 1999) collected a data set with which they were able to ask how well the bursting activity of the cells in the superior colliculus corresponded to saccade onset. In some trials, the saccade was made as early as 100 ms after target onset (termed *express saccades*), whereas in other trials the latency was twice as long. They found that among their cells, there was usually a strong correlation between burst latency and saccade latency: regardless of the latency from target onset to saccade onset, the burst occurred around 20 ms before saccade onset. They suggested that the onset of the burst was the critical event that initiated the saccade.

Work of Neeraj Gandhi and colleagues established that in addition to onset of the saccade, the velocity profile of the saccade was also dependent on the activity of cells in the superior colliculus. Ivan Smalianchuk, Uday Jagadisan, and Neeraj Gandhi (2018) recorded from neurons in the superior colliculus in an overlap task in which after fixation, a target was presented in the response field of the neuron, and after 500–1200 ms, the fixation stimulus was removed, instructing the monkey to make a saccade. Saccade velocity was measured along a vector that was aligned to the preferred movement field of the cell. In some trials, the monkey made a fast saccade (trace 1, figure 4.5): this typically occurred early in the recording session when the animal was thirsty. In other trials, the monkey made a much slower saccade (trace 3, figure 4.5). The authors selected those saccades that had amplitude within ±5% of the mean. Figure 4.5 shows velocity of the mean saccade

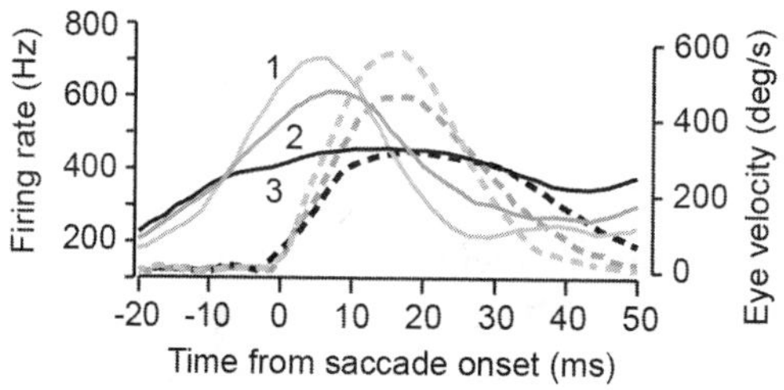

Figure 4.5
Saccade velocity is correlated with activity of cells in the superior colliculus. Firing rate (solid trace) of a visuomotor neuron and the accompanying saccade velocities (dashed trace). Amplitude of the saccades was maintained within ±5% of the mean. Trace 2 is for the mean saccade. Trace 1 is the mean of the fastest 10%, and trace 3 is the mean of the slowest 10% of the saccades. (From Smalianchuk et al., 2018.)

(trace 2), and velocity of the fastest 10% (trace 1) and slowest 10% of the saccades (trace 3). This figure also shows the activity of a typical cell. A low discharge rate in the cell was followed by a slow saccade, whereas a high discharge rate was followed by a fast saccade.

To find correlations between the firing rate and saccade kinematics, the authors removed the mean of each signal and then measured correlation between residuals of the velocity and of the firing rate traces. In a typical cell, after a 9 ms delay, the correlation between the firing-rate residuals and saccade-velocity residuals was roughly r=0.5 throughout the duration of the saccade. Across cells, the optimal delay was around 12 ms. Therefore, the residuals of both signals fluctuated around the mean in a coherent fashion, suggesting that superior colliculus activity influenced saccade velocity throughout the saccade.

4.3 Bursting to Hold Still

In addition to the bursting in the visuomotor cells that encode the saccade vector, there is another critical event that takes place in the superior colliculus during the saccade. A group of cells located in the rostral pole is generally very active during fixation. However, just before the saccade, these cells pause their activity, allowing the saccade to take place.

Doug Munoz and Robert Wurtz (Munoz and Wurtz, 1993a) explored this rostral pole region of the colliculus. In the superficial layers, they found cells with visual response fields around the fovea. In layers beneath the superficial layer (intermediate layers), they found saccade-related cells that burst for small saccades (for example, less than 2° from the fovea). In still deeper layers, they found cells that were active when the eyes were fixating on a rewarding stimulus but paused their activity when the animal made a macrosaccade to a target (with amplitude that was greater than approximately 2°).

An example of a fovea-related neuron is shown in figure 4.6. When the fixation stimulus was present and the animal was looking at it, this cell had a high level of sustained discharge (figure 4.6A). If a target was presented on the ipsilateral side and the animal made a saccade to it, the cell paused its discharge for both microsaccades (<1°) and

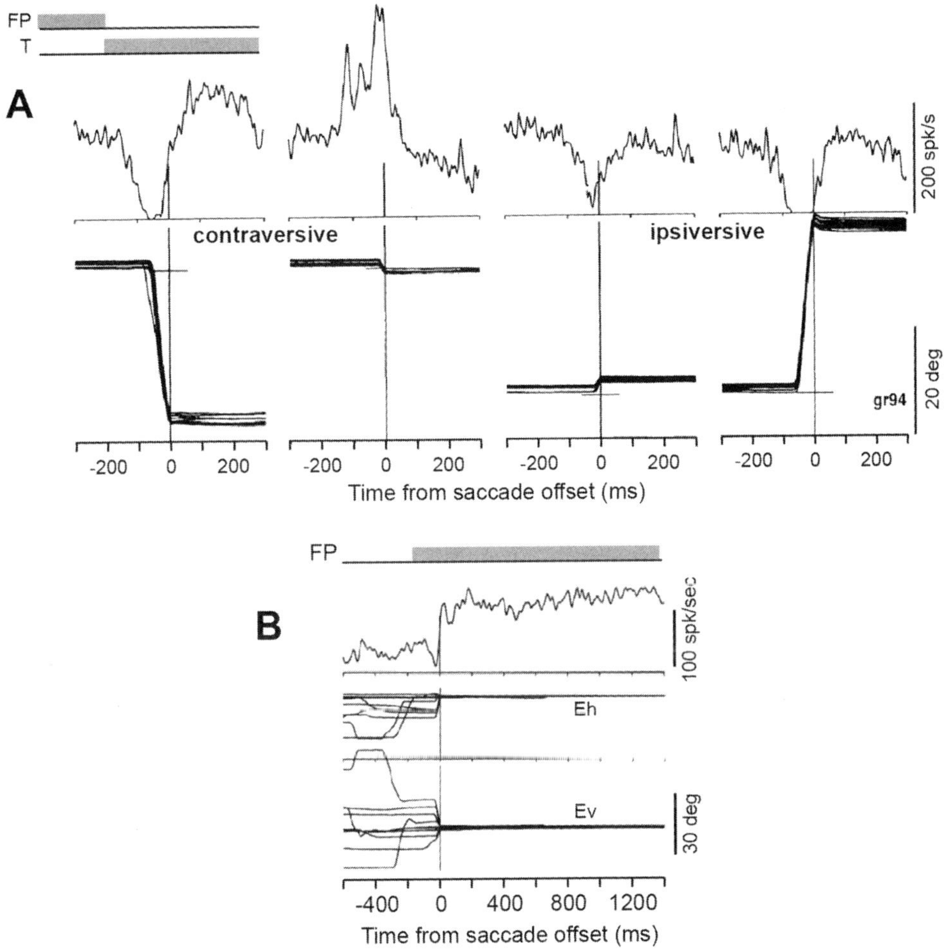

Figure 4.6
Examples from two fovea-related neurons in the rostral superior colliculus. **A**. After fixation-point offset, the animal makes a saccade to the target. The cell exhibits a pause in discharge during the saccade to the target in the ipsilateral direction. The cell exhibits a burst for microsaccades in the contralateral direction and pause for macrosaccades in the contralateral direction. Traces are aligned to saccade end. Targets were at 1.5° or 30°, ipsilateral or contralateral to the recording site. Horizontal eye position is displayed. **B**. Data from another cell in trials in which the fixation point was off for a period of time. The animal maintains period of fixation, but the cell's activity is low. Once the fixation point is displayed and the animal makes a saccade to it, the cell exhibits a high discharge. Ev and Eh refer to vertical and horizontal eye positions. (From Munoz and Wurtz, 1993a.)

macrosaccades. The pause ended near the time of saccade termination, and so the pause duration was shorter for small amplitude saccades and longer for the large amplitude saccades. Immediately after the saccade completed, activity in the cell returned to a level near or slighter higher than the activity that was present before the saccade. In contrast, if a target was presented on the contralateral side, the cell exhibited a burst for a microsaccade but a pause for a macrosaccade. Therefore, like cells in figure 4.4 and figure 4.5, this cell produced a burst, but only for microsaccades in the contralateral direction. Because its response field was near the fovea, it was highly active during fixation. When a macrosaccade took place, the cell's activity paused, regardless of the direction of that saccade.

The fovea-related neurons are located in the rostral pole of the deeper layers of the colliculus, among the layers that house the buildup cells (figure 4.7B). Animals had been trained to fixate on a point. After a certain period of time, the fixation point was removed. After another delay, a target appeared on either contralateral or ipsilateral to the recording site. While the animal fixated on the location where the fixation point had just been removed, the fovea-related neurons reduced their activity (figure 4.7A). These neurons paused their discharge entirely during the period of the saccade.

The fovea-related neurons are located in a region of the colliculus that sends monosynaptic excitatory projections to the omnipause neurons in the brainstem (Buttner-Ennever et al., 1999). The omnipause neurons are active during fixation, and as their name would suggest, pause during saccades of all sizes. Their job is to prevent a saccade from taking place, effectively acting as a threshold for onset of the movement. However, the burst and buildup neurons send projections to the brainstem burst generators (Harting, 1977; Keller et al., 2000), which are directly responsible for engaging motoneurons that move the eyes (figure 4.1). The principal superior colliculus cells that send axons to the burst generators are the burst neurons (Moschovakis et al., 1996). The fovea-related neurons are well situated to prevent a saccade while the brain is attending to a region near the fovea, while the buildup and burst neurons are well situated to move the eyes to their preferred location.

Discharge of fovea-related neurons varies with utility of fixation.

An important property of the fovea-related cells is that their discharge is modulated by the value of the stimulus at the fovea. For example, in figure 4.6B the animal was not instructed to maintain fixation, and no fixation point was available. Nevertheless, there were periods in which the animal held the eyes fairly steady as it looked around the room. However, despite the fixation of the eyes, there was low activity in those cells. Once the trial began and the fixation point was displayed, the animal made a saccade to it, and at saccade offset the discharge of the cell increased and was maintained during fixation. Therefore, it appeared that the fovea-related cells were active during fixation, but the magnitude of that activity was higher if there was a stimulus with high utility near the fovea. (The animal would receive juice for maintaining fixation.) Without this rewarding stimulus, the fovea-related cells had a lower level of activity during fixation.

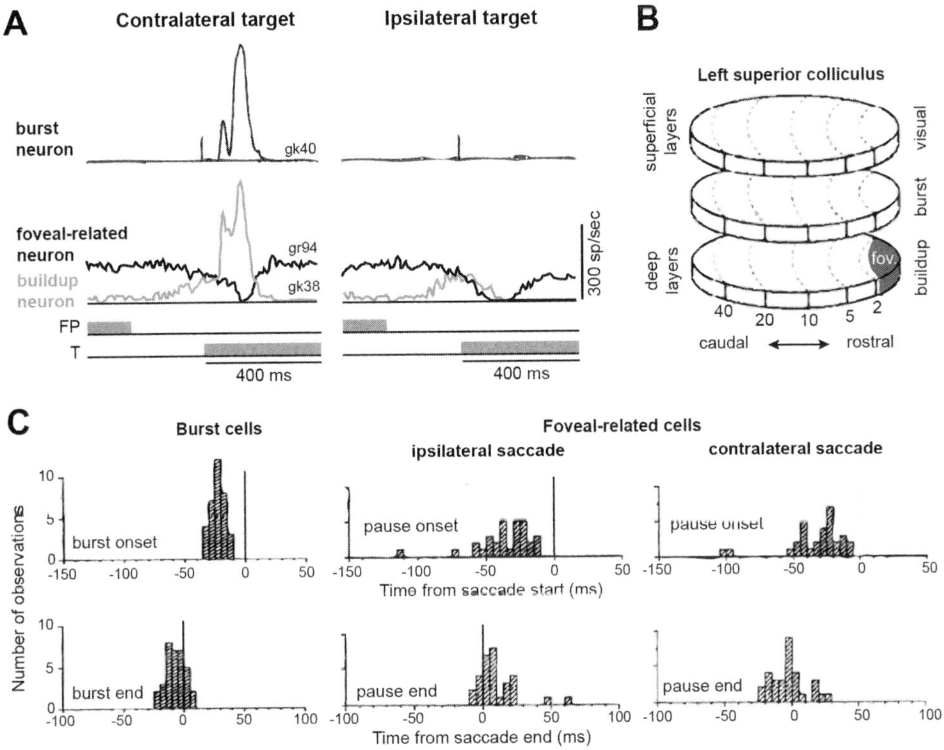

Figure 4.7
Discharge of fixation, buildup, and burst cells in the superior colliculus, and their relative anatomical location. **A**. After the removal of the fixation point, there is a temporal gap before the onset of the target. The target appeared randomly at either the left or the right of the fixation point, and therefore the animal expected the stimulus in one of two locations: ipsilateral to the recording site or contralateral to it. Data are aligned to the onset of the target. The fixation neuron shows a gradual decline in discharge during the gap period, and then pauses during the saccade, regardless of whether the target is ipsilateral or contralateral. The burst neuron shows a rapid increase in discharge near saccade onset when target is on the contralateral side. The buildup neuron shows an increase in discharge, and then shuts off the activity if the target is ipsilateral or bursts if the target is contralateral. **B**. Schematic drawing of visual, burst, and buildup cell layers of monkey superior colliculus. The gradation on each layer indicates visual/motor receptive field in degrees with respect to the fovea, along the horizontal axis. The fixation neurons are noted as part of the buildup layer. (From Munoz and Wurtz, 1995.) **C**. For the burst cells, onset of discharge leads saccade onset by around 23 ms. The burst ends around 7 ms before saccade end. For the fixation-related cells, regardless of saccade direction, the pause onset is around 34 ms before saccade onset. When the saccade is ipsilateral to the fixation cell, pause end is 10 ms after saccade end. When the saccade is contralateral to the fixation cell, pause end is 3 ms before saccade end. (From Munoz and Wurtz, 1993a.)

Munoz and Wurtz (Munoz and Wurtz, 1995) also measured the discharge of the neurons in the caudal regions of the colliculus, further differentiating the visuomotor cells that Sparks had described into burst and buildup neurons. They trained a monkey to fixate on a point, and then presented a target either in the response field of the neuron or in the opposite direction (same amplitude). In one version of the task, the fixation point disappeared, and then after a delay, the target appeared (gap task, as shown in the bottom row of figure 4.7A). They found that just below the superficial layer of the superior colliculus, there was a layer that housed neurons that generally had a burst right before a saccade to the preferred movement field (burst neuron, figure 4.7A). (The preferred movement vector was always contralateral to the recording site.) In a slightly deeper layer, the researchers found cells that had clear buildup of activity in the delay period before the saccade. For example, in the scenario displayed in figure 4.7A, the animal expected a target either in the movement field of the cell (contralateral target), or in a location opposite to that (ipsilateral target). In both cases, after the fixation point disappeared, the cell exhibited a buildup of activity. If the target appeared in a location ipsilateral to the cell, it was outside of its response field, and so the buildup that had started abruptly terminated (the cell stopped firing). However, if the target appeared contralateral to the cell and inside of its response field, the buildup continued and culminated in a burst.

In other words, when the animal was waiting for one of two possible targets, one to the left and the other to the right of fixation, but neither had been presented, the buildup cells that encoded those two specific movements displayed a gradual buildup of activity. When one of the targets was displayed, the buildup cells on the ipsilateral colliculus shut off, while the buildup and burst cells on the contralateral colliculus increased their discharge. The anticipation of a rewarding movement produced activity in the buildup cells that encoded that movement vector.

An intriguing question is whether the sustained activity in the fovea-related neurons is a reflection of the harvest rate: if there is a valuable visual stimulus on the fovea, and the animal is fixating, perhaps the activity of the fovea-related cells reflects the rate of reward that is being harvested or the expectation of this rate. Missing from these data are data from experiments that explicitly modulated stimulus value at the fovea and asked whether activity of the fovea-related cells during fixation was related to value of the stimulus. Such experiments have not yet been performed. The prediction is that a larger stimulus value would be reflected in greater discharge of rostral pole neurons during fixation. A complication is that fovea-related cells also discharge for microsaccades; hence, it may be difficult to dissociate activity due to microsaccades and activity due to expectation of reward.

In summary, anticipation of a rewarding movement to the periphery produced an increase in the discharge of the cells that encoded that movement vector. In addition, the removal of the rewarding stimulus at fixation resulted in a reduction in the discharge of the fovea-related cells. As we will see shortly, both of these events contribute to the latency of the movement.

Dynamics of collicular activity suggest online control of the ongoing saccade. Whereas Sparks had demonstrated a tight correspondence between the onset of the burst and the onset of the saccade (figure 4.5), Munoz and Wurtz (Munoz and Wurtz, 1995) uncovered a tight correspondence between the duration of the burst and duration of the saccade, as well as the duration of the pause and the duration of the saccade (figure 4.7C). The burst onset led saccade onset by around 23 ms, and ended around 7 ms before the saccade ended. The pause onset led saccade onset by around 34 ms, and ended around 3 ms before saccade end for saccades to the contralateral side, and 10 ms after saccade end for saccades to the ipsilateral side. Therefore, the duration of the burst was associated with duration of the saccade.

The duration of the pause was also associated with duration of the saccade, but the fovea-related cells were reactivated somewhat sooner for contralateral saccades than for ipsilateral ones (note the example displayed in figure 4.7A). The precise temporal correspondence between saccade duration and pause duration of the fovea-related neurons is particularly noteworthy because pause offset cannot be due to response of the cell to the arrival of the visual stimulus on the fovea (a delay that usually requires around 70 ms). Rather, this precise timing, particularly among the fovea-related neurons, hints that the activity of the collicular cells might be part of an internal control loop that is monitoring the ongoing movement. Although we will not consider this fascinating topic further in this book, there is a wealth of data arising from experiments that have perturbed an ongoing saccade (for example, via an air puff to the eye) and measured the response in the colliculus (Keller and Edelman, 1994; Soetedjo et al., 2002; Goossens and Van Opstal, 2006; Fleuriet and Goffart, 2012). That data suggest that collicular cells are not simply producing a stereotypical response, but one that is intimately related to real-time control of the ongoing movement.

In summary, while the eyes are holding still, gazing at a rewarding stimulus, the fovea-related cells in the rostral pole of the superior colliculus exhibit sustained discharge. The sustained discharge in the fovea-related cells may be related to the value of the stimulus on the fovea, and not fixation per se, because holding the eyes still without a fixation stimulus produces lower activity in the fovea-related neurons. The removal of the stimulus from fixation results in a gradual decrease in the activity of the fovea-related cells. Presentation of a rewarding stimulus away from the fovea produces a gradual increase in the activity of buildup cells that encode that movement vector. When the saccade takes place, the fovea-related neurons pause their activity, while the burst and buildup neurons that encode that movement vector produce a burst. The onset and offset of the burst are precisely timed with the start and end of the saccade.

Taken together, these results illustrate that during the period of deliberation that precedes shifting a gaze and making a saccade, there is reduction in the activity of the fovea-related neurons in the rostral colliculus, potentially encoding the reduced utility of holding still, and simultaneous buildup of activity in various neurons in the caudal colliculus, each encoding the utility of a particular movement. As we will see, the visual response in the caudal region is sometimes strong enough to produce a saccade, resulting in movements that have extremely short latency (less than 120 ms). These express saccades may reflect movements that are reflexively generated toward salient stimuli, without the ability to compute their utilities. Under normal conditions, however, the visual response to the salient stimulus is followed by some form of suppression, possibly from cortical or basal ganglia regions, preventing the colliculus from starting a movement until the utilities of the various options are considered. The movement that is actually performed is due to comparison of various utilities, reflected in the rate of fall in the fovea-related cells and the rate of rise in the buildup neurons, which accumulate until they trigger a movement.

4.4 Movement Latency and the Activity in the Fovea-Related Cells

In the models that we considered in the previous chapter, the decision variable associated with performing an action increased until it reached a threshold, at which time it triggered production of that action. The latency of the movement was associated with the time to threshold: the slower the rise, the longer the latency. In the superior colliculus, during the period before execution of a macrosaccade, there are two events that are simultaneously unfolding: (1) a reduction in the activity of the fovea-related cells and (2) an increase in the activity of the buildup cells that encode the movement vector. Is reaction time dictated by the rate of change in the presaccadic discharge of the foveal and buildup neurons?

Reaction time depends on the interactions between two neural systems in the colliculus: the fovea-related neurons, encoding utility of staying, and the buildup neurons, encoding utility of moving.

Michael Dorris and Doug Munoz (1995) explored this question in an experiment in which the reaction times exhibited variability. In their experiment, the fixation point was removed, but the saccade target (placed at 10° to the left or right of fixation) was presented only after a variable delay of 0–800 ms (a gap task, as shown in figure 4.8A). The gap was chosen randomly on each trial from the following set: 0 ms, 100 ms, 200 ms, 300 ms, 400 ms, 600 ms, and 800 ms. Therefore, the expected value of the gap period, the period between fixation point offset and target onset, was 343 ms.

When the gap period was 0 ms, the offset of the fixation point coincided with the onset of the saccade target. Under these conditions, the animal displayed a reaction time of around 210 ms (figure 4.8B). If there was a 100 ms gap before target onset, the subsequent reaction time was shorter by around 30 ms. However, as the gap period became longer, the reaction time increased back toward normal. The shortest reaction time occurred when the gap duration was 300 ms. Therefore, the animal reacted fastest on trials in which the gap period was near its expected value.

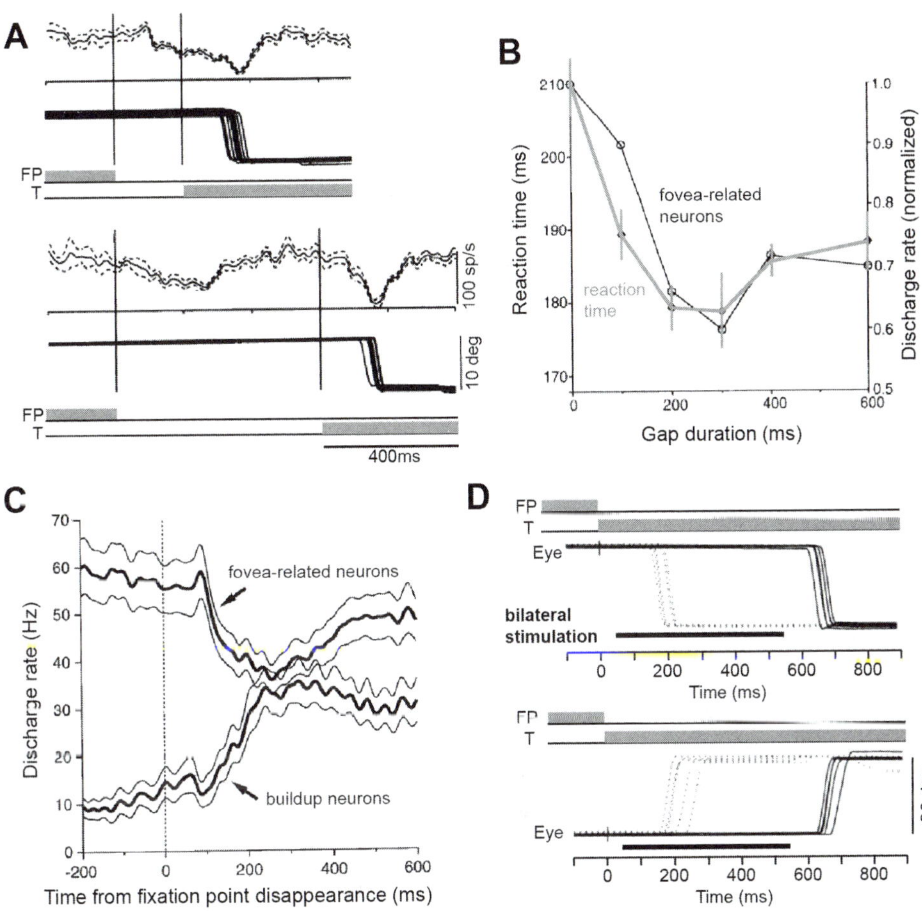

Figure 4.8
Activity in the fovea-related neurons and buildup neurons during a gap task, and the effects of bilateral stimulation of the fovea-related neurons. **A**. The gap period between fixation disappearance and target appearance was randomly selected between 0 ms, 100 ms, 200 ms, 300 ms, 400 ms, 600 ms, and 800 ms in a block of trials. For saccade-related neurons, target position in each trial was randomly selected to be either ipsilateral or contralateral to the side of recording. For fovea-related neurons, target position was randomly selected to be to the left or right of fixation at 10°. Activity of a fixation cell located in the right superior colliculus in the gap paradigm. Target appeared at 200 ms or 600 ms after the removal of the fixation point. Discharge declined after the removal of the fixation point, reached a minimum at around 300 ms, and then rose to near baseline levels by around 500 ms. (From Dorris and Munoz, 1995.) **B**. The time course of the saccade-reaction times appeared to follow the change in discharge of the fovea-related neurons. For each gap duration, the activity of each fovea-related neuron was calculated by dividing spike rate at target appearance by the value at fixation point removal. (From Dorris and Munoz, 1995.). **C**. Population averaged discharge in the fixation and buildup neurons. (From Dorris et al., 1997.) **D**. Effects of bilateral stimulation of the rostral pole of the superior colliculus. Left figure shows a saccade to a 20° target to the left of fixation, and right figure shows a saccade to a 20° target to the right of fixation. Saccades made without stimulation are shown by dotted lines. Saccades made with stimulation are delayed, occurring only after end of stimulation. (From Munoz and Wurtz, 1993b.)

Dorris and colleagues (Dorris and Munoz, 1995; Dorris et al., 1997) recorded from the cells in the rostral and caudal regions of the colliculus. They found that during the gap period, at 200 ms after fixation point offset, the discharge level of fovea-related cells dropped to around 60% with respect to discharge level just before fixation removal (figure 4.8A, top panel). This makes sense because the rewarding stimulus was removed from a place near the fovea and that removal was a cue that a rewarding target in the periphery was about to be displayed. However, if the gap duration was extended, this discharge recovered, increasing to around 75% (figure 4.8A, bottom panel). It is unclear why this recovery took place, as there were no visual stimuli that appeared near the fovea. Regardless, reaction time showed a similar U-shaped pattern. When the gap was 0 ms, reaction time was around 210 ms. As the gap increased to around 200 ms, reaction time fell to around 180 ms, and then increased to 190 ms with a gap of 600 ms. Therefore, it appeared that the reduction in the reaction time coincided with the decrease in the activity of the fovea-related cells, and the increase in reaction time coincided with an increase in the activity of these same cells.

During the gap period, as the activity in the fovea-related cells decreased and then increased, the discharge of the buildup neurons that encoded the movement vector showed an opposite pattern of change, displaying an initial increase and then a decrease (figure 4.8C). During a long gap period, the maximum in the discharge of the buildup neurons and the minimum in the discharge of the fovea-related neurons occurred at around 250–300 ms, about the time of the mean gap period (343 ms). As a result, the shortest reaction times coincided with two events: (1) the activity of the fovea-related neurons reached a local minimum and (2) the activity of the buildup neurons reached a local maximum.

In summary, when the fixation point was removed, activity in the fovea-related neurons decreased in anticipation of an upcoming macrosaccade, but the activity then increased when the saccade target did not materialize. Simultaneously, activity in the buildup neurons that encoded that movement vector increased, reaching a peak at around 300 ms and then decreasing when the target did not materialize. Reaction times showed a similar U-shaped pattern, exhibiting a minimum at around 300 ms into the gap period.

Latency of a movement is affected by the interaction between the utility of staying and the utility of moving.

To establish whether the pause in the activity of the rostral cells was necessary to release fixation and generate a saccade, Munoz and Wurtz (Munoz and Wurtz, 1993b) inserted two stimulating electrodes, one in the left superior colliculus and the other in the right superior colliculus, both in the rostral poles. With the animal looking at the fixation point, the researchers turned off the fixation light and immediately turned on the target. Normally, the animal would make a saccade to the target with a latency of around 200 ms (dotted traces, figure 4.8D). However, when they bilaterally stimulated the rostral pole of the colliculus the animal could not start the saccade, and instead maintained fixation (black traces, figure 4.8D). Once the bilateral stimulation ceased, the saccade started after a reaction time of around 100 ms.

These results suggest that reaction time depends on the interactions between two neural systems in the colliculus: the fovea-related neurons, with activity that perhaps reflects the process of harvesting reward from fixating the stimulus at the fovea, and the buildup neurons, encoding the vector of the desired movement to the next harvest location. A high level of activity in the fovea-related neurons should promote a long reaction time. A fast rate of rise in the buildup neurons should promote a short reaction time. The key step was to design experiments in which there was natural variability in the reaction time, and then test whether this variability coincided with variability in the activity of the fovea-related and buildup cells.

We had noted earlier that one way to modulate the activity of the fovea-related cells is via the presentation of a reward-associated stimulus near the fovea (figure 4.6B). For example, consider the gap and the overlap tasks (figure 4.9A) studied by Stefan Everling, Doug Munoz, and colleagues (Everling et al., 1999). Animals exhibited a shorter reaction time in the gap task than they did in the overlap task (figure 4.9B). The reason for this is the difference in the activity of the fovea-related cells. To explain this, consider the gap task: the fixation point is a stimulus associated with a reward. When the stimulus registers on the animal's fovea, the stimulus engages the fovea-related cells of the colliculus. When the fixation point is removed, the removal of this salient stimulus from near the fovea coincides with a reduction in the activity of the fovea-related cells (figure 4.9C). Coincident with the reduced activity in the fovea-related neurons is an increase of activity in the buildup cells, because the animal expects the saccade target to appear in one of two possible locations, one of which is the response field of the buildup cell (gray trace, figure 4.9D). Onset of a reward-associated target in the response field quickly brings the activity of the buildup cells to threshold. As a result of the reduced activity in the fovea-related neurons, the gap task produces relatively fast reaction times (top row of figure 4.9B).

In contrast, in the overlap task, the fixation point remains present throughout the trial. The presence of this salient stimulus constantly engages the fovea-related neurons. This makes it so that when the target is presented, it takes time to reduce the high level of activity in the fovea-related neurons (black trace, figure 4.9C). Coincident with the high activity in these neurons is the low-level or lack of activity in the buildup cells (black trace, figure 4.9D). The onset of the reward-associated target in their response field brings them to threshold slightly later. As a result, the overlap task produces slower reaction times (bottom row of figure 4.9B).

One of the factors that affect the latency of a movement is the utility of the stimulus that is currently engaging the subject (i.e., rate of current harvest). A second factor that affects the latency of a movement is the utility of the stimulus that is the target of the movement. As a result, the time it takes to disengage from looking at something and moving to look at something else depends on both the utility of the stimulus that currently falls on the fovea and the utility of the stimulus that turns the subject's attention away from the prior

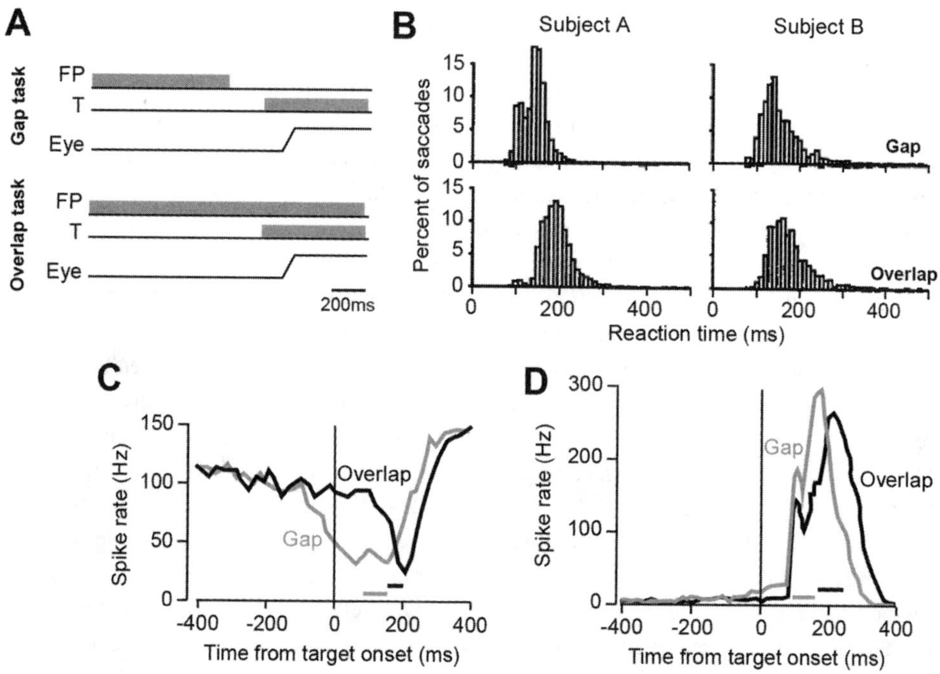

Figure 4.9
Activity of cells in the superior colliculus during the gap and overlap saccade tasks. **A**. The gap and overlap trials were randomly placed within a block of trials. (The gap period was fixed at 200 ms). The position of the target was randomly selected from one of two fixed locations within each block. **B**. Distribution of reaction times of two monkeys in the two types of trials. **C**. Activity of a fovea-related neuron in the two tasks. The horizontal bars indicate the range of saccade onsets. **D**. Activity of a buildup neuron in the two tasks. The horizontal bars indicate the range of saccade onsets. (From Everling et al., 1999.)

object of focus. The former is likely reflected in the discharge of fovea-related neurons of the colliculus, whereas the latter is likely reflected in the discharge of buildup neurons.

4.5 Movement Latency and the Rate of Rise in the Buildup Cells

Is the rate of increase in the buildup cells faster in the trials in which the animal exhibits a shorter reaction time? This is a difficult question to answer because the buildup cells often have a visual response to the onset of the stimulus, followed by buildup of activity that culminates in a burst. It would be more convenient if the location of the visual stimulus where decision-related information is presented were distinct from the location of the stimulus toward which movement is directed. In that scenario, the onset of the visual target would not produce a visual response in the buildup cell, allowing one to focus on the rate of buildup of activity and its relationship to movement latency.

Reaction time is related to the rate of rise in the activity of buildup neurons. Roger Ratcliff, Anil Cherian, and Mark Segraves (2003) presented a fixation point and then displayed an array of seven lights that had only one element turned on. The monkey had to decide whether the distance of the light to the fixation point was short or long. If the monkey decided that the distance was short, it made a saccade to the target labeled S. If the monkey decided that the distance was long, a saccade to the target labeled L target. (The letters were not presented on the screen.) In the example shown in figure 4.10A, the monkey was very likely to decide that the distance was short, making a saccade to the S target.

The researchers placed an electrode in the intermediate layers of the monkey's superior colliculus and then displayed the S and L targets so that one of the two targets was in the response field of the cell. This setup ensured that the decision of the animal was based on visual information that was not in the response field of the cell. In this way, activity in the cell could not be related to its visual response to the stimulus. Rather, any activity was likely related to decision-making and/or movement generation.

Of particular significance were the trials in which the light was in the middle of the array, thereby presenting a situation in which the distance of the light to the fixation point was neither short nor long, but in the middle. In this case, in about half the trials the animal decided that the distance was short (and made a saccade to S) and in the other trials it decided that the distance was long (and made a saccade to L). The average latency of the saccade was roughly the same regardless of the decision of the animal (around 250 ms in one subject, and 270 ms in another subject). On the basis of saccade latency, the authors divided the trials into three bins: fast, medium, and slow. Figure 4.10B illustrates the average activity of 28 neurons (recorded in two monkeys) when the L target was in the response field of the neuron and the animals decided that the distance was long. The cells showed no visual response to the onset of the array (no activity at around 70–90 ms). In fact, on average, the cells showed a dip in activity at around 90 ms after array onset. However, at 130 ms, the cells showed a buildup of activity. The rate of buildup was faster in the trials in which the reaction time was fast and slowest in the trials in which the reaction time was slow.

In some cases, the L target was in the response field of the neuron, but the animals decided that the distance was short (figure 4.10C) and therefore made a saccade to the S target. In these trials, the cells again showed a dip around at around 90 ms and a buildup that started around 130 ms after array onset. Notably, the activity rose faster in the trials with fast reaction times. However, the buildup did not continue; it declined after the initial rise.

These results suggest that when there are two potential actions, the neurons that encode these actions show buildup of activity. In trials in which the reaction time is fast, the buildup accumulates at a faster rate. Selection of one action over the other appears to coincide with continued buildup in the activity of neurons that encode the selected action and a simultaneous decline in the activity of neurons that encode the alternative movement.

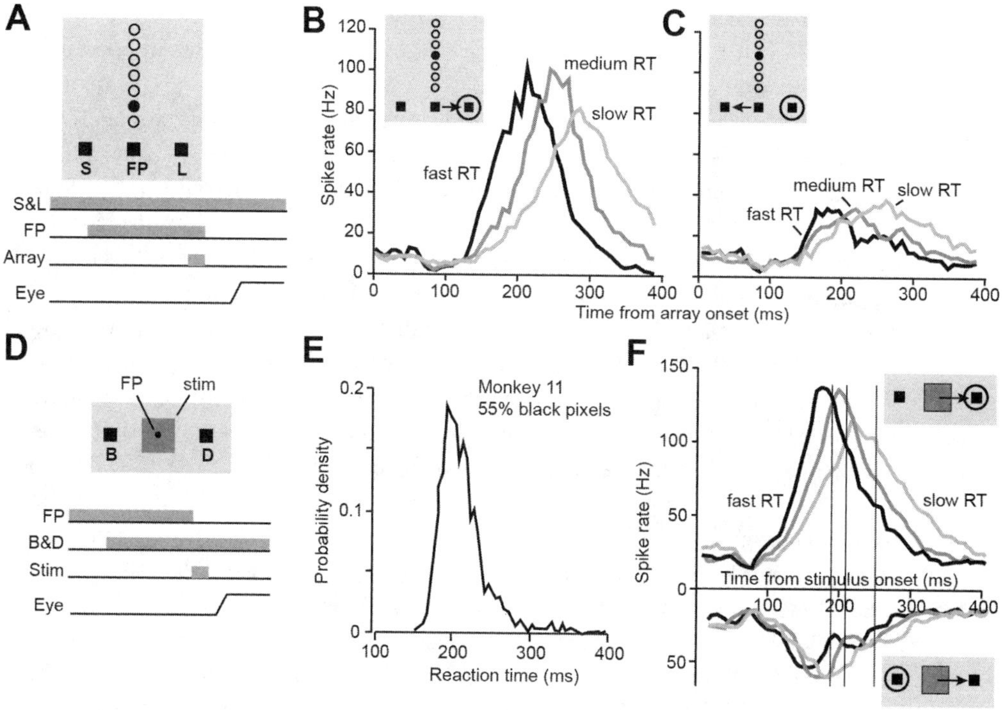

Figure 4.10
Variability in reaction time and the activity in the intermediate layers of the superior colliculus. **A**. Experiment design. One light among a possible seven lights was lit, and the animal decided whether the distance of that light with respect to the fixation point was short or long. Upon offset of the fixation point and the array, the animal made a saccade to the S or L target, depending on whether its decision was a short distance or a long distance. In this case, the likely decision is S. **B**. Activity of cells in the superior colliculus when middle light in the array was displayed, the L target was the in the response field of the neuron, and the decision was a saccade to that target. The responses are binned on the basis of the reaction time of the saccade (fast, medium, or slow). **C**. Activity of cells when middle light in the array was displayed, the L target was in the response field of the neuron, but the decision was a saccade away from that target. (From Ratcliff et al., 2003.) **D**. Experiment design. Monkey maintained fixation while a stimulus consisting of 2%–98% bright pixels appeared at center. The task for the monkey was to saccade to the target on the right if the image was dark (<50% bright pixels), and to the left if the image was bright (>50% bright pixels). **E**. Distribution of reaction times for one monkey in the 55% black pixel condition. **F**. Averaged activity of n=82 neurons in the superior colliculus of the same monkey and the same stimulus as in part E. Top part of the plot shows activity when the dark-associated target was in the response field of the cell, and the animal decided that the stimulus was dark. The vertical lines show median of reaction times for fast, medium, and slow bins. The bottom part of the plot shows activity when the bright-associated target was in the response field of the cell, but the animal decided that the stimulus was dark (increased activity is shown as downward moving). (From Ratcliff et al., 2007.)

The strength of the data in figure 4.10B is that it demonstrates that activity of buildup cells increases faster when the reaction times are shorter. However, the weakness is that the data are from two monkeys, with one monkey having longer reaction times than the other. When the data were sorted on the basis of reaction time, it was possible that the sorting also separated to some extent the recordings from the two monkeys. What we would like to know is whether within the same monkey, trials in which the animal showed a faster reaction time coincided with faster rise in the activity of the buildup cells of the colliculus.

In a follow-up experiment, Ratcliff et al. (2007) explored this question. At the fixation point, they presented an image that was composed of a certain percentage of bright pixels, ranging from 2% (very dark) to 98% (very bright). The animal made a saccade to the right if it judged the stimulus to be dark and to the left if it judged the stimulus to be bright (D and B in figure 4.10D). (There were no B or D labels displayed on the screen, we use them here as a way to label the two targets.) As in their previous experiment, the response field of the cell was centered at the target labeled D or B, and the decision of the animal was based on the visual information, which was on the fovea, not in the response field of the cell.

Figure 4.10E displays the reaction times for a single monkey in trials in which the stimulus had 55% black pixels. The reactions times are skewed, with the median being smaller than the mean. The authors focused on trials in which the animal decided that the stimulus was dark (i.e., a correct decision) and therefore made a saccade to the target labeled D. They split these trials into three groups by reaction time: fast, medium, and slow. The researchers then sorted the neurophysiological data on the basis of the reaction times. When the D target was in the response field of the cell, it showed a buildup of activity that started at around 80 ms after stimulus display (upper plot, figure 4.10F). This buildup increased faster in trials in which the monkey had a shorter reaction time. Therefore, the rate of rise was faster in trials in which the movement took place earlier (median of movement latency for each bin is indicated by the vertical line). However, when the animal decided to make a saccade to the D target even though the cell's response field encoded a movement to the B target, that cell also increased its response starting from around 80 ms after stimulus display, but at a much lower rate than the cell that encoded the choice for the D target (lower plot, figure 4.10F).

Note that in figure 4.10B, as well as in figure 4.10F, the peaks of activity are approximately the same for the fast and medium reaction time trials, but significantly smaller for the slow reaction time trials. Yet in all cases, a saccade is generated when the activity reaches its peak. Therefore, we see that the activity of buildup and saccade-related cells in the superior colliculus need not reach a constant threshold before a saccade is triggered. That is, based on these data, the threshold for the decision variable does not appear to be described uniquely by activity of buildup neurons in the superior colliculus. In our next chapter we will suggest a potential resolution to this problem when we consider activity of cells in the frontal eye field.

In summary, during the period before onset of a saccade, activity increases in the buildup neurons that encode the saccade vector. This rate of increase is related to the reaction time: the faster the rate of rise, the earlier the onset of the movement.

Rate of rise in the buildup cells appears to reflect an urgency signal, not accumulation of evidence.

There are two broad classes of models regarding the period of decision-making: one in which the brain acquires evidence for each potential action and acts on it when the evidence for one action reaches a threshold, and another in which the momentary evidence is weighted by a time-varying urgency signal. In the urgency model, a single urgency signal is shared by all potential actions: the decision is reached because passage of time increases the weight of the most recent evidence (Cisek et al., 2009). The data in figure 4.10 provide us with an opportunity to contrast these two models.

In the trials with fast reaction times, activity increased rapidly both when the decision was toward the preferred stimulus of the cell and when the decision was away from it. For example, in figure 4.10B, the cells encoded a saccade to the right. In trials in which the reaction time was fast, activity in these cells increased rapidly, and in trials in which the reaction time was slow, activity increased gradually. Notably, these patterns occurred regardless of whether the ultimate decision was to go right or left.

In a scenario in which evidence is accumulated, fast trials entail rapid accumulation of evidence for the movement that is eventually produced, but not for the movement that is not selected. In contrast, in a scenario in which local evidence is amplified by an urgency signal, fast trials entail rapid accumulation of evidence for all potential movements. The data in figure 4.10B and figure 4.10C suggest that when the decision is reached rapidly, the collicular neurons that encode evidence for the eventual winning movement and the neurons that encode evidence against that movement both show a rapid rise. This coincidence in the rate of rise for and against the movement appears more consistent with a common drive to all cells. This common drive may be an urgency signal, possibly mediated through the basal ganglia (Thura and Cisek, 2017).

4.6 Reaction Time as an Interaction between Utility of Holding versus Utility of Moving

Earlier, we saw that subjects exhibited a shorter saccade latency in the gap trials than they did in the overlap trials (figure 4.9B). This pattern suggests that reaction time is a function of not only the utility of the stimulus that elicits the movement, but also the utility of the stimulus that is currently being fixated on, encouraging harvesting of reward. This framework would predict that a high utility stimulus at fixation should result in longer reaction times, just as a high utility stimulus at the periphery should result in shorter reaction times.

To consider the interaction between utilities of the current and future actions, we can imagine that there is a decision variable x that accumulates the difference between the utilities for the option of moving and the option of holding still. This is a version of the drift diffusion model (Ratcliff, 1978) we discussed in the previous chapter, in which the value of the decision variable associated with moving is the difference between the evidence for moving and the evidence for staying (evidence for moving minus evidence for staying).

Suppose that the utility of the current action (for example, the stimulus at fixation) is represented by J_0 and the utility of the new action is J_1. We set the instantaneous evidence in support of each action as a normal random variable with a mean that is proportional to the utility of that action:

$$\begin{aligned} r_o &\sim N(kJ_o, \sigma_o^2) \\ r_1 &\sim N(kJ_1, \sigma_1^2) \end{aligned} \tag{4.1}$$

The decision variable x indicates the merits of moving. It integrates the evidence for and against moving.

$$dx = (r_1 - r_o)\, dt \tag{4.2}$$

The decision to move is made when $x(t)$ reaches threshold x^*. (This formulation can also be used in the urgency model, in which case the decision variable rises as a consequence of the difference between the instantaneous evidence, multiplied by a rising urgency signal.) Reaction time is specified by the following ratio:

$$t = \frac{x^*}{r_1 - r_o} \tag{4.3}$$

If we set $r = r_1 - r_o$, then $t = x^*/r$. Because r_1 and r_o are normally distributed independent variables with mean $\bar{r}_1 = kJ_1$ and $\bar{r}_o = kJ_o$, the probability density of r is as follows:

$$\begin{aligned} &r \sim N(\bar{r}_1 - \bar{r}_o, \sigma_o^2 + \sigma_1^2) \\ &p_r(r) = \frac{1}{(\sigma_o^2 + \sigma_1^2)^{1/2}\sqrt{2\pi}} \exp\left(-\frac{(r - (\bar{r}_1 - \bar{r}_o))^2}{2(\sigma_o^2 + \sigma_1^2)} \right) \end{aligned} \tag{4.4}$$

The probability density function for the random variable t (reaction time) becomes the following:

$$p_t(t) = \frac{x^*}{t^2} p_r\left(\frac{x^*}{t} \right) \tag{4.5}$$

The mean of the reaction time is this:

$$E[t] = \frac{x^*}{\bar{r}_1 - \bar{r}_o} \tag{4.6}$$

This result implies that if one is fixating on a central stimulus, as the utility of the peripheral stimulus increases with respect to the utility of the central stimulus, mean reaction time will decrease hyperbolically as a function of the difference in utilities. Similarly, increases in the utility of the central stimulus will tend to increase the reaction time for movements toward the peripheral stimulus.

Manipulating activity of fovea-related cells alters reaction time and velocity of saccades. One way to causally relate the activity in the fovea-related cells of the colliculus to reaction time is via artificial manipulation of that activity. During the period before execution of a macrosaccade, activity in the fovea-related cells declines. Presumably, the rate with which this decline takes place affects the reaction time of the saccade, as illustrated by reaction time differences in the gap and overlap tasks.

Doug Munoz and Robert Wurtz (Munoz and Wurtz, 1993b) injected muscimol (a GABA agonist that increases efficacy of inhibitory neurons) in the rostral pole region of the superior colliculus and recorded behavior in visually guided and memory guided saccades.

In a single block, the animal would experience various kinds of trials: visually guided trials (fixation spot removed, target appeared), memory-guided trials (fixation remained, target was flashed, fixation removed), overlap trials (fixation and target would overlap in time), and gap trials (fixation removed, a gap followed, then target appeared). This richness of trial types made reaction times much longer than in a typical experiment in which only one type of trial was considered.

In the case of visually guided trials, under normal conditions the animal had a typical reaction time of around 300 ms. Muscimol injection altered this reaction time: if the target was located contralateral to the side of injection (for example, injection in the right colliculus, target to the left of fixation), injection reduced saccade latency (figure 4.11A). It also increased saccade velocity and made it slightly hypermetric (low plots of figure 4.11A). Latency of saccades to the ipsilateral direction was also decreased, but by a smaller amount. In the case of memory-guided saccades, injection of muscimol made it difficult for the animal to perform the task: when fixation spot was still present and the target was flashed contralateral to the side of the injection, the animal broke fixation and made a saccade to the visual stimulus (figure 4.11B). Velocity of these saccades was also higher than normal, and the saccades were again somewhat hypermetric.

One way to view these results is to imagine that injection of muscimol is resulting in reduced activity in the fovea-related cells, artificially reducing the utility of the stimulus at fixation. If the presentation of a visual target in the periphery is to trigger a movement toward it, its utility must compete with the utility of the stimulus at fixation. When activity in the fovea-related neurons is artificially reduced, presentation of the peripheral stimulus leads to faster reaction times and greater vigor. The change in the reaction time is particularly evident in the memory-guided task, in which the animal finds it difficult to withhold the saccade until the fixation spot has been removed. Instead, upon presentation

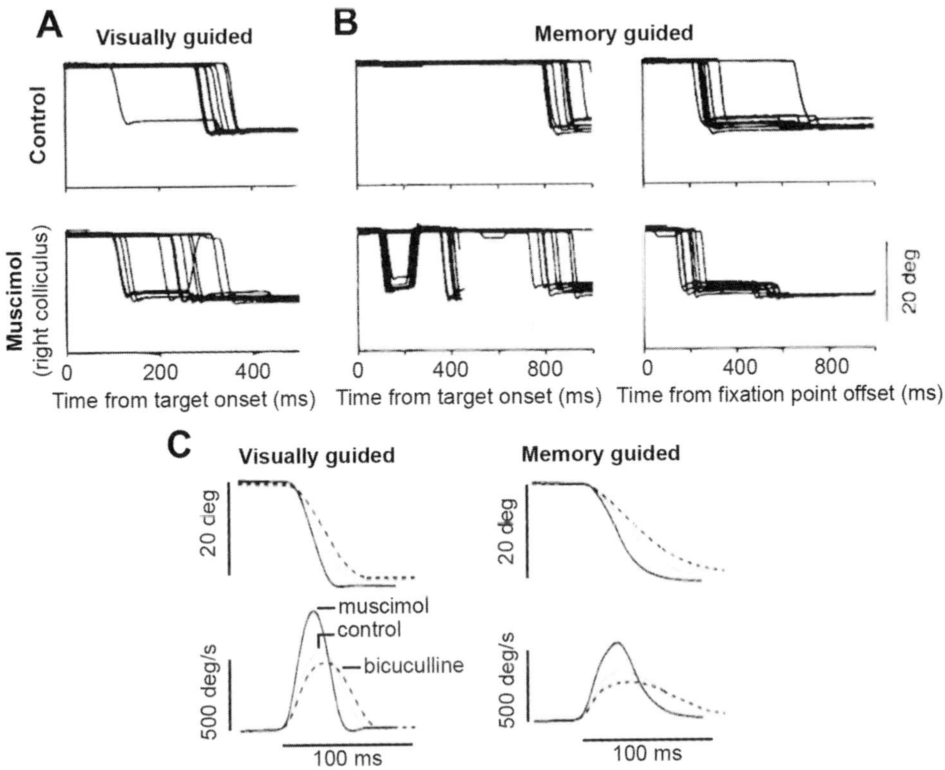

Figure 4.11
Disruption of the foveal region of the superior colliculus reduces reaction time and increases saccade vigor. Muscimol was injected into the rostral pole region of the right superior colliculus. **A**. Visually guided saccades. Leftward saccades are shown. The effect of muscimol injection is to reduce latency of rightward saccades and increase their peak velocity. Note that there is little or no change in saccade amplitude. **B**. Memory-guided saccades. The effect is increased break of fixation and early saccade to the target. **C**. The effect of muscimol and bicuculline (a GABA antagonist) on saccade position and velocity. (From Munoz and Wurtz, 1993b.)

of the peripheral stimulus, the animal immediately makes a saccade toward it. Notably, the effect on saccade amplitude (hypermetria during muscimol injection) suggests that the fovea-related cells may be participating in control of the ongoing saccade via the timing of their activity as the pause period ends and activity returns to high levels (figure 4.6A).

Because manipulating the activity of fovea-related cells in the colliculus alters the reaction time and vigor of the macrosaccade, it seems that the colliculus, despite its proximity to the saccade-generating motor circuitry, is not simply responding to an upstream decision-making system by following orders and generating a movement. The activity of the colliculus indicates that it is influencing the timing as well as the vigor of the movement. We will see further evidence for this shortly as we consider what happens when the colliculus is artificially stimulated.

4.7 Salience versus Utility

Our movements, especially saccades, are affected by not just the utility of the stimulus, but also its salience. In chapter 1, we saw examples of this. When a distractor stimulus appears about the same time that subjects need to make a saccade to task-relevant stimulus, subjects occasionally make an incorrect saccade to the distractor. A novel but irrelevant stimulus that happens to have high salience will capture our attention and direct a movement toward it, even when another stimulus that has been associated with reward is present. In figure 4.3 we saw that collicular neurons can have a visual response to the stimulus at around 80 ms and then a second response at about 100 ms later. Is there a fundamental difference between these two responses? Here, we consider the possibility that the short-latency response is one that largely reflects stimulus salience, not utility. However, the second response, the one that precedes the movement, is likely a reflection of stimulus utility.

The earliest collicular response reflects stimulus salience, not utility.
In the natural environment, there are many potential stimuli that can serve as the goal of a saccade. There are generally multiple options, each with its own utility and salience, attracting our gaze. Robert McPeek and Ed Keller (McPeek and Keller, 2002), along with Byounghoon Kim and Michele Basso (Kim and Basso, 2010) considered a task in which after the presentation of a fixation spot, four visual stimuli appeared, three of which were the same color, all serving as distractors, and one which was a different color, serving as the target. The animal had to decide which stimulus was unique and then saccade to that stimulus in order to receive reward. Importantly, the appearance of the array of stimuli was a highly salient event, making each stimulus stand out locally and potentially engaging collicular neurons that encoded that spatial location.

Kim and Basso (2010) inserted four electrodes, two in the left colliculus and two in the right colliculus, which afforded them the ability to simultaneously record activity of four neurons as the stimulus array was presented. Each stimulus was located in such a way that its position was inside the response field of one of the neurons. The location of the unique target was random in each trial. They found that around 70 ms after the presentation of the stimulus array, all four neurons responded with a small burst (figure 4.12A). Therefore, the initial response to the presentation of the stimuli did not distinguish the target. However, this visual response was short-lived. Three of the four neurons had a distractor in their response field and not the target of the movement; the response of those neurons was followed by a decline back toward baseline. In the fourth neuron, which had the target in its response field, the visual response was followed by a buildup and an eventual burst, leading to production of a saccade.

The observation that the response at 70 ms was the same regardless of whether the stimulus was a distractor or the target is further illustrated in the activity of two neurons recorded by Robert McPeek and Edward Keller (2002) in figure 4.12B. After a short

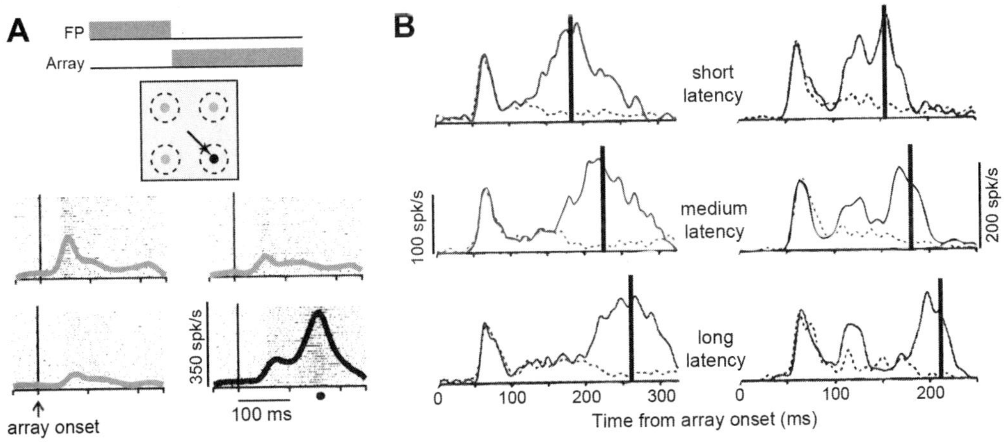

Figure 4.12
Activity in the superior colliculus during a visual search task in which the animal had to select the unique stimulus among four stimuli. **A**. Activity in four simultaneously recorded collicular neurons. Each stimulus was positioned inside the response field of one neuron. Upon onset of the visual array, all neurons responded at a latency of around 80 ms. However, activity in only one of the neurons (that which encoded the unique stimulus) was sustained and then increased, resulting in a saccade. Dot at bottom right figure indicates average time of saccade. (From Kim and Basso, 2010.) **B**. Each column displays response of one neuron. The response to distractor stimulus in the response field is shown with the dashed line, and the response of the same cell to the target stimulus is shown by the solid line. The initial burst, occurring at around 70 ms, signaled occurrence of a salient event in the response field of the neuron, but did not signal the utility of that stimulus. Time of saccade is shown with the heavy black line. (From McPeek and Keller, 2002.)

period, the neurons signaled that a salient visual event had occurred. However, this response did not indicate the utility of that event. If the stimulus in the response field was the target, activity increased at a rate that appeared to foretell the latency to movement. In these cells, utility was reflected in the activity that began to rise around 100 ms or later following stimulus onset. Indeed, there was diversity among the cells, with some showing a second peak, before a third and final rise to threshold. The final burst, which appeared to be largest, was the predictor of the saccade latency.

4.8 Effect of Reward on the Colliculus

Is the rate of rise in the activity of the buildup cells related to the utility of the ensuing movement? One way with which utility of a movement may be modulated is via reward magnitude. Increased reward increases the utility of the action, and this coincides with reduced reaction time and increased vigor of the movement. For example, Reiko Kawagoe and colleagues (1998) trained monkeys to look at a fixation point and then flashed a target (for 100 ms) at one of four locations. In a given block of trials, only one of the four locations was paired with a reward. Once the fixation point was removed, the monkey made a saccade to the remembered location of the target (memory-guided saccade task).

Saccades that were made to the rewarding target had lower latency and higher velocity than saccades that were made to the same target in blocks in which it was not paired with reward (figure 3.5A and figure 3.5B).

The rate of rise in the buildup cells is modulated by utility of the stimulus. Takuro Ikeda and Okihide Hikosaka (2003) recorded from the caudal regions of the superior colliculus while the animal performed a version of this memory-guided saccade task (top row, figure 4.13A). The trial began with a fixation point. After 1 s, a target was flashed for 100 ms in either the cell's response field or outside the response field in the mirror symmetric position relative to the fixation point. The fixation point turned off 1.0–1.5 s after target presentation, and this served as the cue that instructed the animal to make a saccade to the remembered location of the target.

As the animal looked at the fixation point and waited for the appearance of the target, many cells showed activity buildup. Because there were only two possible targets, there was 50% chance that the target would appear in the response field of the neuron. We

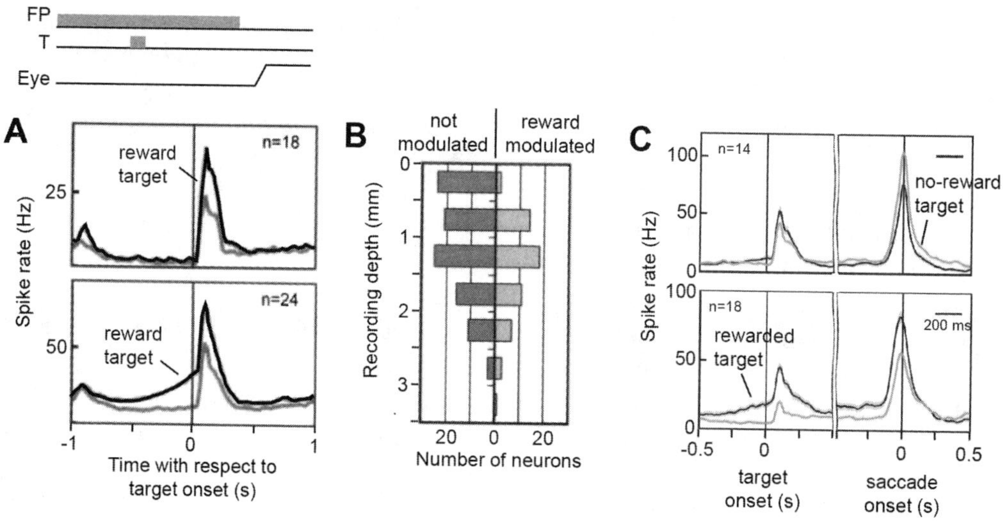

Figure 4.13
Reward modulates response of cells in the superior colliculus. Monkeys watched a screen that displayed a fixation point. At 1 s after onset of fixation, the target was flashed for 100 ms. Saccade to the remembered location of the target took place after offset of the fixation point. **A**. Activity of neurons that not only responded when the target appeared in their response field, but also modulated this response as a function of reward. All modulated neurons responded more strongly to the rewarded target. (From Ikeda and Hikosaka, 2003.) **B**. Depth of the neuron and whether its discharge in response to target onset was modulated by reward. Most cells that were modulated by reward were in the intermediate layer. (From Ikeda and Hikosaka, 2003.) **C**. Activity of neurons that responded near saccade onset and modulated this response by reward value. Some neurons with saccade-related activity showed negative reward-related modulation (reduced activity with reward), while a majority showed positive reward-related modulation (increased activity with reward). (From Ikeda and Hikosaka, 2007.)

would expect that in anticipation of target presentation, many cells should show buildup of activity (bottom row, figure 4.13A), while some cells may not show this buildup (top row, figure 4.13A). If the target was associated with a reward, in some cells the buildup of activity rose faster than if the same target was not associated with a reward (bottom row, figure 4.13A). Once the target was presented in the cell's response field, there was a short delay before some of the cells showed a burst of activity, exhibiting a visual response. Of the 156 cells that exhibited a visual response, 56 (36%) showed greater activity in response to presentation of the rewarding target.

Ikeda and Hikosaka (2003) noted that while only 1/3 of the cells recorded in the superior colliculus were modulated at target onset by reward, these cells were generally located in the intermediate and deeper layers, and not in the superficial layers (figure 4.13B). The intermediate and deep layers of the superior colliculus receive inputs from the basal ganglia, the frontal eye fields, and the LIP area in the parietal cortex (figure 4.1). The activity in these cortical and subcortical areas is modulated by reward and could potentially drive the buildup of activity in the colliculus.

Ikeda and Hikosaka (2007) also described the activity of the same cells around the time of the saccade. They focused on the cells that exhibited significant saccade-related activity (101 neurons) and found that 39% of these neurons showed modulation by reward. Most of these neurons showed higher activity when the rewarding target appeared in the response field of the cell, as illustrated in the bottom row of figure 4.13C. These cells had exhibited a higher buildup of activity in anticipation of the reward-target presentation and maintained their higher activity in anticipation of making a saccade after the removal of the fixation point. Indeed, just before saccade onset, these cells produced a burst after the rewarding target was presented.

However, in addition to these positive reward-coding neurons, there were also a substantial number of negative reward-coding neurons. At saccade onset, the negative reward-coding neurons showed a burst of activity that decreased in intensity if the saccade was to a rewarded target (top row, figure 4.13C). The negative reward-coding neurons tended not to exhibit a buildup of activity in anticipation of the saccade. Interestingly, the recording sites were deeper within the layers of the superior colliculus for the positive reward-coding neurons than for the negative reward-coding neurons.

In summary, expectation of reward increased utility of the action and resulted in movements that occurred after a shorter latency period and were performed with higher vigor. Before the presentation of the visual stimulus, some of the buildup cells in the superior colliculus exhibited a greater rate of increase in their activity in anticipation of making a rewarding movement. Therefore, the modulation of buildup rate appeared consistent with the idea that utility of the movement was reflected in the rate of increase in the buildup activity. When a visual stimulus that was the goal of the movement fell in the response field of the cells, after a delay, some cells responded by discharging (the visual response). In many, cells the magnitude of this visual response was larger if that stimulus was associated with reward. If the target was paired with a reward, as the animal waited for the fixation point to disappear, there was greater activity among the buildup cells and a larger burst at saccade onset. Other cells that showed little or no buildup of activity before the saccade still produced a burst near saccade onset, but this peak was smaller when the target was paired with a reward.

Probability of reward modulates activity of collicular cells.
A different way to alter utility is to introduce uncertainty as to whether a stimulus will be rewarded. This uncertainty can be represented as a probability. For example, if a stimulus is paired with reward α with probability p, and zero reward with probably $1-p$, then the stimulus has an expected value of $\alpha p+(0)(1-p)$, or simply αp. If the reward is harvested during period t_h, the utility of the movement toward that stimulus is a function of the expected value of the reward (and effort):

$$J=\frac{f(\alpha p,t_h)-g(t_m)}{t_h+t_m} \tag{4.7}$$

The effect of the probability of acquiring a reward is similar to the effect of the magnitude of the reward: an increase in probability of reward decreases saccade duration (producing a faster saccade) and increases the utility of the reward, as shown in figure 3.4A. As a result, an increase in probability of reward should increase movement vigor and reduce latency.

One way to modulate the probability of acquiring a reward is to present a certain number of potential targets, and only later specify which one is the actual target of the movement. If we imagine that the cells of the colliculus produce activity that relates to the utility of their movement vector, then a reduction in probability of selecting a movement also reduces the utility of that movement and, in turn, should result in a reduction in the discharge of the cells that encode that movement vector.

Michele Basso and Robert Wurtz (1997) performed an experiment (figure 4.14A) in which the fixation point was displayed for about 1 s (0.8–1.2 s), then an array of potential saccade targets were displayed. During a given trial, the array had 1, 2, 4, or 8 targets. The array stayed visible for about 1 s, and then one of the targets on the array was dimmed. After another delay of about 1 s, the fixation point was removed, signaling the animal to make a saccade to the dimmed target. In all trials, the saccade made to the dimmed target was rewarded. However, at array onset, the animal did not know which stimulus would be selected. The probability of selection was 1, 1/2, 1/4, or 1/8. Did the response to array onset take into account this probability?

Basso and Wurtz (1997) first found the optimum movement field of the cell by recording its saccade-related response when the animal made saccades to a single target. They then ensured that one of the targets in the array was always located at the position that produced the maximum saccade-related response in the cell. The response of a

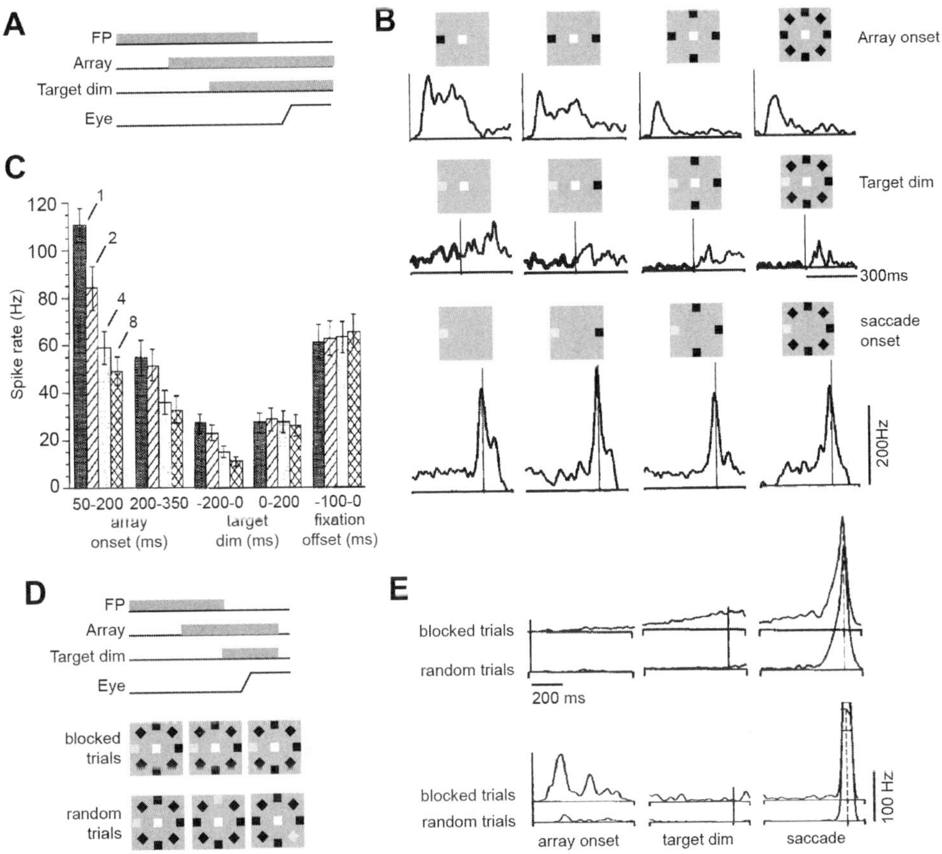

Figure 4.14
Effect of target uncertainty on activity of cells in the superior colliculus. **A**. Experimental protocol. **B**. Activity in one cell in response to array onset, target dim, and saccade onset. The array consisted of one to eight targets. The target that was selected was always in the response field of the neuron. **C**. Average spike rate across all neurons during various time periods. Uncertainty, as reflected by the number of targets in the array, reduced the response to array onset and the response during the period before target was selected. **D**. Protocol for a second experiment. The array always consisted of eight targets. In the blocked trials, a single target was repeatedly selected. **E**. Discharge of two cells during the blocked and random trials. When the stimulus in the response field of the cell is certain to be selected, the cell shows a larger buildup of activity or a stronger response to array onset. (From Basso and Wurtz, 1997.)

representative cell is shown in figure 4.14B. For this cell, the response field was for a target to the left of fixation. At array onset, presence of a single target produced the largest response, and presence of eight targets produced the smallest response (50–200 ms after array onset, figure 4.14C). Before one of the targets was dimmed, there was buildup of activity that anticipated selection of the target in the response field. As shown in figure 4.14C, this buildup was largest for the one-target array (200 ms before target onset) and smallest for the eight-target array (0 ms before target onset). Once the target was dimmed, the cell's activity level in the one-target trial no longer differed from the response in the eight-target trial. Therefore, both the response that followed array onset and the buildup of activity that preceded selection of the target reflected the probability of whether the stimulus in the cell's response field would be selected for a rewarded movement.

A potentially confounding variable with this experiment is the luminance of the visual display, which may have changed with the number of targets in the array and thus may have affected the response to the array onset. To account for this possible situation, Basso and Wurtz (1997) designed a second experiment in which the trial always included an array of eight targets (figure 4.14D). About 1 s after array onset, one of the targets was dimmed and simultaneously the fixation point was removed, signaling the animal to make a saccade to the identified target. To modulate probability of selecting a particular target, in one group of trials the stimulus in the response field of the neuron was always the target that was selected. These were termed blocked trials (figure 4.14D). In another group of trials, the target was randomly selected for each trial (random trials, figure 4.14D). As a result, in the block trials, the stimulus in the response field of the cell had 100% chance of being selected. In the random trials, that stimulus had a one in eight chance of being selected.

Responses recorded from two representative cells are presented in figure 4.14E. In the block trials, one cell did not have a response to the visual input at array onset but exhibited buildup of activity in anticipation of target selection (top row, figure 4.14E). In the random trials, the same cell showed little or no buildup of activity. Therefore, this cell's rate of buildup increased in trials in which the movement that the cell encoded was likely to be paired with a reward. A second cell's activity is illustrated in the bottom row of figure 4.14E. This cell had a strong response to the visual input at array onset during the block trials but produced only a weak response in the random trials. Therefore, this cell showed modulation in its response to the visual input, exhibiting greater activity in the case in which a saccade to the stimulus in its response field was likely to result in a reward.

In summary, when the location of a visual stimulus serves as the goal of a movement, the reward associated with that movement modulates the response of superior colliculus cells: cells increased their activity in anticipation of the onset of the rewarding stimulus (figure 4.13A). When there is variance in the probability that a movement to a particular location will result in reward, the discharge of neurons that encode movement to that location is modulated by this probability (figure 4.14B, target dim subfigure). It seems possible that the reward-induced reduction in latency may be associated with the increased rate of buildup before the saccade, reaching threshold sooner. However, it is unclear whether the reward-dependent changes in saccade vigor are associated with the modulation of the burst of activity that takes place near saccade onset in the colliculus.

4.9 Reacting Much Earlier

The activity of the buildup cells in the colliculus to some extent reflects the decision variable, rising from a baseline toward to a threshold and ultimately triggering a saccade to a specific location. The rate of rise depends on the utility of the stimulus; activity builds faster when a greater utility is at stake and rises more slowly when the utility is uncertain. In addition, the activity of these cells at target onset affects the reaction time, which is shorter when there is greater baseline activity and longer when the baseline activity is lower. If we take this argument to its conclusion, we would guess that at target onset, if the activity of the buildup cells that encode a macrosaccade is high enough, the arrival of the visual stimulus should bring the buildup cells to threshold, making them burst immediately after the detection of the visual stimulus. In this scenario, the shortest reaction times will be composed of the time it takes for the visual stimulus to reach the superior colliculus and the time it takes for activity of the burst cells of the colliculus to start that saccade. It turns out that the brain can produce these extremely short reaction times in certain conditions, and the critical structure that makes it possible is the superior colliculus.

It is possible to make movements without computing its utility: express saccades. Express saccades are movements that follow target onset with a latency of 120 ms or less. The fact that such short-latency movements can be generated is important because it forces us to confront the question as to why normally it takes so much longer to start a movement. We will suggest that under normal circumstances, a movement takes about 200 ms to start because the brain structures that sit atop the superior colliculus usually act to prevent it from producing a reflexive response to the visual stimulus: they evaluate that stimulus, and on the basis of its utility, decide whether to move toward it or not. Computing a utility takes time, a reflection the price that we pay for flexibility.

The behavioral study of express saccades suggests that when reward is associated with a stimulus that repeatedly occurs at a particular location, the healthy brain can produce movements in response to that stimulus with extremely short reaction times. The study of the neural basis of express saccades demonstrates that these movements occur because the buildup cells that encode the repeatedly rewarded movement are primed to reach threshold, exhibiting a high baseline activity in anticipation of the arrival of the visual stimulus in their response field. In certain pathological conditions, this priming occurs routinely, resulting in the inability to withhold movements in response to the visual stimulus. The study of express saccades in disease, something that we will take up in the next chapter, raises the possibility that a function of the cortex and the basal ganglia is to inhibit the tendency of the colliculus to reflexively react to visual stimuli, and instead to replace it with a flexible response that depends on its utility.

To see how an express saccade may be generated, we will reconsider the data in figure 4.3. The presentation of the visual stimulus (labeled "A") in the response field of the cell in the superficial layer results in a burst of activity with a latency of around 80 ms. This same stimulus also produces activity in the neuron in a deeper layer, but this activity is not large enough to produce a saccade. The saccade takes place later when the neuron in the deeper layer (visuomotor neuron) produces a larger burst, about 100 ms after its initial response to the visual stimulus. Another example is illustrated in figure 4.7A. In the first column of figure 4.7A, we have a condition in which the buildup cell is presented with a target in the contralateral direction. The cell responds with an initial peak in discharge at around 80 ms after target onset (this is the visual response), and then a second but larger peak just before saccade onset. As we will see, express saccades occur when the baseline activity of the cell is high enough so that the onset of the target in its response field brings its activity to threshold, thus generating a saccade. That is, an express saccade occurs when the first burst (due to a visual response) and the second burst (sending a motor signal to generate a saccade) merge into one, producing a saccade at the time of the visual response.

Martin Pare and Doug Munoz (Pare and Munoz, 1996) trained monkeys in a task in which gap and no-gap trials occurred in random order (figure 4.15A). The target always appeared to the right of fixation. Therefore, the animal knew precisely the location of the target on each trial, but did not know the timing of the target onset. The distribution of saccade latencies in the two types of trials for one monkey in one session (session 2) are displayed in figure 4.15B. The distribution had two modes: express saccades that had latencies of less than 120 ms and regular saccades that had longer latencies. With each training day, the portion of express saccades in each type of trial increased. Importantly, express saccades were always more likely in gap trials than no-gap trials.

Michael Dorris, Martin Pare, and Doug Munoz (Dorris et al., 1997) hypothesized that these extremely fast reaction times were possible because with training, the inputs to the superior colliculus from the basal ganglia and the cerebral cortex had changed, making it possible for the visual response of the colliculus to bring activity of buildup and burst neurons to threshold, resulting in a saccade. To explore this hypothesis, they recorded from the superior colliculus and found that during the period before presentation of the target, the fovea-related neurons showed a reduced discharge in anticipation of the fixation point offset, but this reduction did not differ between express and regular saccades (figure 4.15C, part 1). After the fixation point offset, there was a pause during the saccade (figure 4.15D, part 1). The activity of these cells did not appear to differ between trials in which the animal exhibited an express saccade and those in which the animal expressed a regular saccade. Therefore, the shorter reaction time in the express saccades did not appear to arise from a faster rate of fall in the activity of the fovea-related neurons.

In contrast, express saccades occurred in trials in which the buildup neurons had a greater activity at target onset (figure 4.15C, part 2). Interestingly, when an express saccade did not take place, the buildup neurons still had a burst in response to target onset, but this burst was too small to drive a saccade, as it was followed by a second burst at a delay of 100 ms (figure 4.15C, part 2). This second burst coincided with the onset of a regular saccade. Finally, in cells that did not exhibit buildup activity, presentation of the visual stimulus produced bursts in trials in which the animal generated an express saccade (figure 4.15C, parts 3 and 4)

In summary, saccades with regular reaction times (200 ms) occurred because the visual response in the buildup cells was normally not large enough to drive them to threshold.

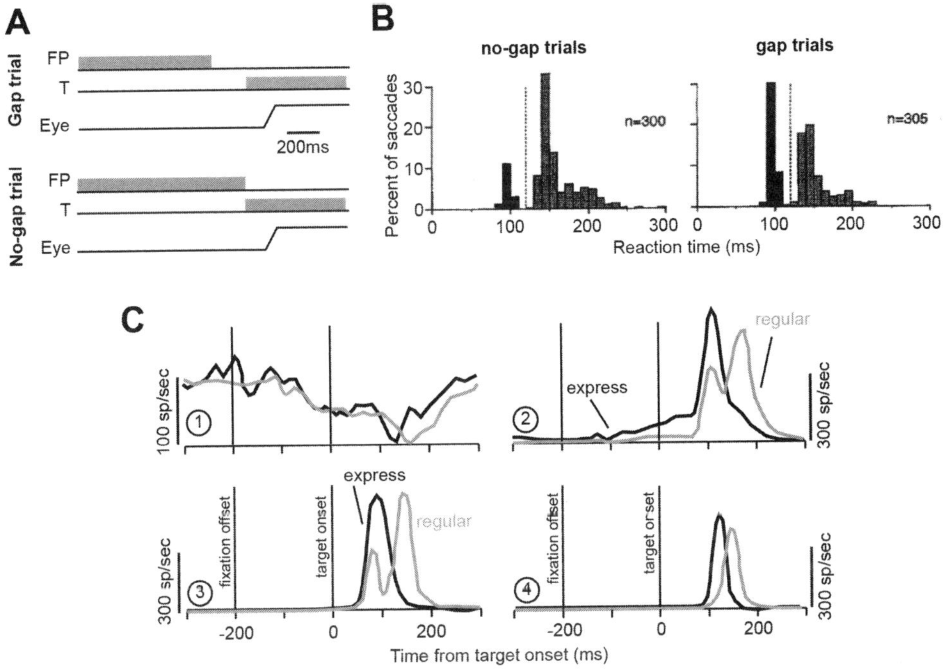

Figure 4.15
Express saccades. **A**. In the task, gap and no-gap trials appeared in random order. The target was always at the same location with respect to fixation. As a result, the animal was certain of target location but uncertain of target onset. **B**. Performance of a monkey during the two types of trials in session 2 of training. Express saccades are separated from regular saccades by the dashed line. **C**. Recording from four cells in the superior colliculus during a gap task with a 200 ms gap period. The task had only gap trials, but the target position was randomly placed in the cell's response field or polar opposite. Express saccades occurred when the visual response to target onset was large enough to generate a burst. (1, fovea-related neuron; 2, buildup neuron; 3 and 4, burst neurons) (From Pare and Munoz, 1996; Dorris et al., 1997.)

After a transient visual response, activity of many cells typically returned to near baseline, then had to buildup to threshold. However, with training, it was possible for a healthy brain to produce reaction times that were near the limit of sensory-to-motor information transmission, around 100 ms. These express saccades occurred because the buildup cells of the superior colliculus were primed to receive the excitation generated by the visual input. That visual input took the cells to threshold, triggering a burst that resulted in a saccade.

Express saccades rely on a network that resides in the colliculus.

Is the principal circuit responsible for generating an express saccade in the visuomotor network within the superior colliculus? The answer appears to be yes, because unilateral damage to the superior colliculus completely eliminates express saccades to the opposite side, but leaves the other, slower reaction time saccades roughly unimpaired.

Peter Schiller and colleagues (Schiller et al., 1987) tested the role of the superior colliculus in generating express saccades by training monkeys in a detection task in which after fixation offset and a gap of time (fixed within block to 16 ms), a target appeared at one of two locations. The animal made saccades to the target with reaction times that showed a bimodal distribution: express saccades with short reaction times, and normal saccades with long reaction times (figure 4.16B, top panel). Three days after lesion to the left superior colliculus, express saccades to the left increased in frequency, but those to the right were completely eliminated (figure 4.16C). The distribution of reaction times for rightward saccades was much slower than before, illustrating that soon after damage to the colliculus, contralateral saccades were possible but greatly delayed. After 70 days, leftward saccades showed a normal distribution of latencies, but no express saccades were made to the right.

Schiller and colleagues (Schiller et al., 1987) also trained the animals in a discrimination task in which the target was now a green stimulus among three to five red distractors. The position of the target was randomly changed in each trial. The animals never produced express saccades during this task (figure 4.16B, lower panel). The lesion to the left superior colliculus increased saccade latency in this task, particularly for rightward saccades (figure 4.16D).

These results demonstrated that a fundamental and enduring effect of damage to the superior colliculus was the complete elimination of express saccades to the contralateral side. Therefore, the circuitry for generating the fastest possible reaction times is within the colliculus. This circuitry is normally inhibited by the contralateral colliculus, as evidenced by the fact that at 3 days after the unilateral colliculus lesion, express saccades became more frequent to the side contralateral to the intact colliculus (figure 4.16C, second row).

Arm muscle activity suggests a possible non-cortical pathway for visual stimuli to produce a reach.

Are short reaction times also possible for arm movements, and if so, is there is a pathway in the brainstem to activate arm muscles in response to a visual stimulus, long before the cortex has had a chance to evaluate that stimulus? Recent results hint that like saccades, there may be a brainstem pathway for rapid arm responses to visual stimuli. Chao Gu, Daniel Wood, Paul Gribble, and Brian Corneil (Gu et al., 2016) recorded from the pectoralis muscle (a shoulder flexor) while subjects made movements toward or away from a target (figure 4.17). During each trial, the color of the fixation point provided instructions regarding whether the movement should be toward or away from the upcoming target. The targets appeared randomly to either the left or the right of the fixation point, and subjects were instructed to move as quickly as possible in response to the presentation of the target.

The authors found that regardless of the instruction, there was a small but robust activation (or inhibition) of the pectoralis muscle at 80–120 ms after target onset. This activation appeared to be instruction-independent. For example, regardless of whether the

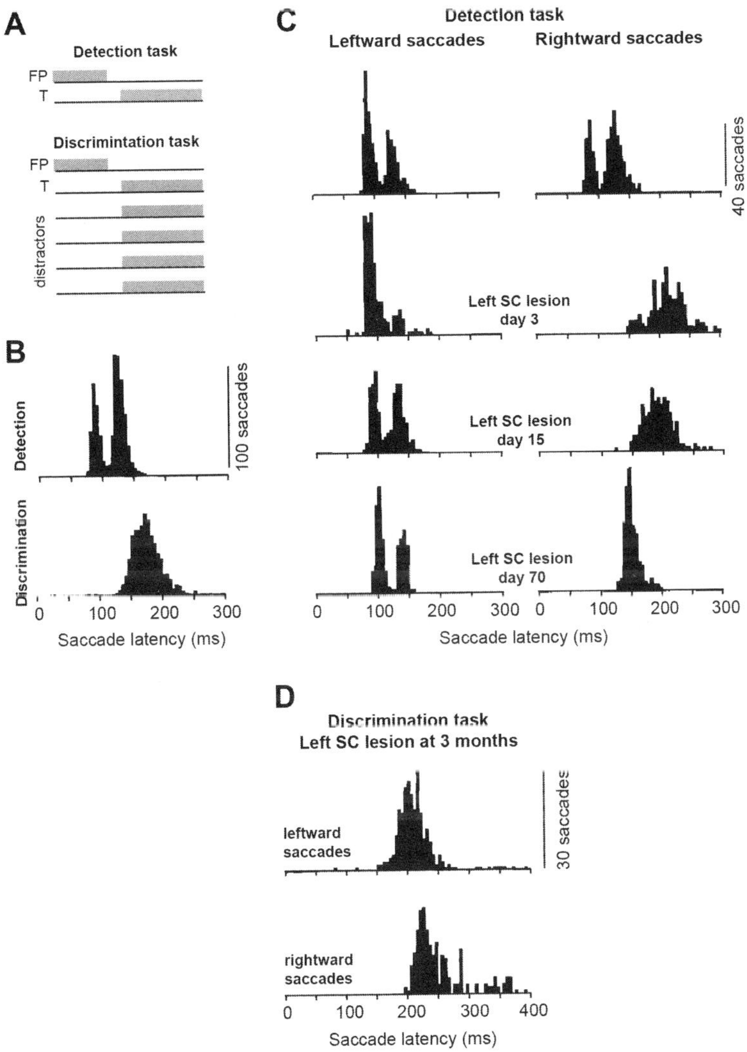

Figure 4.16
Damage to the colliculus permanently removes the possibility of express saccades. **A**. In the detection task, after fixation offset, the target appeared after a gap of 16 ms in one of two locations, randomly selected on each trial. In the discrimination task, the target was a green stimulus among three to five red distractors. The positions of the stimuli were random on each trial. The tasks were examined on separate days. **B**. Reaction times were bimodal in the detection task: the animal produced express saccades as well as regular saccades. In contrast, reaction times were unimodal in the discrimination task, and the animal never produced express saccades. **C**. Effect of damage to the left superior colliculus (SC). At 3 days after the lesion, express saccades to the right had disappeared. In contrast, express saccades to the left had increased. With the passage of time, the number of express saccades to the left returned near normal, but no express saccades were made to the right again. This suggests that the colliculus contains the essential circuitry for express saccades. **D**. In the discrimination task, damage to the colliculus increases the latency for contralateral saccades. (From Schiller et al., 1987.)

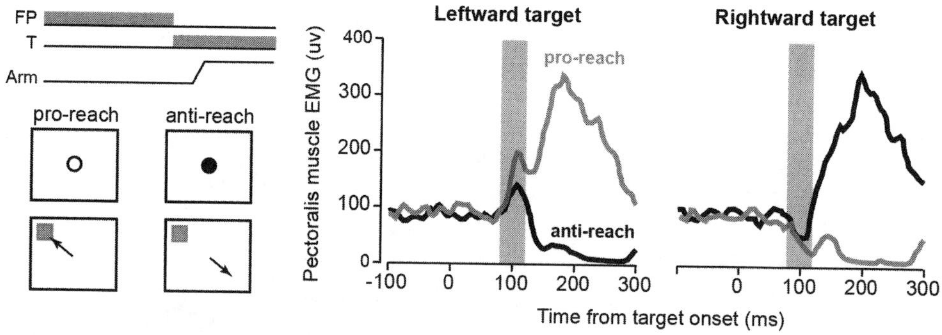

Figure 4.17
Early activity in the pectoralis muscle before onset of a reaching movement appears to be dependent on the stimulus location, not the instruction. On each trial, subjects were instructed to either reach toward (pro-reach) or reach away (anti-reach) from the visual target. When the target was toward the left, target onset resulted in activation of pectoralis muscle at 100 ms regardless of the instruction. (EMG: electromyogram) The reach typically started at around 200 ms. Similarly, when the target was toward the right, target onset resulted in inhibition of the pectoralis, regardless of the instruction. Data are from a single subject. (From Gu et al., 2016.)

instruction was to move toward or away from the target, if the target was located on the left side (a target for which a reach would require activation of the pectoralis), then the pectoralis experienced a small activation at around 100 ms. However, if the target was toward the right, then regardless of the instruction, the pectoralis experienced a small inhibition at around 100 ms. In contrast to this early response, a later response in the muscle (around 180–250 ms) was strongly dependent on the instruction, resulting in a movement toward or away from the target location. Reach reaction times toward the target were around 200 ms. Therefore, the instruction-independent response at 100 ms is akin to an express reach. Because this early activity is instruction independent, it has been termed the *stimulus locked response* (Pruszynski et al., 2010). Electrical stimulation of the superior colliculus can evoke activity in the limb muscles (Philipp and Hoffmann, 2014). Indeed, the authors (Gu et al., 2016) speculated that this early response was generated by the superior colliculus responding to the visual stimulus and activating neck and arm muscles after extremely short periods through the brainstem to spinal cord pathways.

Overall, it appears that under normal conditions, the onset of a visual stimulus produces activity in the buildup cells of the superior colliculus, but this visually triggered activity is not large enough to generate a saccade. After the visually driven response, there is a separate buildup of activity that, upon reaching the threshold, results in a saccade to the stimulus. Very short reaction times are possible when the anticipation of the stimulus results in buildup that precedes stimulus onset. In this case, an express saccade is generated because the visually induced response in the collicular cells adds with the already present buildup of activity to produce a combined response that reaches threshold, generating a saccade that has reduced the normal reaction time by half. Recent results suggest that under certain conditions, arm muscles are activated 100 ms after stimulus presentation. Because this activation appears to be independent of the instruction for that trial, its presence has suggested that the colliculus may also have a role to play via brainstem pathways in generating early responses in the arms. Much, however, remains unknown regarding the short-latency visuomotor pathways in control of the arm.

4.10 Influencing Decisions with Manipulation of the Colliculus

Although there is some evidence that activity in the neurons of the superior colliculus may be related to computation of a movement utility, the actual test of this hypothesis requires one to examine whether manipulation of this activity biases the process of decision-making. Is the colliculus part of the network that computes the utility of the movement, participating in the decision-making process, or is it simply acting as a low-level surrogate of an upstream overlord, controlling the motor commands that are needed to produce the movement?

Inactivation of a region of the colliculus makes it hard to direct attention to the corresponding region of the visual space.

Christopher Carello and Richard Krauzlis (Carello and Krauzlis, 2004) lowered their electrode to the intermediate or deep layers of the colliculus and then stimulated that site, which resulted in a saccade vector indicating the response field of the cells at the stimulation site (indicated by the cross-hatched circle in figure 4.18). They then presented the monkey with a fixation spot. After fixation, either a gray or a white cue appeared, one of which was inside the response field of the stimulation site. This was followed by reappearance of the fixation spot, and then presentation of either a gray or white target. In the saccade version of the task, the monkey was rewarded if it made a saccade toward the target that was previously cued. In the pursuit version of the task, both targets moved toward the fixation point, but the monkey was rewarded if it pursued the target that was previously cued. In half of the trials, starting at around 100 ms before the target+distractor onset, the researchers stimulated the colliculus site at a subthreshold level and continued the stimulation for around 400 ms. Therefore, the stimulation occurred during the period when the animal would be deciding which was the target and which was the distractor. If collicular activity influenced decision-making, then two things should happen. In the saccade task, stimulation should bias the decision-making process by generating movements toward the object in the response field of the stimulated cells, regardless of whether the object was the cued target. In the pursuit task, stimulation should bias the decision-making process by generating pursuit of the object that was located in the response field of the stimulated cells.

In the trials without stimulation, the animal's decisions were usually, but not always, toward the correct target. In the saccade trials, the monkey's performance was 93%

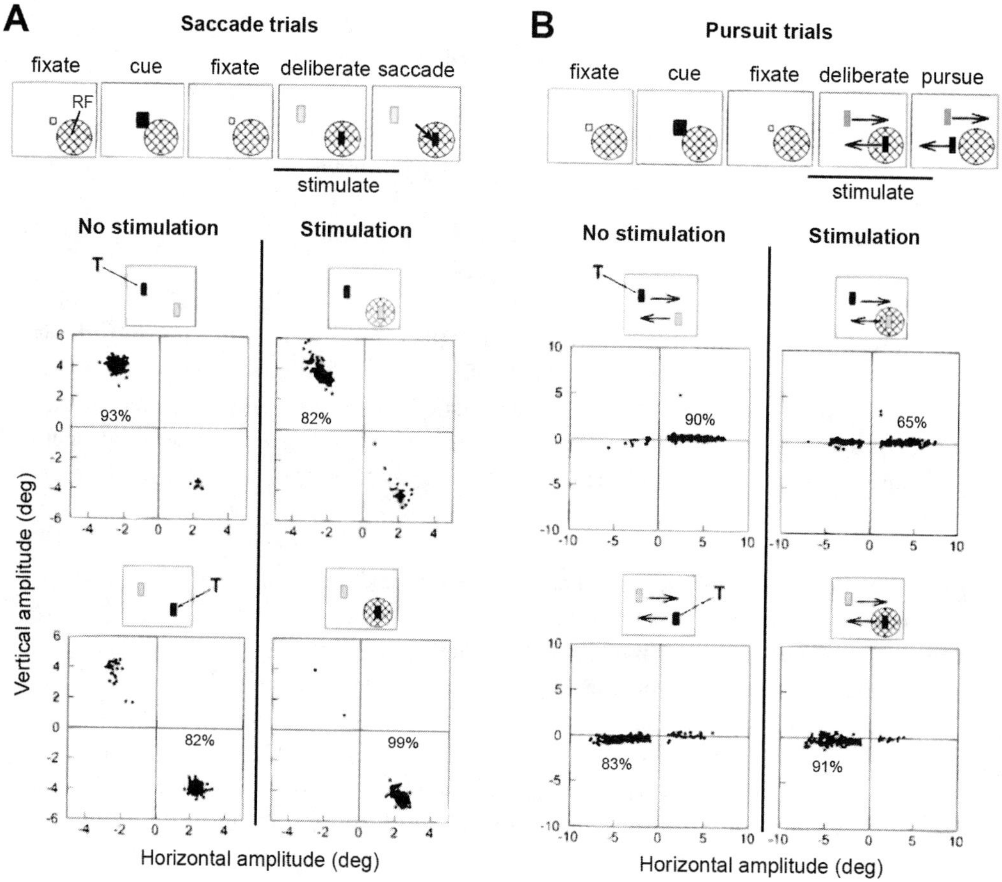

Figure 4.18
Evidence for a causal role for the superior colliculus in deciding which action to perform. **A**. Saccade task. The monkey performed a delayed sample to match task. The trial began with a fixation spot, which was then followed by a white or a gray cue. Following refixation, a target and a distractor were displayed. The monkey was rewarded for making a saccade to the target. The cross-hatched circle indicates the saccade that was generated when a part of the superior colliculus was stimulated at above threshold (RF, response field of the stimulated site). In the no-stimulation condition, the monkey made the correct choice (saccade to the target) in 93% and 82% of the trials, depending on target location. In some trials, stimulation started at 100 ms before presentation of the target and distractors. The stimulation biased the decision, making the animal more likely to choose the object that appeared in the response field of the stimulation site. **B**. Pursuit task. The trial began with fixation and presentation of a cue. After refixation, a target and a distractor were displayed and both began moving toward the midline of the screen. The animal was reward for pursuing the target, not the distractor. Stimulation made the animal more likely to pursue the object that had appeared in the response field of the neurons. (From Carello and Krauzlis, 2004.)

correct when the target was on the left side and 82% correct when the target was on the right (left column, figure 4.18A). Sub-threshold stimulation biased this performance. When the distractor was in the response field (right column, middle row, figure 4.18A), the stimulation biased the probability of saccades away from the target and toward the distractor; the monkey's performance was correct only 82% of the time. However, when the target was in the response field, stimulation biased the probability toward the target, raising the probability of success from 82% to 99% (right column, bottom row, figure 4.18A). Stimulation also affected the latency of the saccades by reducing the latency when the target was inside the response field and increasing the latency when the distractor was inside the response field.

The stimulation was subthreshold and by itself did not generate a saccade. However, the results may be hard to interpret because stimulation of the site, if strong enough, could induce a saccade that was in the same direction as the bias induced in the decision. The pursuit task allowed for dissociation between the decision of what to do and the direction of movement that resulted from suprathreshold stimulation of the colliculus.

In the pursuit trials without stimulation, the performance was 90% correct when the target moved to the right and 83% correct when the target moved to the left (left column, figure 4.18B). Sub-threshold stimulation biased this performance. When the distractor was in the response field, which was located on the right side, pursuit was biased toward the distractor; the animal had a greater tendency to pursue motion of the distractor (right column, middle row, figure 4.18B). Notably, when the animal chose the distractor, the resulting movement was away from the point where the distractor first appeared, not toward it. When the target was in the response field (right column, bottom row, figure 4.18B), pursuit was biased toward the target. Once again, the movement itself was away from the location where the target first appeared. Therefore, the authors found that stimulation of the region of the superior colliculus that corresponded to the location where the cue had originally appeared biased the decision of which target to pursue, as well as the latency of the pursuit. These results show that activity in the colliculus is participating in the choosing the objective for the next movement, not just generating that movement.

In a follow-up study, Samuel Nummela and Richard Krauzlis (Nummela and Krauzlis, 2010) re-examined the performance of monkeys in the same task, but manipulated activity of the colliculus through deactivation via muscimol injection. In addition, they considered a reach task in which the same two cues appeared, but the monkey expressed its decision via pressing one of two keys in a response box. The day before each experiment, the researchers identified an injection site using single-unit recordings and microstimulation. They then injected muscimol and noted reductions in peak velocity of visually guided saccades, delineating the retinotopic space affected by the injection. On the next day, they injected muscimol again and found that the animal was less likely to select the target when it appeared in the affected region; this pattern held true during both the saccade and the pursuit trials. The key-pressing task was also affected, although mildly so.

Importantly, the researchers also interleaved trials in which only the target appeared (no distractor). In this case, they found no effect on the probability of correct selection. The monkeys could pursue or make saccades when the target appeared in the affected region, but not if the same visual stimulus competed with a distractor that appeared outside the affected area. Therefore, in both the saccade and pursuit trials, inactivation of a region of the colliculus biased selection away from the visual stimulus that appeared in the affected visual field, particularly if the selection involved a comparison with a competing stimulus outside the affected area.

These results illustrate that stimulation or inactivation of a region of the colliculus alters the allocation of attention, especially when there are multiple visual cues that compete to identify the goal of the movement. In addition to receiving input from the cortical decision-making system, the colliculus itself actively manipulates the decision-making process by weighing sensory information. In the next chapter, we will see further data illustrating that inactivation of a part of the colliculus not only makes movements less likely to the region controlled by that part, but it also effectively produces neglect-like symptoms by making the subject discount the visual information that appears in the affected part of the space.

Limitations

Largely because of the discoveries of David Sparks, Robert Wurtz, and their students, over the past 30 years, the superior colliculus has become arguably the single most well studied and best understood sensorimotor region of the brain. However, important puzzles still remain.

The function of the rostral region of the colliculus was thought to be related to the generation of motor commands that are needed to hold the eyes still during fixation. Indeed, the neurons in this area were initially referred to as fixation neurons. However, disruption of this area does not affect the ability to hold still, but rather alters the equilibrium position of the eyes. When the rostral pole region is injected with muscimol, the animal can still fixate on a spot of light. In fact, fixation becomes more stable because microsaccades become less frequent (Goffart et al., 2012). The principal effect of the disruption is to bias the precise position of the eyes, shifting them slightly off target. For example, if the disruption is to the right colliculus, during fixation the eyes shift slightly to the right of the target. The function of the neurons in this region of the colliculus still remains to be better understood.

The work of Richard Krauzlis and colleagues (Krauzlis et al., 2000) has demonstrated that in the rostral pole, cells burst during microsaccades and increase their discharge during pursuit. Their discharge appears to depend on the distance of the valuable visual stimulus with respect to the fovea. Here, we viewed this activity in terms of the utility of the visual stimulus near the fovea. Our inference was based on the fact that the fovea-related neurons are anatomically adjacent to the buildup cells that appear to encode value of a distant visual object (figure 4.7B). Therefore, we imagined that value encoding was a defining feature of these cells, but experimental evidence for this idea is indirect. Experiments are needed that alter value of the stimulus near the fovea and measure response of rostral cells in the colliculus.

A function of the cells in the caudal regions of the colliculus is clearly related to encoding the utility of the movement vector, as disruption of this region alters the decision of what to do. However, we have not considered the role of these cells in actually generating a saccade. This control depends on coordination of activity between the superior colliculus, brainstem burst generators, and the cerebellum, transforming decisions into control policies that accurately bring the eyes to the intended target. The specific details of this topic lie outside the scope of our book.

An important note of caution is required in interpreting neural data for which utility is manipulated via reward: when the reward value associated with visual stimuli increases, so does the attention that is allocated to those stimuli. To test whether the neural changes are associated with utility, one needs to consider not only positive utility events that produce gains, but also negative utility events that produce loss. For example, a stimulus that is associated with a large loss is important and will gather attention, as will a stimulus that is associated with a large gain. However, the utility associated with a loss is lower than one that is associated with a gain. Decision-making under time pressure is influenced by salience, but without time pressure, decision-making is largely driven by utility. In the next chapter, we will see examples of experiments that have attempted to disentangle neural activity that encodes salience from the neural events that encode utility.

Summary

The neurons in the superior colliculus are largely responsible for controlling fixation and saccades. In the rostral pole, collicular activity reflects the utility of staying still, maintaining fixation. In the caudal regions, collicular activity reflects utility of moving, generating a saccade. During the reaction time period, as we deliberate between our various options for movement, there is a reduction in the activity of the fovea-related cells in the rostral pole and a rise in the activity of movement-related cells in the caudal regions. The rate of change in these activity patterns reflects the difference between the utility of the stimulus that encourages us to hold still and the utility of the stimulus that encourages movement. The rate of rise in the activity of the movement-related cells is a reflection of the reward promised for completing the movement, and this rate of rise affects the reaction time and the vigor of the ensuing saccade. The earliest response of the movement-related collicular cells is not a reflection of the stimulus's utility, but rather its salience. Under normal conditions, it takes time for the brain to compute stimulus utility, and this information is provided to the colliculus by the cortex and the basal ganglia. However, under some conditions, the colliculus is allowed to bypass this computation. In this case, the colliculus directs movements that occur after extremely short reaction times, movements that are induced because of stimulus salience, not utility.

Computing the utility of a stimulus is complicated, partly because it involves subjective evaluation of reward and effort, and partly because choices and actions need to consider the history of experiences. When there is damage to a cortical area, it can affect the patterns of both decisions and movements. In our next chapter, we will consider how various regions of the cortex participate in computing utility and how this cortical activity leads to control of reaction time and movement speed.

5

Cortical Computation of Utility

Life is the sum of all your choices.
—Albert Camus

Habituation puts to sleep the eye of our judgment.
—Michel de Montaigne

The superior colliculus receives direct projections from the retina and responds to the onset of visual stimuli as early as 70 ms. Through its downstream projections to the pons, the colliculus has privileged access to the saccade generating machinery; without the colliculus, the cerebral cortex is not able to generate a saccade (Hanes and Wurtz, 2001). As a result, after the sudden appearance of a visual input on the retina, and absent of visual inputs that would encourage fixation, occasionally the superior colliculus issues a command for an express saccade, which occurs at a latency of around 120 ms. However, express saccades are not the normal state of affairs. Usually, another 80 ms pass before a movement is made. Why do we need this extended time to make a movement?

The reason for this delay is that the colliculus is not trusted with choosing between various targets of action. Instead, its privileged access to the gaze control machinery is controlled via the supervision that it receives from the cerebral cortex and the basal ganglia. The cerebral cortex provides the excitation that the colliculus needs to generate the rewarding movement, while at the same time the output nucleus of the basal ganglia removes the inhibition that it has been imposing, thereby allowing release from fixation. During the 100 ms delay between the arrival of visual input to the colliculus and the decision to move, these structures compute the utility of the new stimulus, deciding whether that utility warrants a movement, and if so, the vigor with which it will be expressed. In this chapter, we will consider the neural basis of this utility formulation and supervisory control.

We will begin with some of the data regarding effects of damage to the cerebral cortex and the basal ganglia on the control of gaze. Although the retina does provide the superior colliculus with direct inputs, the vast majority of retinal projections are destined to the cerebral cortex. In primates, more than a million fibers per eye project to the thalamus (dorsal lateral geniculate nucleus) on their way to the visual cortex, whereas about 150,000 fibers project to the superior colliculus. If there is damage to the primary visual cortex (V1), the subject becomes blind to the corresponding region of the contralateral visual space. Despite the fact that the original insult is limited to the visual cortex, thalamic neurons that project to the damaged cortical area degenerate, effectively deafferenting much of the cerebral cortex. However, with passage of time, the remaining regions of the cortex become privy to some of the visual information that the colliculus receives (conveyed from the colliculus to the thalamus, then to the cortex). Although subjects remain blind to the visual information in the affected part of their visual field and will state that they cannot see anything there, when forced to act, they will make saccades (and even point) to the stimulus that they cannot see.

This illustrates that conscious perception of visual stimuli depends on the cerebral cortex, not the colliculus. However, with training, people and other animals can rely on information received by the superior colliculus to direct their gaze. Therefore, with a damaged visual cortex we cannot see, but we can move our eyes (and arms) toward the unseen.

Because the superior colliculus has direct access to the gaze control machinery, partial damage to the colliculus can affect the ability to make saccades to the corresponding region of visual space. However, the effect of this damage goes beyond altering kinematics of a movement. Rather, the reduced ability to make a saccade of a given displacement vector also produces a cognitive deficit: the subject can no longer divert attention toward that region of space. Thus, with temporary damage to the colliculus and the ensuing disability in performing a particular saccade vector, the brain acquires a form of neglect, an inability to direct attention to the affected region of visual space. This implies that when we are paying attention to a particular location of the visual space (away from the fovea), we are covertly engaging the collicular machinery that controls moving our eyes there. If that collicular machinery has been partly disabled, requiring more activity than normal to generate a saccade, we become biased in how we direct our attention by avoiding the visual region to which directing our gaze requires greater effort. Thus, effort in the motor system affects the cognitive control of attention.

Damage to the frontal eye field (FEF), a cortical area that sends strong projections to the superior colliculus, returns the system to a state in which the colliculus has greater autonomy in choosing the action. The colliculus now responds by directing the eyes toward the most salient visual stimulus, rather than one that is associated with greater utility. To understand how the FEF may contribute to the computation of utility, we will consider the activity of neurons there and find that like the colliculus, the FEF houses neurons that are active when the eyes are fixating on a rewarding stimulus. These fovea-related cortical neurons pause when a saccade takes place. The FEF also houses neurons that build up their activity and burst around the time of a saccade. The firing rate of these saccade-related neurons reflects the utility of the task: if the destination is valuable, the saccade is more vigorous and is accompanied by greater discharge of these neurons. Indeed, the rate of rise in the activity of FEF cells corresponds to the reaction time of the movement, suggesting that these cells are playing a critical role in determining which movement will be performed.

However, a puzzle arises when we consider the relationship between the discharge of cortical cells and the timing of the movement: saccades that occur after a shorter latency tend to not only coincide with a faster rise of activity in the FEF cells, but also with a higher level of activity in these cells. The higher level of activity implies that contrary to some models, activity in the FEF and the colliculus does not need to reach a fixed threshold before a saccade is initiated. Therefore, if we imagine that a movement starts when the neural activity that represents a decision variable reaches threshold, that threshold is not set by the neurons in the FEF or the superior colliculus. Rather, it may be that the threshold for initiating a movement is set deep in the brainstem, in a nucleus that receives converging inputs from both the colliculus and the FEF and potentially integrates the activity in both regions before permitting the saccade to start.

5.1 Moving without Computation of Utility

In 1973, Ernst Poppel, Richard Held, and Douglas Frost (Poppel et al., 1973) examined four patients who had suffered cortical damage to their occipital lobe. The patients were unable to detect a flash of light in a part of the visual field contralateral to the damaged cortical hemisphere. That is, they had no light perception within their cortical scotoma, as evidenced by their inability to verbally report occurrence of the flash. The authors examined the patients in a forced-choice task in which during fixation of a central point, a target was flashed for 100 ms within the region of the cortical scotoma, and the subject had to saccade to it. Because the patients could not detect the flash, a sound was simultaneously played via a speaker at a central location. Therefore, the sound signaled occurrence of the visual event, but did not localize it. Indeed, the task was puzzling for the patients. One commented: "How can I look at something that I haven't seen?" Despite this, the patients produced a saccade, and remarkably, the saccades were not random. Rather, there was a small but significant increase in saccade amplitude with increasing target eccentricity. That is, although the patients were not conscious of seeing the targets, they nevertheless made saccades that hinted that some spatial information about the stimulus was in fact available to them, a phenomenon called *blindsight*.

Damage to the visual cortex removes the conscious ability to see, but preserves the ability to move toward the unseen.

To detect blindsight, a typical approach is to have the patient fixate on a central location, and then present a visual stimulus at a location in the cortical scotoma region during one of two intervals. The subject is then asked to guess in which of the two intervals the stimulus appeared. As stimulus contrast increases, the probability of picking the correct interval goes from chance to near 100% correct. However, although the patients are able to point or even reach out and grasp the object, they do not have a conscious perception of the visual stimulus.

This loss of conscious perception is not solely because the V1 has been damaged, but because that damage initiates further large scale degeneration: the neurons that project to the damaged V1 from the lateral geniculate nucleus of the thalamus undergo degeneration (Leopold, 2012). This effectively produces a visually deafferented cortical hemisphere. The cortical deafferentation, and not the damage to V1 alone, is likely the basis for the loss of conscious awareness of the visual stimulus in the scotoma.

In their original report on blindsight, Poppel et al. (1973) had speculated that "perhaps this effect is due to the midbrain in these patients." Indeed, despite a damaged V1, the retinal projections to the colliculus are preserved. Is this the pathway that allows the brain to localize the visual stimulus and direct actions toward it?

To test for this, Rikako Kato's team (Kato et al., 2011) trained monkeys to perform a visually guided saccade task. They presented the animals with a fixation point. After a random duration, the fixation point was removed and the target appeared at one of eight possible locations. A saccade to the target within 1 s of its appearance resulted in a drop of juice. Once the animals had learned the task, the authors removed much of V1 in one hemisphere. One week after the surgery, the ability of one monkey to discriminate (by making a saccade) between two possible targets in the affected visual hemifield dropped to chance. However, performance gradually improved through training, reaching 90% accuracy within 2 months. In another (younger) monkey, performance was at 90% at 3 weeks after the surgery.

To quantify how well the animals could identify the location of a visual stimulus, many months after the surgery, the authors measured the minimum luminance that was required for an accurate saccade. They found that although the animals could make accurate saccades to the contralateral hemifield, the threshold for luminance contrast for many targets had more than doubled (with respect to before V1 damage); the animals needed much more contrast to be able to localize the visual stimulus in the region of their cortical scotoma.

To test the role of superior colliculus in this localization, the authors lowered an electrode to the ipsilesional colliculus (the same side as the lesioned V1) and after finding a saccade-related neuron in the intermediate layers, they injected muscimol, effectively disabling that region. They found that the animal could no longer make visually guided saccades to the target that was within the response field of the collicular neuron. However, this could have been because collicular access to the saccade generating machinery had been impaired. To test for this, they recorded the behavior of the monkey in the dark and found that it occasionally made saccades to the disabled response field of the colliculus, but with significantly reduced velocities. Therefore, the partial collicular inactivation had not completely removed the ability to make saccades to the affected region, suggesting that after the V1 lesion, the ability of the animal to detect and make saccades to the affected V1 visual field was likely due to the retained function of the colliculus.

It appears that after V1 damage and the ensuing degeneration of neurons to the cortex, the superior colliculus can provide the cortex (likely via the thalamus) information about the location and perhaps other properties of the visual stimuli. Remarkably, this information is sufficient to allow the subject to orient toward the stimulus in a forced-choice trial, but is insufficient to allow the brain to become consciously aware of that stimulus.

5.2 Attention and the Effort Cost of Movement

Injecting muscimol into a small region of the colliculus raises the amount of excitation that the cells in that region need in order to reach threshold. This increases the cost of activating the cells, an effect that we might interpret as an increase in the cost of making a saccade to that region of space. However, the inactivation of a small collicular region has behavioral consequences that are broader than reduced saccade velocities. Impairment in the circuitry that directs movements to a region of space also affects the ability to direct attention to that region. Essentially, an increase in the effort needed to make a saccade to a part of the visual space makes the subject exhibit neglect-like symptoms toward the sensory stimuli that are placed there.

Directing attention to a region of visual space is impaired when the corresponding region in the colliculus is inactivated.

Recall that in the experiments shown in figure 4.18, the monkey needed to attend to two stimuli: one in the response field of the affected cells in the colliculus, and another in the opposite quadrant. In the deliberation period, the monkey decided which of the two stimuli matched an earlier cue, and then made a saccade toward it (or pursued it). When the colliculus was partially disabled, the monkey largely ignored the information in the affected part of the visual space. That is, when the cost of making a saccade to that region was increased (in the sense that cortical input to the colliculus needed to be greater to overcome the effect of muscimol), the animal appeared less willing to direct attention to that part of the visual space, and instead became more reliant on the information that it gathered from the unaffected visual space.

We have seen hints of this possible link between cost of moving and control of attention in our earlier discussion in humans (figure 3.12). In that experiment, a visual motion discrimination task, people had to decide whether dots were moving mostly toward the left or toward the right. They then signaled that choice by moving their left or right arm. Their perceptual ability was affected by the cost of moving the associated arm; they were less likely to determine that the dots were moving leftward when that choice required moving the left arm against a heavier load than was required by the right arm. It appeared that during deliberation, the increased effort cost for the movement that signaled leftward motion reduced the weight of the evidence associated with that decision, making it easier for evidence for rightward motion to reach threshold and trigger the less effortful movement. Therefore, the effort involved in the movement that signaled the choice affected the way sensory evidence was evaluated.

To account for this result, we imagined that at each moment of time during deliberation, the brain considered the motion evidence r_1 for rightward motion. This evidence was a random variable drawn from a normal distribution $N(\mu_1, \sigma^2)$, in which μ_1 was proportional to the coherence of the rightward motion. This random variable was integrated until it reached an upper bound, at which point the choice was that the random dots were moving rightward (or reached a lower bound, at which point the choice was leftward). When the right arm had to move against a load to indicate this choice, the cost of reporting a rightward motion had increased. The choices that people made suggested that the utility of the movement that reported the choice biased the evidence for that stimulus:

$$r_1 \sim N(\mu_1 + kJ_1, \sigma^2) \tag{5.1}$$

In equation 5.1, evidence r_1 for rightward motion is drawn from a distribution that depends not just on the (objective) sensory information $N(\mu_1, \sigma^2)$, but also on the utility of the movement J_1 that will report the decision. As a result, increased effort in the movement that signals the decision weighs the evidence for that choice.

Lee Lovejoy and Richard Krauzlis (2010) examined the effects of temporary collicular inactivation on the ability to direct attention and accumulate evidence from a region of the visual space. Before their work, it was thought that control of attention, particularly covert attention (when we look at one location but are attending to another location) was a purely cortical phenomenon. Imagine a situation in which you are speaking with someone and looking at that person's face, but are actually interested in someone else, who happens to be in another part of your visual space. As a result, despite the fact that you are looking at the person in front of you, your attention is focused elsewhere. As you direct your attention covertly (without moving your eyes), cortical cells that encode that part of the visual space will discharge more than they would have if your attention were not directed toward the actual person of interest, even if the visual information would have been the same in both cases. Remarkably, Lovejoy and Krauzlis found that the ability to make decisions based on information in a given region of space was affected if the part of the colliculus that corresponded to making movements to that region was temporarily disabled.

In their task (figure 5.1A), trials began with a fixation point. While the animal maintained fixation, four rings appeared, one of which was a unique color. The uniquely colored ring cued the animal to covertly attend to this region of the visual space. As the animal fixated on the center location, dots appeared in each of four regions and began moving in random directions. At cue offset, the rings disappeared and the dots in the cued ring and the location opposite to it (foil region) began moving coherently in two different directions (motion onset). After this period, the moving dots disappeared and the monkey

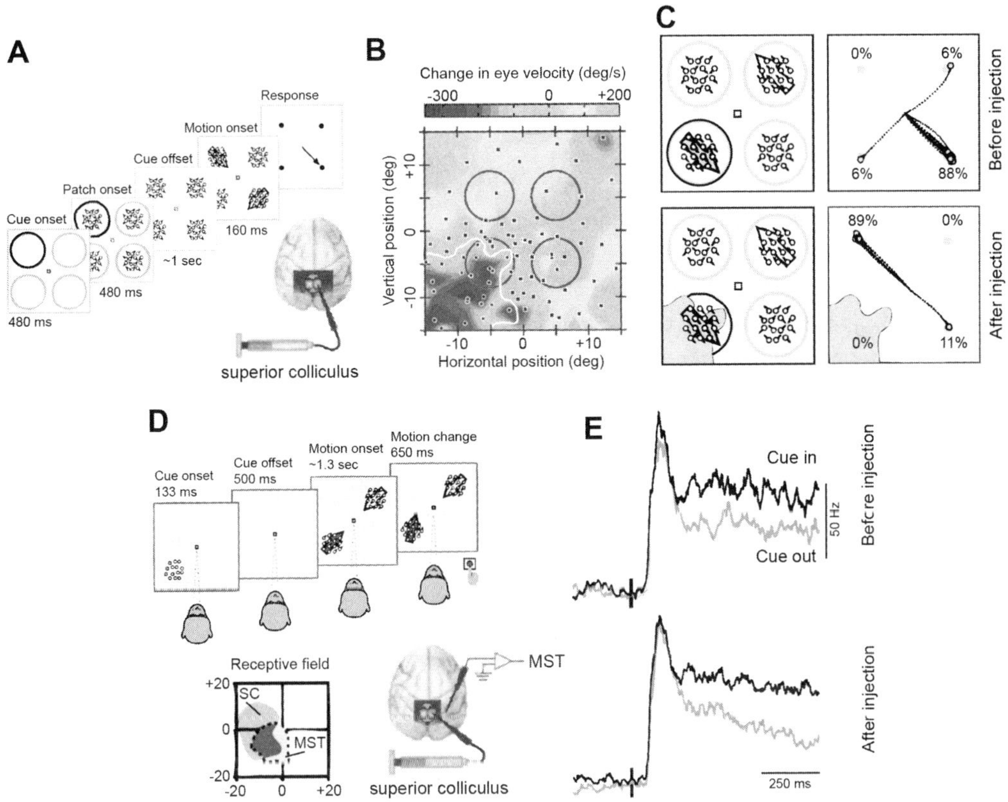

Figure 5.1
Effect of collicular inactivation on control of attention. **A**. Experimental protocol. The animal had to detect direction of motion in the cued location, and ignore the direction of motion in the foil location (in the quadrant opposite to the cued location). **B**. Local inactivation of superior colliculus (SC) resulted in substantially reduced saccade peak velocities to a region of the visual space (dark shadings). **C**. Example of single trials. Before the injection, motion in the cued location (dark circle) was toward −45°, and motion in the foil location was toward 135°. Motion in the other two locations was random. In 88% of the trials, the animal correctly made a saccade toward −45°. After collicular inactivation in a spatial region that included the cued location, in 89% of the trials, saccades were toward 135°. That is, the animal was strongly biased away from the cued location. **D**. In this task, the animal had to detect change in motion in the cued location, and ignore the motion in the foil location. It signaled its choice by pressing a button at the time it detected change. Recordings were made from MST (medial superior temporal) neurons while a small region of the colliculus was disabled. Receptive field of the affected collicular neurons and the recorded MST neurons are shown. **E**. Response of an MST neuron to motion onset (dark vertical bar). In all cases, the motion was in the response field of the neuron. However, when the animal was instructed to attend to that motion (motion was in the cued location, cue in), the cell's response was sustained at a higher level than when the animal was instructed to not attend to it (same motion was in the foil location, cue out). This modulation was unaffected by collicular inactivation. Despite this, performance dropped significantly after injection. (From Lovejoy and Krauzlis, 2010; Zenon and Krauzlis, 2012.)

had to make a saccade to one of four dots. The direction of the saccade indicated the direction of the coherent dot motion in the cued ring. Therefore, the direction of saccade was unrelated to the position of the cued ring with respect to fixation. Rather, the task was to detect direction of coherent motion of the stimulus in the cued ring and ignore the motion in the foil region.

In 88% of the trials, the monkeys made the correct decision: a saccade based on the motion in the cued ring. However, in 12% of the trials, the saccade was directed on the basis of the information in the foil region (figure 5.1C, before injection). The authors injected muscimol into the intermediate and deep layers of the superior colliculus, thus inactivating the location that corresponded to one of the rings (left column of figure 5.1C, after injection). This substantially reduced the peak velocity of saccades to that location while slightly increasing peak velocities for saccades to other locations (figure 5.1B). The researchers then presented the cued ring either in the affected region or in the opposite quadrant. Collicular inactivation made the animals largely ignore the motion in the cued region when that region overlapped with the inactivated part of the space. After inactivation, the animal appeared to have made its decision on the basis of the signal in the foil region: performance switched to 89% based on information in the foil region and 11% based on the cued region (right column of figure 5.1C, after injection).

This apparent reluctance to ignore the foil region and pay attention to the cued region persisted even when the choice was signaled via a manual response (button press). The researchers considered the possibility that collicular inactivation impaired motion discrimination, and they found that this was indeed the case. To compensate for the potential effect of collicular inactivation on motion discrimination, in their final experiment they substantially increased coherence of the cued region with respect to the foil region. Despite the increase in cued-signal strength, the animals still largely ignored the cue signal and instead based their decision on the information available in the foil region. Importantly, this tendency to ignore information from the cued/affected region was not present if there was no competing foil region. That is, the neglect-like symptoms were present only when there were at least two regions that provided information, but only one of which was to be attended.

These results show that if subjects have a hard time directing gaze toward a region of visual space (as evidenced by reduced saccade velocity for those movements), they discount the weight of the visual information there. This internal reduction in the weight of the evidence from one part of the visual space allows evidence from another part (foil region) to accumulate faster, thus suggesting that the effort required by the motor system affects the cognitive control of attention.

Increased effort to direct gaze to a part of the visual space affects decisions that depend on information from that region.

Two cortical regions in the visual cortex that are highly sensitive to motion are the middle temporal cortex (MT area), and the medial superior temporal cortex (MST area). Cells in these regions have receptive fields centered in a part of the contralateral visual space and respond by firing when there is motion in a preferred direction in that region of the space. During fixation, when there is motion in a region of space and the animal is covertly attending to it, the cortical cells respond with greater discharge than when the animal is not attending there. This enhanced response to information is thought to be the normal way that the cortex covertly directs attention. However, we just noted that collicular inactivation affects the control of attention. Did this inactivation produce changes in the discharge of motion-selective cells in the visual cortex? That is, when the colliculus was impaired, did it affect the ability of cells in the visual cortex to selectively acquire information from the affected region?

Alexandre Zenon and Richard Krauzlis (2012) considered a task in which during the entire trial, the monkey had to maintain fixation on the center dot. The trial began with a cue (a small collection of static dots) appearing in the response field of the neurons that were being recorded in MST or MT (figure 5.1D). The cue disappeared and was replaced with moving dots in two patches, one corresponding to the cue location and the other in the opposite quadrant. The dots moved in opposite directions, one of which was the preferred direction of the cortical neuron that was being recorded. The direction of motion remained constant for around 1.3 s and then changed in one of the patches by 16°–20°. The animal had to detect if the change occurred in the cued location and indicate this detection by pressing a button. The paradigm required the animal to actively ignore the stimuli in the foil location, covertly focusing only on the information from the cued location. Under the control conditions, animals were able to do this (press the button correctly) in 50%–60% of the trials.

Zenon and Krauzlis (2012) placed the cue in a location that overlapped with the region of visual space that corresponded to collicular inactivation and also corresponded to the receptive field of the cell that was recording in the MST or MT (receptive field, figure 5.1D). Under the control conditions, the neurons in the MST and MT showed higher discharge rates when the motion stimulus appeared in their receptive field and was in the cued location (figure 5.1E). Therefore, when covert attention was directed to the region of space covered by the receptive field of the cortical cell, that cell responded more vigorously to the motion information. During collicular inactivation, this attention-driven modulation in the visual cortex was intact (figure 5.1E, after injection). However, the animal's performance dropped to 10%–15% correct.

In summary, when muscimol was injected into a small region of the colliculus, saccade velocity to stimuli located in that region of space decreased substantially. We may interpret this as an increase in the effort needed to move toward that region of space, brought about by the increased excitatory input that is needed to generate a movement. When the animal had to make a decision based on motion information in that part of space, neurons in the motion sensitive regions of the visual cortex responded as before, with increased discharge when the motion was being attended. This suggests that attention, as defined by modulation of the visual cortex neurons with respect to the motion information, was unaltered. However, the decision-making process was strongly affected: the weight assigned to the information acquired from the region of space with reduced movement utility was downregulated.

Therefore, decision-making that depends solely on visual information nevertheless is strongly affected by the state of the colliculus, which affects whether the eyes can effectively move to the part of the space where that information appears. When the cost of moving to a region of space is increased, the decision-making system, presumably residing in the cortex, largely ignores the information that is available in that part of the visual space.

5.3 Inhibiting Reflexive Behavior

Although the colliculus receives direct retinal projections, allowing some of its cells to respond at around 70 ms after the appearance of a stimulus, that response indicates occurrence of a salient event, but not the utility of whether one should make a movement toward it. For example, consider the data in figure 4.8, gathered from the animal that fixated on a center spot while four stimuli appeared at various locations. Three of the stimuli were the same, and a fourth, the target, was the unique. The task was to identify the target and make a saccade toward it; saccades made to the target were rewarded. The response at 80 ms in the colliculus was the same regardless of the identity of the stimulus in the response field. After the early response, activity declined toward baseline, but then rose at around 100 ms later if the stimulus was indeed the one associated with the reward. Utility is reflected in the delayed response, not the initial response (similar to the response in the MST neurons in figure 5.1E). Is the information about stimulus utility computed elsewhere and then sent to the colliculus? To consider this question, let us begin by examining behavior of patients who have cortical or basal ganglia deficits.

After basal ganglia or FEF damage, patients have a hard time inhibiting the habitual response that the visual stimulus elicits.

Florence Chan, Doug Munoz, and colleagues (Chan et al., 2005) examined patients with Parkinson's disease (PD) who were on their regular schedule of medication. The experiment started with one block of prosaccades and then two blocks of antisaccades. In the prosaccade block, the instructions were to look at the target, whereas in the antisaccade block, the instructions were to look in the opposite direction. Each block contained both gap and overlap trials in a random order, and the target appeared randomly at 20° to either the left or the right of fixation.

In the prosaccade block, the PD patients produced more short-latency saccades than did the control group: in figure 5.2 (prosaccades), the cumulative probability of saccade latency in correct trials rises faster in PD than in the control group. Importantly, in both the gap and overlap trials, the patients produced more express saccades than did the control group. (Express saccades were defined by latencies that were between 90 ms and 140 ms.)

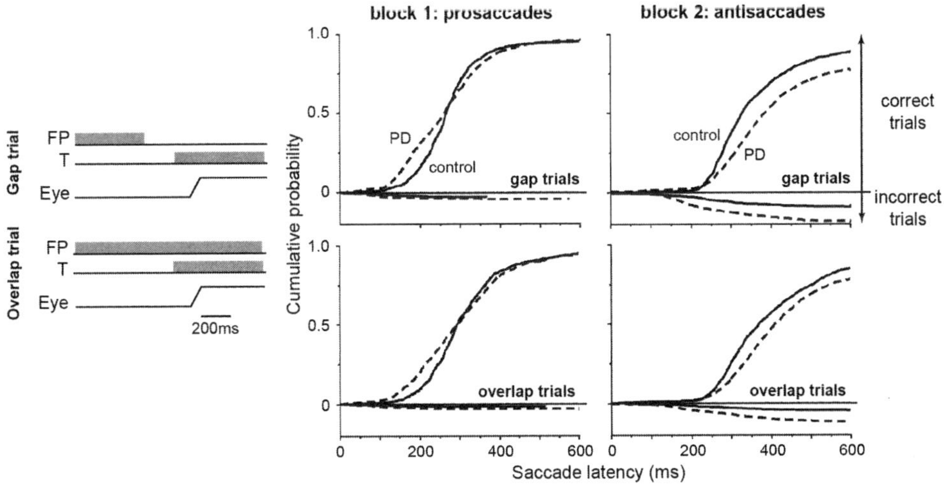

Figure 5.2
Parkinson's disease (PD) patients exhibit a reduced ability to inhibit reflexive saccades. In separate blocks of trials, the instructions were to make prosaccades or antisaccades. In each block, the gap and overlap trials occurred in random, and the target was at 20° to the left or right of fixation. In the prosaccade block, PD patients exhibited faster reaction time and a greater number of express saccades. In the antisaccade block, for trials in which the movement was correctly selected (opposite to the direction of the target, plotted as a probability above the 0 line), the patients required a longer reaction time. (From Chan et al., 2005.)

Surprisingly, saccade vigor (peak velocity) was greater in the PD than in the control group, a finding which may reflect the greater number of express saccades and the fact that short latency movements typically are associated with greater vigor.

In the antisaccade block, the instructions were to look in the opposite direction of the visual stimulus. When the PD patients made the correct movement, their reaction times were much longer than those of the control: in figure 5.2 (antisaccades), the cumulative probability of saccade latency in correct trials (in both gap and overlap trials) rises more slowly in the PD group than in the control group. When they made the incorrect movement, their reaction times were shorter than those of the control (data below the zero-axis in right column of figure 5.2). Indeed, in the antisaccade task, the PD patients made many more errors than did the controls.

These data suggest that in PD there is an impairment in the ability to inhibit the reflexive (or habitual) response that the visual stimulus elicits. The basal ganglia inhibit the superior colliculus through the projections from the substantia nigra reticulata (SNr). It is possible that in PD, this inhibition level is lower than normal, and hence there is an impairment in the ability to inhibit the superior colliculus's response to the visual stimulus. To generate an antisaccade, the visually induced response in the collicular cells must not be allowed to accumulate to a burst. Rather, cortical structures must remap that target, activating the opposite colliculus at a location where no visual stimulus exists. This remapping takes time, during which the colliculus needs to be prevented from reflexively responding to the visual stimulus. Perhaps in PD, the inhibition of the reflexive response is weaker than normal.

This lower ability to inhibit a reflexive saccade is also present in patients with FEF damage, but not patients with damage to the parietal cortex (lateral intraparietal sulcus region, LIP). Liana Machado and Robert Rafal (Machado and Rafal, 2004) examined patients with FEF damage and a separate group of patients with damage to LIP. The task was to make antisaccades to visual targets. A center fixation point was presented along with two open circles at 6° to either the left or the right of fixation. At 500 ms, either a unidirectional arrow appeared, informing the subject of the future target's location (informative), or a bidirectional arrow appeared (uninformative). After 200 ms, one of the two targets was filled. The subject had to make an antisaccade. In some trials, the center cue disappeared with target onset (non-overlap condition). In other trials, the center cue remained visible at target onset (overlap condition).

The data shown in figure 5.3 are collapsed across all conditions but divided into trials in which the target was ipsilateral or contralateral to the side of cortical damage. The

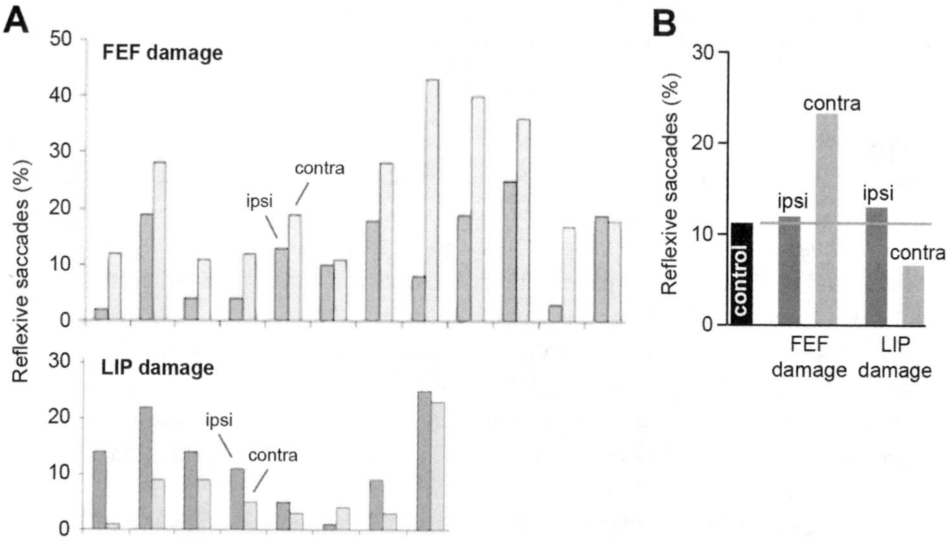

Figure 5.3
Probability of erroneous saccades in an antisaccade task in patients with frontal eye field (FEF) or parietal cortex (lateral intraparietal sulcus, LIP) damage. **A**. Data for individual subjects. Each pair of bars represents a single subject. **B**. Data averaged across all subjects. FEF patients had trouble inhibiting reflexive saccades when the target was contralateral to the side of the lesion. In contrast, their performance was like healthy controls when the target was ipsilateral to the side of the lesion. Unlike FEF patients, LIP patients were better than healthy controls in their ability to inhibit reflexive saccades when the target was contralateral to the side of the lesion. (From Machado and Rafal, 2004.)

patients with FEF damage made more errors: they were less able to inhibit the reflexive saccade when the target was contralateral to the lesion. However, their behavior was normal when the target was ipsilateral to the lesion. This suggested that like the PD patients, the patients with FEF damage had trouble inhibiting reflexive saccades, but only if the target was contralateral to the side of cortical damage.

In contrast, the patients with parietal cortex damage made errors less frequently than the healthy controls when the target was contralateral to the lesion, and about the same number of errors as controls when the target was ipsilateral to the lesion. This suggested that with parietal damage, the utility of the action toward the contralateral side was less than normal, a property consistent with neglect. As a result, these patients did better than healthy controls in inhibiting their reflexive saccades toward the contralateral side.

The reduced ability of the patients with FEF damage to inhibit the reflexive response toward the stimulus in the antisaccade task was confirmed in later experiments (Hodgson et al., 2007; Van der Stigchel et al., 2012). Robert Rafal and colleagues (Van der Stigchel et al., 2012) considered a task in which six green dots were arranged about an imaginary circle. After a fixation period, all except one of the dots turned red. At the same time, a new red dot appeared in a location polar opposite to the remaining green dot. The objective for the subject was to saccade to the green dot. This was difficult because a new visual stimulus (the new red dot) had appeared in a location that was opposite to the goal. The subjects had to inhibit the tendency to look at the new stimulus, and instead direct their action toward the only stimulus that had not changed. Four patients with FEF damage were tested. The authors found that when the distractor (the newly appearing red dot) was in the visual space contralateral to the lesioned hemisphere, the patients made more saccades toward it (in error) than when the same distractor was in the visual space ipsilateral to the lesioned hemisphere.

In the antisaccade task, the increased prevalence of prosaccades to the contralateral field exhibited by patients with FEF damage are an example of what John Hughlings Jackson, a 19th century neurologist, called *positive symptoms*. He postulated that damage to the cortex produced both negative symptoms (paralysis that follows a stroke), and positive symptoms (increased strength of spinal reflexes). Writing about epilepsy, he noted: "The condition after the paroxysm is duplex: 1. there is a loss or defect of consciousness and there is 2. mental automatism. In other words, there is 1. loss of control permitting, 2. increased automatic action." (The quote is from Berrios, 1985.) The weaker ability to inhibit reflexive saccades, perhaps resulting in greater autonomy of the colliculus, is an example of increased automatic action.

In summary, damage to the basal ganglia or the FEF impair the ability of humans to direct attention and saccades away from the irrelevant visual stimuli that suddenly appear in the contralateral visual space. In contrast, damage to the posterior parietal cortex impairs the ability to direct attention and saccades toward stimuli that appear in the contralateral visual space.

5.4 Activity in the FEF during the Reaction Time Period

Like the superior colliculus, the FEF contains a motor map; this is the reason why stimulation produces saccades that have a contralateral horizontal component. In the FEF, stimulation of the dorsomedial region produces larger saccades and stimulation of the ventrolateral region produces smaller saccades (Bruce et al., 1985). Indeed, projections of the FEF neurons onto the superior colliculus are topographic: lateral parts of the FEF project to the rostral colliculus (fovea-related region, producing microsaccades), and medial parts of it project to caudal colliculus (producing macrosaccades).

Although there are direct projections from the FEF to the brainstem oculomotor centers, the FEF needs the superior colliculus to generate a saccade. Doug Hanes and Robert Wurtz (2001) stimulated a region in the left FEF and then inhibited the corresponding region in the superior colliculus. For instance, they stimulated a region in the left FEF and noted that the result was an 18° rightward saccade. They then identified a region corresponding to this saccade in the superior colliculus. In this case, stimulation of a region in the colliculus produced a 29° rightward saccade. (The match is not perfect, but this is due to the difficulty in placing electrodes precisely in a deep brain region.) They then injected muscimol in that region of the colliculus, thereby inhibiting it along with nearby neurons. They found that when the monkey was provided with a visual stimulus at 17°, it made a 14.5° saccade. The latency of the saccade was much longer (+50 ms) and had a lower peak velocity. Next, they stimulated the FEF site and found that it no longer generated a saccade. Therefore, the direct pathway from the FEF to the brainstem was insufficient to generate saccades in the absence of the pathway that projected to the superior colliculus.

FEF cell types resemble those present in the superior colliculus, but without the tight correspondence between discharge and timing of saccade onset and offset.

Emilio Bizzi (1968) was among the pioneers who first recorded from the FEF. He recorded cortical activity while the animal made spontaneous saccades in a lit room but without any explicit rewards. He found some cells that exhibited a burst of activity around the time of the saccade but not during fixation, and other cells that exhibited sustained activity during fixation but reduced activity during saccades.

Charles Bruce and Mickey Goldberg (Bruce and Goldberg, 1985) followed up on Bizzi's work but in a more controlled setting in which the monkey was rewarded for maintaining fixation and for making saccades to specific targets. Their work described four broad groups of cells in the FEF. One group consisted of cells that responded to a visual stimulus in their response field but not to saccades that followed, as shown in the top trace in figure 5.4A. The activity in these cells resembled that of the visual neurons in the superficial layers of the colliculus. Another group of cells responded to the visual stimulus and then showed buildup of activity in anticipation of the saccade, as shown in the

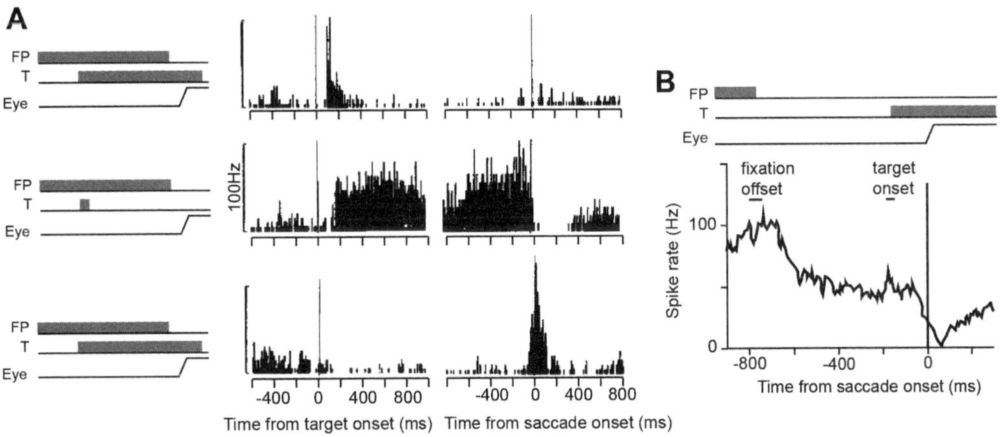

Figure 5.4
Activity of cells in the frontal eye field (FEF) during saccades. **A**. The top row illustrates activity of a cell with a response to the onset of the visual stimulus. Middle row illustrates activity of a cell with buildup of activity that precedes onset of a saccade. Bottom row illustrates activity a cell that exhibits a burst at saccade onset. (From Bruce and Goldberg, 1985.) **B**. Example of a cell with fovea-related activity. The cell fired strongly while the animal maintained fixation and the fixation point was illuminated. Upon removal of the fixation point, the animal maintained fixation but the discharge diminished. The cell exhibited a minimum in discharge around the time of the saccade. (From Hanes et al., 1998.)

middle trace in figure 5.4A. The activity in these cells did not require the presence of the target stimulus. Rather, the activity resembled that of the buildup cells in the deep layers of the colliculus. A third group of cells did not show a response to visual stimuli but had a burst near saccade onset, as shown in the bottom trace in figure 5.4A. The activity in these cells resembled the activity of the burst cells in the intermediate layers of the colliculus. A final group of cells were active while the animal was maintaining fixation on a rewarding visual stimulus.

An example of a cell with fovea-related activity is shown in figure 5.4B (taken from data in Hanes et al., 1998). When the fixation point was on, the fovea-related FEF cell exhibited a high discharge rate. When the fixation point was turned off, the animal maintained fixation but the cell's discharge rate declined. When the target was displayed, activity declined further and reached a minimum around the time of the saccade. After the saccade, activity rebounded toward presaccade levels. However, note that in the example shown here, the activity in the fovea-related cell in the FEF did not exhibit a sharp pause that was tightly associated with saccade duration, as was the case for the fovea-related cells of the superior colliculus. Therefore, the cell types present in the FEF do resemble those present in the superior colliculus, but without the tight correspondence between discharge and timing of saccade onset and offset.

The early response in FEF appears to reflect salience, not utility.

To further illustrate this correspondence, let us revisit the experiment in which the subjects were tasked with finding the unique dot (the target) when an array of dots appeared on the screen. All of the dots had the same color except for the target. The objective was to find the unique stimulus and make a saccade toward it. Jeffrey Schall and Doug Hanes (Schall and Hanes, 1993) placed either a distractor or the target in the response field of a FEF neuron. They found that at around 80 ms after target presentation, the cell responded to the stimulus in its response field. As in the colliculus experiments, the target and the distractors both elicited similar early FEF responses; the firing rates were similar for both types of stimuli (figure 5.5). In the ensuing 100 ms or so, the activity of the cell changed, returning close to the baseline if the stimulus was a distractor but staying elevated if the stimulus was the target.

Utility of holding still is reflected in the activity of fovea-related cells in FEF.

To gain a better understanding of the role of the FEF in controlling the colliculus, Stefan Everling and Doug Munoz (2000), along with Jeffrey Schall (Munoz and Schall, 2004) compared the activity of the cortical cells with the collicular cells in the prosaccade and antisaccade task. Everling and Munoz (2000) recorded from FEF neurons that projected directly onto the superior colliculus. (They identified the cortical cells by stimulating the colliculus and noting antidromic activity in the cortex.) During fixation, a target would appear in the receptive field of the cortical cell, and its color would instruct the animal as to whether the trial was prosaccade or antisaccade.

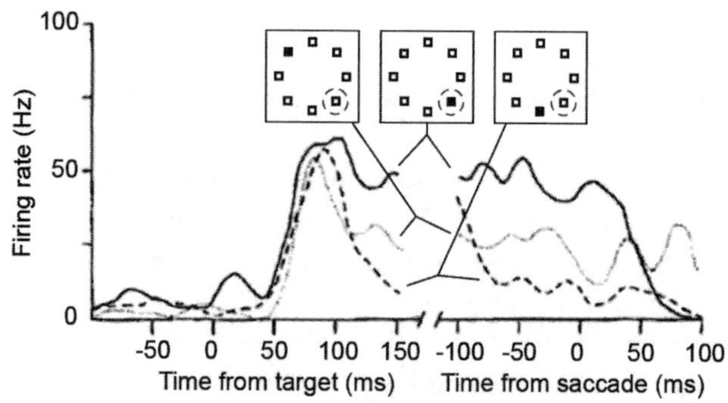

Figure 5.5
Activity of a frontal eye field (FEF) neuron during a visual search task. In the task, a stimulus appears in the response field of the cell, but it is one of eight stimuli that are displayed; the target is green, but the seven distractors are red. The animal is rewarded if makes a saccade to the target. As a result, there is always a stimulus in the cell's receptive field. Regardless of stimulus color, the cell discharges 80 ms after stimulus onset. If the stimulus in the response field is not instructing a saccade, the visually evoked burst shows a decline to near baseline. If the stimulus in response field is instructing a saccade, the visually evoked burst is sustained until the saccade time. (From Schall and Hanes, 1993.)

When the animal was holding fixation and awaiting the arrival of the target, there was sustained activity in the fovea-related cells of the FEF and the colliculus. However, if the instructions indicated an antisaccade trial, this activity was higher than it was when the instructions indicated a prosaccade trial (figure 5.6, part 1). That is, in the hard task (antisaccade), in which the animal needed to ignore the (upcoming) visual stimulus and instead saccade to the polar opposite location, there was greater activity in the fovea-related neurons. Because fovea-related neurons inhibit the buildup and burst neurons that respond to visual stimuli associated with macrosaccades, the sustained activity in the fovea-related neurons is likely the key factor that allows the animal to hold still and not reflexively respond to presentation of a visual stimulus. By holding still, the brain buys time, allowing it to remap the target, thereby making a saccade to the polar opposite location.

As the animal maintained fixation, it expected the visual stimulus to appear in one of two locations on the screen: one of the target locations was in the response field of the buildup or burst neuron that was being recorded, and the other target location was in the polar opposite field. The animal did not know which of the two locations the target would appear in, but the locations were consistent across trials, resulting in anticipation. This anticipation could be inferred from the activity of the neurons: FEF buildup neurons that had their response field centered on one of the upcoming targets showed elevated activity during the fixation period (figure 5.6A, part 2). To a lesser extent, this elevated activity was also present in the collicular buildup cells.

Importantly, the instructions for the trial induced large differences in the activity of the buildup neurons of the FEF: for antisaccade trials, the anticipatory activity was lower than for the prosaccade trials. This difference appeared more muted in the colliculus. Therefore, although the visual stimulus was as likely to appear in the response field of the neuron in the prosaccade trials as it was in the antisaccade trials, the anticipatory activity in these neurons was lower when the instruction indicated an antisaccade trial. This likely reflected the increased activity in the fovea-related neurons during the fixation period, effectively reducing the utility of the stimulus that was about to appear in one of two locations and thus preventing that neural activity from increasing enough to generate a movement.

After the fixation point disappeared, there was nothing on the screen. This is a vulnerable period of time for the colliculus because onset of a visual stimulus is now likely to result in an express saccade, a movement that occurs because visual response of the colliculus to the stimulus is large enough to drive its burst neurons. Indeed, the collicular cell shown in figure 5.6A (part 2) shows a strong burst soon after target onset. In the antisaccade trial, this burst is smaller than in the prosaccade trial. In the prosaccade trial, the initial burst is followed by a larger, second burst, which coincides with saccade onset. In contrast, in the antisaccade trial, the initial burst is followed by suppression of discharge. Therefore, if the trial was for prosaccades and the target appeared in the response field of the neuron, then during the gap period, the elevated activity in the FEF as well as that in the colliculus rose to higher levels, reaching a peak around time of the saccade. However,

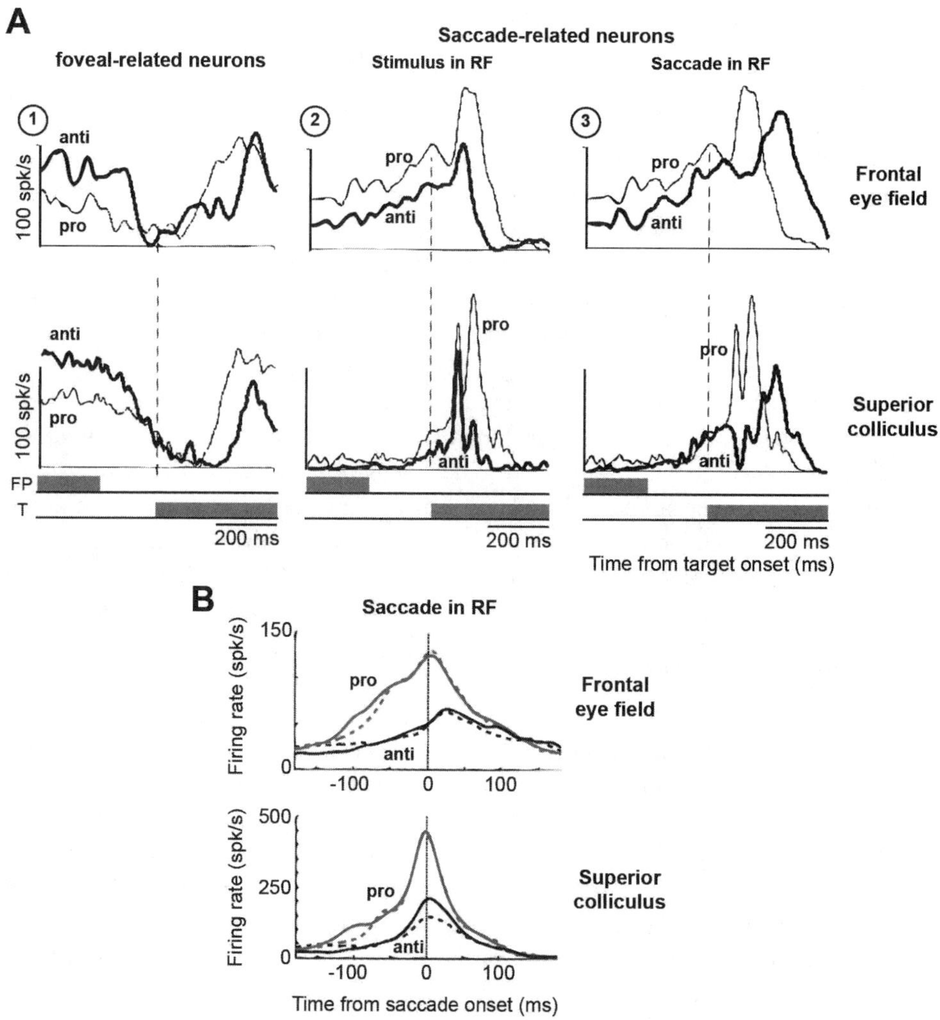

Figure 5.6
Activity of frontal eye field (FEF) and superior colliculus neurons in an antisaccade task. **A**. Activity of fovea-related neurons (part 1) in the FEF and superior colliculus. Data are aligned to target onset. In the antisaccade trial, the discharge was higher during the period of fixation and remained higher during the gap period. Parts 2 and 3 show activity of saccade-related neurons. The target either appeared in the response field (RF) of the neuron or in the polar opposite location. Similarly, the saccade was either toward the response field of the neuron or in the opposite direction. In the antisaccade trial, activity was lower in the saccade-related cells both during the gap period and after target onset. (From Munoz and Schall, 2004.) **B**. Saccade-related activity in the FEF and superior colliculus neurons was generally lower at saccade onset. Solid and dashed lines are data from overlap and gap trials, respectively. (From Jantz et al., 2013.)

if the trial was for antisaccades, the onset of the target led to a smaller burst, followed by a return to baseline. The reduced activity during the gap period of the antisaccade trial may have been the key in buying time following onset of the stimulus, preventing the reflexive response.

Indeed, the movements in the antisaccade trials occurred after a longer reaction time, as illustrated by the fact that the activity levels of the FEF and collicular cells reached their peaks later (figure 5.6A, part 3). Remarkably, the burst of activity in the cortical and collicular neurons (figure 5.6B) was much lower in the antisaccade trials than in the prosaccade ones. (Saccade peak velocity was also generally lower in the antisaccade trials, but not by the large margin of difference in the discharge of these neurons.) This illustrates the important point that activity in the FEF and the colliculus does not need to reach a fixed threshold before a saccade is initiated.

In summary, sometimes we need to hold still and buy time before we make a movement. In order to hold still and not respond to a visual stimulus, fovea-related cells in the FEF need to maintain a high level of discharge. Indeed, a critical difference between discharge patterns in the prosaccade and antisaccade tasks was in the fixation period, during which the objective of the task was identified (via color of the fixation point). In the harder task (antisaccade), there was greater activity in the fovea-related neurons, and smaller activity in the buildup cells. The greater activity in the fovea-related neurons may indicate the greater utility associated with holding still and not moving. This results in lower activity in the buildup cells during fixation in the antisaccade trials. In the ensuing gap period, the buildup in anticipation of stimulus arrival is at a lower level, making it less likely that arrival of the visual stimulus in the cell's response field will be sufficient to produce a burst. One of the keys in the control of this behavior, dampening the reflexive response to the visual stimulus, may be the activity of the fovea-related neurons of the FEF.

5.5 FEF and Deciding Where to Direct Gaze

To quantify activity of cells in the FEF during the deliberation period of a decision, Doug Hanes and Jeffrey Schall (1996) considered a task in which there was substantial variability in the reaction times.

During deliberation, the decision to move coincides with reduced activity in the fovea-related neurons and increased activity in the movement-related neurons.

In their task (called "countermanding"), an instruction was provided to the subject to make a movement, and there was some probability that, after a period of time, the instruction would change to canceling the request for the movement (identical to the task described in figure 3.20A). If the second instruction had occurred and the animal had waited, it would be rewarded. However, in some cases, the second instruction never came, and the animal was rewarded if it made the movement in response to the first instruction.

The task began with the monkey viewing a fixation point (figure 5.7A). After a variable interval (250–350 ms), a target appeared at one of two locations, either in the movement field of the FEF neuron or in the polar opposite location. At the time the fixation point was offset, the target was displayed. In 25%–50% of the trials, after a variable delay, the fixation point reappeared. These trials were called stop-signal trials. On the remaining trials, the target remained on (these trials were called no-stop-signal trials). The task introduced considerable variability in the reaction time of saccades in the no-stop-signal trials, providing a rich opportunity to ask whether changes in movement latency corresponded to changes in activity of the fovea-related and buildup cells of the FEF.

After the target appeared, the animal deliberated on whether to make a saccade or to maintain fixation. As it waited, there was some probability that the fixation point would reappear, in which case it should not make a saccade. However, it was possible that this was a no-stop-signal trial, in which case the fixation point would not reappear. In this case, it should make a saccade. This ambiguity produced variability in behavior, which allowed for understanding the neural basis of that variability in the FEF.

Consider the stop-signal trials, in which the animal waited and did not make a saccade. Activity of a cell whose response field includes the target is shown in the top panel of figure 5.6B. This cell exhibited a response to the target onset with a latency of about 100 ms, followed by buildup of activity. The stop-signal then arrived (fixation point reappeared); after 100 ms, cell activity exhibited a sharp decline (black trace, top panel of figure 5.7B). In these trials, the animal successfully maintained fixation. Now consider the activity of the same cell in the no-stop-signal trials, in which the fixation point never reappeared (gray trace, top row of figure 5.7B). In these trials, activity showed a buildup that continued to rise until a saccade was made (horizontal bar, top row of figure 5.7B). It appeared that the cell's activity needed to cross a threshold to trigger a saccade.

We can see evidence for this conjecture in trials in which the stop-signal arrived late and the animal was unable to wait, resulting in a saccade. Data for the same cell in these not-canceled trials are shown in the lower panel of figure 5.7B. In these stop-signal trials, the reappearance of the fixation point arrived later than the animal was willing to wait. The activity of the cell continued to rise, and the animal made a saccade to the target (black trace, bottom panel of figure 5.7B). In trials in which the animal maintained fixation, the buildup had not reached threshold and was reversed by arrival of the stop-signal, which canceled the saccade. In trials in which the animal was unable to wait, the buildup had crossed the threshold and the arrival of the fixation point was too late to stop the saccade.

Whereas the cell in figure 5.7B had its response field around the target, the cell in figure 5.7C had its response field around the fixation point. The black trace in the top panel of figure 5.7C shows activity of this fovea-related cell in stop-signal trials in which the animal successfully waited and did not make a saccade (successfully canceled trials). In contrast, the black trace in the bottom panel of figure 5.7C shows activity of the same cell in

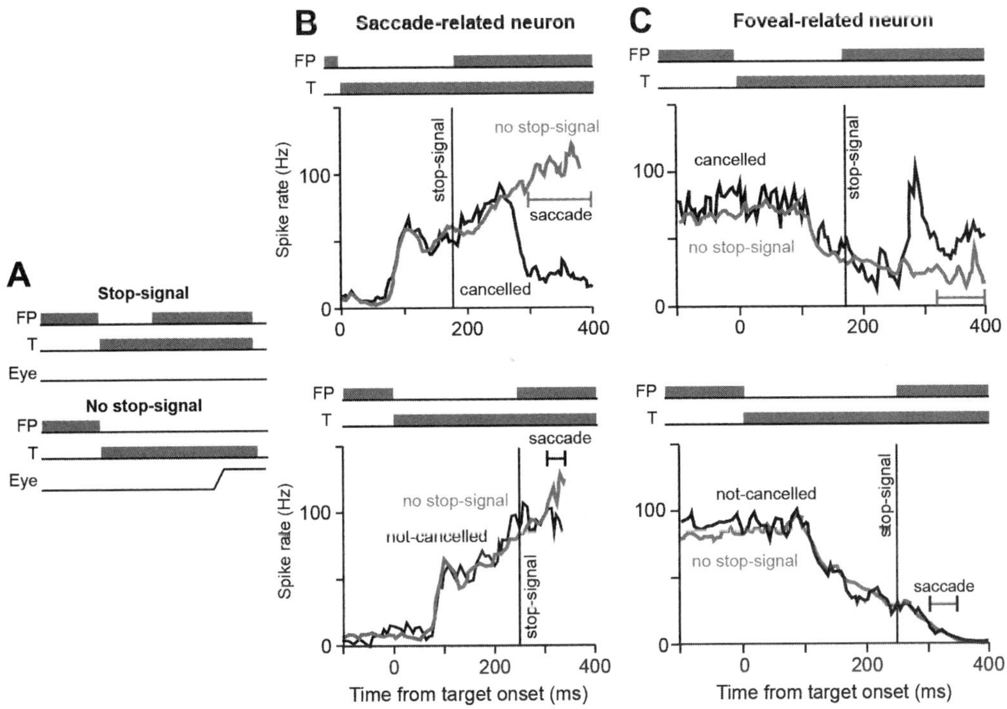

Figure 5.7
Activity of frontal eye field (FEF) cells during deliberation. **A**. The countermanding task. The animal is rewarded by maintaining fixation in the stop-signal trials and is also rewarded for making a saccade to the target in the no-stop-signal trials. However, the animal does not know whether a trial has a stop signal or not. **B**. Activity of a cell that has a response field that includes the target. Top panel: black trace shows activity of the cell in stop-signal trials in which the animal successfully maintained fixation. The gray trace shows activity in no-stop-signal trials in which the animal made a saccade to the target. (Time of saccade is indicated by the horizontal bar.) Bottom panel: activity of the same cell in stop-signal trials (black trace) in which the animal did not wait long enough and made a saccade despite a stop-signal. The gray trace shows activity of the cell in no-stop-signal trials with similar reaction times to the not-cancelled stop-signal trials. **C**. Activity of a cell that has a response field near the fovea. Top panel: black trace shows activity in stop-signal trials in which the animal successfully maintained fixation. Bottom panel: black trace shows activity in stop-signal trials in which the animal did not maintain fixation. In both panels, gray trace shows activity in no-stop-signal trials with reaction times matched to the stop-signal trials. (From Hanes et al., 1998.)

stop-signal trials in which the animal could not wait and did make a saccade (not-canceled trials). In both cases, 100 ms after the offset of fixation, the activity of the cell declined. However, in the case in which the animal successfully waited and maintained fixation, 100 ms after the stop-signal arrived and the fixation point reappeared, the cell produced a strong burst. In the case in which the animal was unable to wait, the saccade took place soon after the stop signal. As a result, the arrival of the stop-signal did not produce a burst in the cell because by the time the information at fixation would have engaged these fovea-related cells, the eyes had moved away and were now fixating the peripheral target.

The variability in reaction time is largely due to variability in the rate of buildup in the activity of cells that encode the upcoming movement.

It is possible that once the activity in the buildup cells crosses a threshold, deliberation has ended and a movement will be performed. However, an alternate possibility is that longer reaction times coincide with a higher threshold. That is, in principle, reaction time may be modulated in two ways: (1) the rate of buildup of activity in the cells that encode the movement may be modulated and (2) the threshold that is required to trigger a movement may be modulated. Is variability in reaction time mainly due to the variability in the rate of rise? Or, do changes to the threshold level lead to the reaction time variability?

Doug Hanes and Jeff Schall (1996) considered this question by focusing on the no-stop-signal trials, aligning activity of cells to saccade onset, as shown in figure 5.8A. This figure shows activity of a cell in trials in which the reaction time ranged from 225–250 ms, and in trials in which the reaction time ranged from 300–325 ms. Approximate time of target onset is indicated by the gray and black arrows. When the reaction time was long (gray trace), the animal was willing to wait. In this case, after target presentation (gray arrow), the cell showed little change in activity. However, at 150 ms after target onset, the cell began to show a slow rate of buildup. In contrast, when the reaction time was short (black trace), at 150 ms after target onset, the cell showed a rapid buildup of activity. To define a possible threshold for triggering a saccade, the authors considered the cell activity at 10–20 ms before saccade onset. The threshold of activity in the cell was invariant with respect to reaction time (top plot, figure 5.8B). However, the rate of rise in the buildup was

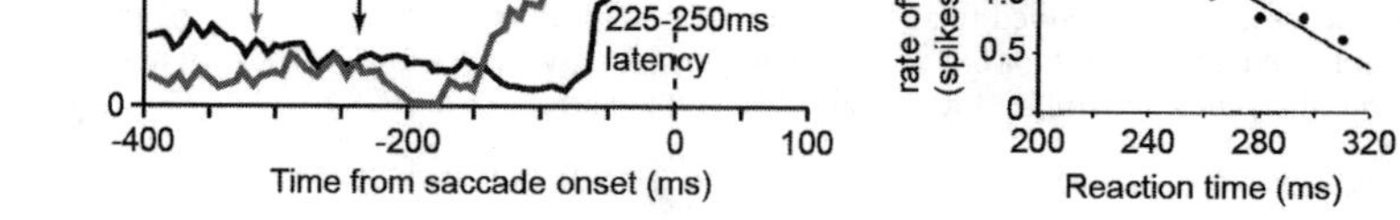

Figure 5.8
Change in reaction time coincided with change in the rate of buildup of activity among some cells in the frontal eye field (FEF). **A**. Experiment set up. In the stop-signal trials, the animal succeeded if it maintained fixation despite presentation of the target. In the no-stop-signal trials, the animal succeeded if it made a saccade to the target. The trials were randomly interleaved and the duration of the target presentation in the stop-signal trials was variable. **B**. Activity of one cell in the FEF in the no-stop-signal trials in which the animal made a saccade, but with variable reaction time: short reaction time (black trace) and long reaction time (gray trace). Target onset is approximately noted by the gray and black arrows. Buildup is slower in trials with long reaction time. **C**. Spike rate of the cell at 10–20 ms before saccade onset did not vary with reaction time. However, the rate of rise in the buildup of activity was greater in trials with shorter reaction time. (From Hanes and Schall, 1996.)

faster for shorter reaction time trials, and slower for longer reaction time trials (bottom plot, figure 5.8B). In 22 of 25 cells, the rate of rise declined with increasing reaction time.

The results suggested that in the FEF, the variability in reaction time was largely due to variability in the rate of buildup in the activity of cells that encoded the upcoming movement, and not variability in the threshold of activity that may be necessary to generate that movement. However, a critical point to note is that while the concept of a fixed threshold may be true in this specific task, in general it is unlikely to be true across tasks. For example, we noted that prosaccades take place with much greater activity in the FEF and colliculus than do antisaccades (figure 5.6B). Therefore, if there is a fixed threshold that specifies the onset of a movement, that threshold may not be defined by the activity of movement-related neurons in the FEF or the colliculus.

In summary, during the period in which the animal is considering making a movement, there is reduced activity in the fovea-related cells of the FEF, reflecting the disappearance of the rewarding stimulus around the fovea. Simultaneously, there is increased activity in the buildup cells that encode the movement vector to the visual stimulus that appeared in the periphery. The decision to make a movement is reached when the buildup cells that encode the saccade have a firing rate that approximately crosses a threshold. The variability in reaction time is largely due to the rate of increase in the activity of the buildup cells of the FEF.

Saccades to more rewarding stimuli are accompanied with greater activity in the movement-related neurons of FEF.

While saccades have a higher peak velocity when the destination promises to be rewarding, the firing rates of the saccade-related cells of the superior colliculus are not generally greater around the time of the saccade (figure 4.13C). The utility of the movement does not appear to be strongly reflected around movement time in the rate of discharge of the collicular cells. Is this utility reflected in the discharge of saccade-related cells of the FEF?

Joshua Glaser, Mark Segraves, and colleagues (Glaser et al., 2016) trained monkeys to search a natural scene for a small item (a star, as shown in figure 5.9A). Once the subject found the rewarding stimulus, it had to maintain fixation for 200 ms, after which it would receive a reward. The trial would reset if the target had not been found after 25 saccades. The authors classified each saccade as landing on the target (T+) or not (T-). T+ saccades were generally followed by the acquisition of a reward. These saccades occurred after a shorter latency and had a higher peak velocity (about +5%), as shown in figure 5.9B. These behavioral characteristics imply that for T+ saccades, before movement onset, the brain was aware that the movement would allow the image of the target to be placed on the fovea and, upon the completion of this, the search would conclude and the reward would be acquired..

The authors measured activity of each saccade-related neuron and identified its movement field. The cells generally had a high discharge around 300 ms before saccade onset,

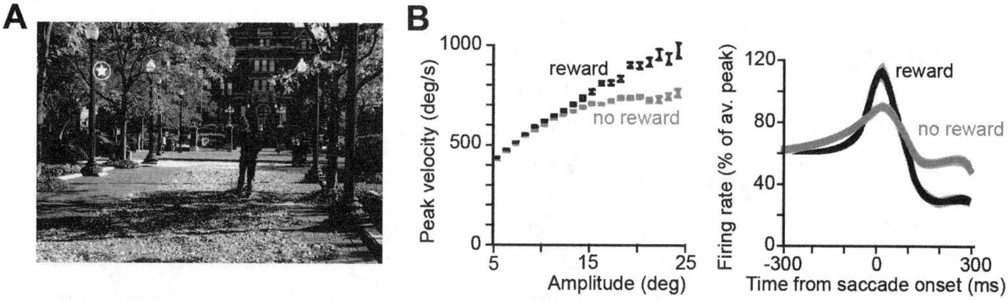

Figure 5.9
Effect of reward on saccade vigor and activity of neurons in the frontal eye field (FEF). **A**. The task was to find a target item (in this case, a star) in a natural scene image. **B**. Saccades that were made to the rewarding target had a higher peak velocity than those made to non-rewarding stimuli. Data are also shown for FEF saccade-related neurons. The saccades were to stimuli that were in the response field of the neurons. The neurons had greater peak discharge near saccade onset in the case in which the target of the saccade was expected to be rewarding. Furthermore, if the target of the saccade was rewarding, the activity of the neuron dropped sharply after saccade completion. (From Glaser et al., 2016.)

and then showed a burst near saccade onset (figure 5.9B). This burst had a higher magnitude when the target was expected to be rewarding. That is, T+ saccades had a higher peak velocity and were accompanied with a higher discharge rate in the FEF cells that encoded that saccade. Therefore, the expectation of reward was readily encoded in the peak rate of discharge of the FEF cells near the time of the saccade.

Interestingly, after the saccade in the T+ trials, the neurons showed a sharp decrease in their activity. However, in the T- trials, this activity remained elevated. We can speculate as to why activity in the saccade-related neurons dropped sharply once the animal found the rewarding stimulus. It seems likely that in T+ trials, after saccade termination, as the rewarding stimulus was fixated upon, fovea-related cells discharged at a higher rate than they did in the T- trials, in which finding and fixating upon the stimulus was not rewarding. This greater activity in the fovea-related cells resulted in greater inhibition of the saccade-related cells, which would then translate into a drop in their activity after saccade termination. In a sense, the utility of the stimulus on the fovea, as reflected in the rate of discharge in the fovea-related cells, would translate into a greater likelihood of staying still. That likelihood would then translate into the acquisition of a reward. (The trial was rewarded when the animal found and continued fixating on the target for at least 200 ms).

Glaser and colleagues trained three monkeys in this task. Two of them behaved as expected: greater vigor for saccades that landed on the rewarding target. However, the third monkey showed a rather weak reward-related modulation of vigor. In that monkey, FEF activity could not strongly dissociate between rewarding and non-rewarding saccades. In our next chapter we will see another example of such paradoxical behavior, and find that in that monkey, the dopamine system is not functioning normally.

In summary, expectation of reward increased saccade vigor, and this behavior was reflected in the rate of discharge of FEF neurons near saccade onset. Activity in the saccade-related neurons dropped more sharply after rewarded saccades than after unrewarded ones. In addition to increasing the activity of saccade-related FEF neurons around movement onset, the expectation of a reward also shut down that activity sooner after the movement culminated at the reward location.

The threshold of activity needed to commit to a decision may not be determined solely by activity in the FEF.

There is a puzzle regarding precisely what variable is encoded by the activity of the FEF and collicular cells. On the one hand, it seems clear that the utility of the movement is at least partly reflected in the rate of discharge of these cells. In the fovea-related cells, the rate of discharge is greater when the eyes are fixating on a rewarding target and falls faster when the reaction time of the movement is shorter. In the saccade-related cells, the rate of discharge rises faster when the reaction time is shorter and reaches a higher peak rate when the movement peak velocity is higher. However, if the rate of discharge during the deliberation period partly reflects the utility of the upcoming movement, the discharge does not reach a fixed threshold before a movement is initiated (figure 5.6B); the peak rate of discharge can be twice as large during prosaccades than during antisaccades. Thus, the threshold for committing to the decision and starting the movement appears to not be reflected solely by the rate of discharge in the saccade-related neurons of the FEF and the colliculus. This idea is further demonstrated in a task in which the animal is under time constraint, requiring it to arrive at a decision before the clock runs out.

Richard Heitz and Jeff Schall (2012) designed a task in which the objective was to detect the letter T among a collection of Ls (figure 5.10A). The experiment was organized in blocks in which the center fixation color indicated how much time the monkey had to find the letter T and make a saccade to it. In the fast condition block, the reaction time for the saccade had to be less than ~375 ms. In the accurate condition block, the reaction time for the saccade had to be greater than ~460 ms. In the neutral condition block, the reaction time had no limit. On average, the animals had a reaction time that was around 600 ms for the accurate condition, and around 300 ms for both the neutral and fast conditions. The error rate was highest for the fast condition block and lowest for the accurate condition block. Therefore, the animal had the greatest time pressure during the fast condition block and made the most number of errors during that time.

Drift diffusion models of decision-making (see chapter 3) explain speed-accuracy tradeoff via a change in the threshold (or equivalently, a change in the baseline). In these models, decision accuracy drops when the threshold is lowered, allowing for a shorter deliberation period (shorter reaction time), but also a smaller accumulation of evidence (greater likelihood of error). As a result, such models predict that under time pressure, the increased error rate results from a reduction in the threshold needed to commit to a decision.

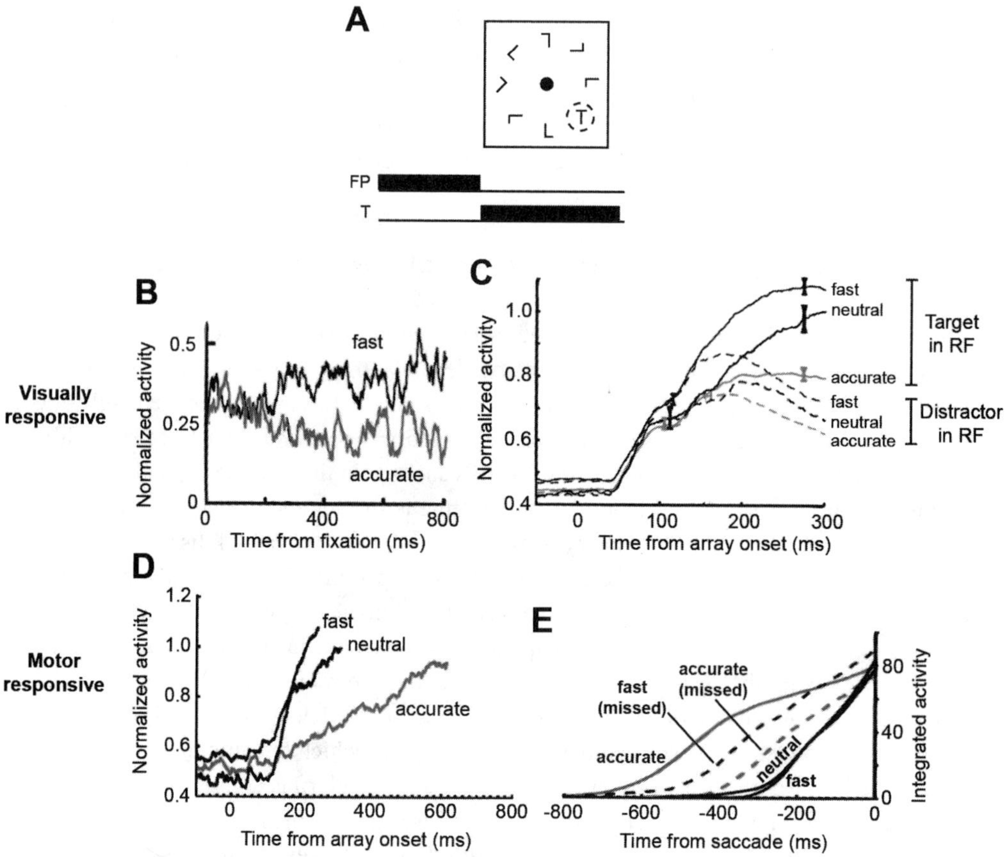

Figure 5.10
Activity in the FEF during a task that varied in speed and accuracy requirements of performance. **A**. The task required the animal to find the target (T) among distractors in a limited period of time. The dashed circle indicates the response field of the neuron. **B**. Activity of a visually responsive cell (cell that, in a memory-guided task, responded to the visual stimulus but did not have a saccade-related response). This cell had a response field that included a potential target location. During the fixation period, before the array of targets were presented, activity in the cell was higher in the fast trials than in the accurate ones. The reaction time was around 300 ms for both the fast and neutral trials. **C**. Normalized activity of all visually responsive cells. Activity increased after presentation of the target array, and the rate of increase was faster if the stimulus in the response field was the rewarding target. **D**. Activity of a motor responsive cell (cell that, in a memory-guided task, did not respond to the visual stimulus but did respond to the saccade). At time of the saccade, the discharge rate reached a higher level in the fast trials than in the accurate ones. **E**. Integrated discharge of the motor responsive neurons (with a time constant of 100 ms). Error trials are labeled as missed. (From Heitz and Schall, 2012.)

To test this model, Heitz and Schall (2012) labeled three broad groups of cells in their collection of FEF neurons: visually responsive, visuomotor responsive, and motor responsive. They made this classification in a memory-guided saccade task. Visual neurons increased their discharge immediately after target presentation but had no saccade-related modulation. Motor responsive neurons increased their discharge before saccade initiation but had no visual response. Visuomotor neurons exhibited responses during both periods of time.

In some trials, the visually responsive neurons had a response field that included the location of the rewarding target. As a result, these cells increased their firing rate when the salient item appeared in their response field. Importantly, the fixation cue, which instructed the type of the trial, induced a shift in the baseline firing rate preceding array presentation (figure 5.10B). In the accurate trials, the baseline activity of these cells was lower than in the fast trials. The authors did not report data from fovea-related cells, but it seems likely that similar differences might also be present among those neurons (with greater activity during the accurate trials).

After the presentation of the target array, the activity in the visually responsive neurons increased, and the rate of increase became faster if the stimulus in the response field was the rewarding target than if it was a distractor (figure 5.10C). Although the reaction time was around 300 ms for both the fast and the neutral trials, activity in these cells around time of saccade was generally lower in the neutral trials. Therefore, in the visually responsive cells, the trial context (fast versus accurate) in the period before the presentation of the target array affected baseline discharge. The cells were more active in the fast trials. As a result, in anticipation of having only a brief period to make a decision, cells whose response fields potentially included the location of the rewarding target increased their discharge before the target was displayed. Furthermore, in the fast trials, the rate at which these cells discharged increased faster in the period before the saccade and appeared to reach a higher level around time of the saccade.

The motor responsive neurons (loosely akin to the burst neurons of the colliculus), which comprised only a small fraction of the cells (about 10%), were identified on the basis of their lack of visual response in a memory-guided saccade task (i.e., they only had saccade-related responses). When the target was in their response field, the rate of rise was faster during the fast trials than during the accurate trials (figure 5.10D). At time of saccade, discharge rate of the motor neurons had reached a higher level during the fast trials than during the accurate trials.

These observations are exactly opposite of what a typical drift diffusion model would predict. Rather than finding a lower threshold for saccade generation in the fast trials, thereby allowing for higher error rates but shorter reaction times, these cells exhibited a faster rise as well as a higher level of peak discharge. Therefore, if there was a threshold for generating a movement, that threshold was not set by the activity in the FEF neurons.

To explain their results, Heitz and Schall (2012) noted that within the FEF, the anatomies of the visual and motor responsive cells differed. Among the various groups of neurons, only the motor responsive cells were likely to project to a region of the pons where a gate-keeper group of neurons, called omnipause neurons (OPNs), regulated the origination of saccades (Segraves, 1992). Omnipause neurons are active during fixation and pause briefly during saccades. It is possible that these neurons act as the threshold that must be crossed before a saccade is allowed.

Heitz and Schall (2012) imagined that the OPNs integrated the activity that they received from the FEF motor responsive neurons. To test for this, they mathematically integrated the discharge of the motor responsive neurons (with a time constant of 100 ms). They found that when the target appeared in the response field of the motor responsive neurons, this leaky integration of the discharge from the array initiation to saccade initiation resulted in activity that reached the same threshold regardless of whether the trial was fast or accurate and regardless of whether the movement was correct or in error (figure 5.10E). Therefore, they raised the possibility that there existed a constant threshold for generating a saccade, but that threshold was not in the FEF. Rather, it was determined in the brainstem via the OPNs, which integrated the converging inputs from the motor responsive cells of the FEF.

5.6 Setting a Threshold to Move: Omnipause Neurons

If decision-making is a process of integration of utility to a threshold, and cells in the FEF and perhaps the colliculus participate in this integration, then what is the neural basis of the threshold?

Omnipause neurons show insensitivity to utility of the stimulus.

One clue is in the brainstem, in the activity of OPNs, a group of inhibitory neurons that reside in a midline region of the pontine reticular formation. They receive projections from the FEF, the superior colliculus, and the cerebellum and send projections to the burst generators, neurons that are one synapse away from the motor neurons that move the eyes (figure 5.11A). OPNs have high sustained discharge during fixation, and then pause when a saccade takes place. Importantly, their pause is present regardless of the direction of the saccade (figure 5.11C).

Stefan Everling, Doug Munoz, and colleagues recorded from the OPNs as well as the fovea-related cells of the colliculus in a gap task (Everling et al., 1998). In the task, the fixation spot was removed and following a 200 ms gap period, a stimulus appeared at ±10° along the horizontal axis with respect to fixation (figure 5.11B). As the gap period commenced, collicular fovea-related neurons reduced their discharge in anticipation of stimulus arrival. However, during the gap period, there was little or no change in the activity of OPNs. Rather, activity in the OPNs changed abruptly just before saccade onset, exhibiting a halt in discharge, with a pause onset that preceded saccade onset by about 10–15 ms. This abrupt halt in OPN activity also ended precisely at saccade offset. In comparison,

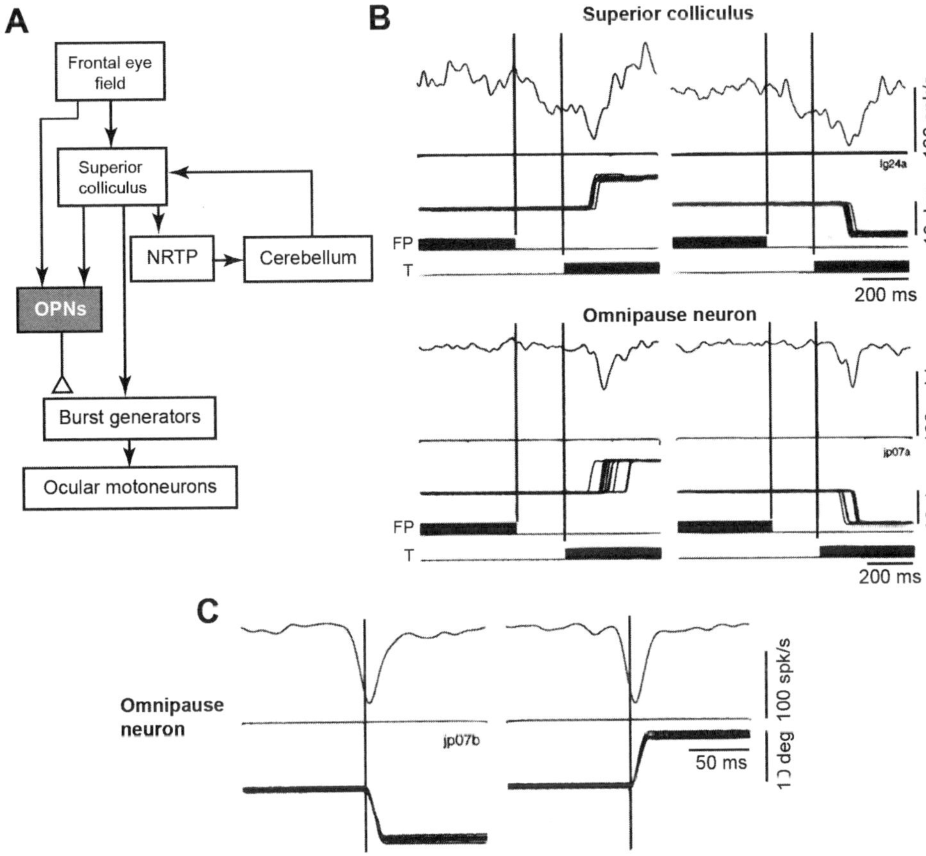

Figure 5.11
Activity of omnipause neurons (OPNs) compared with activity of fovea-related neurons of the superior colliculus. **A**. Schematic of anatomy. **B**. The task for the monkey was to look at a fixation spot and then after a 200 ms gap period, to look at the target. The plot shows activity in a fovea-related cell of the superior colliculus, as well as that of an omnipause neuron. Whereas the collicular neuron exhibited a gradual decay of activity during the gap period, activity in the omnipause neuron remained steady during the gap period. Both neurons showed a reduction in their activity levels during the saccade. Data are aligned to fixation point offset. **C**. To better illustrate the tight coupling between saccade onset and change in the activity of the OPN, data are aligned to saccade onset. NRTP: nucleus reticularis tegmenti pontis. (From Everling et al., 1998.)

fovea-related superior colliculus neurons showed a return to activity with a rather wide distribution of timing that followed saccade offset with a mean of around 20–30 ms.

Therefore, at a gross level, the activity of the OPN resembles that of the fovea-related neurons of the FEF and the colliculus. However, in the FEF and in the colliculus, the discharge of the fovea-related cells is sensitive to utility of the stimulus. Furthermore, the fovea-related cells reduce their firing gradually before onset of the movement at a rate that is related to the reaction time of the saccade. In contrast, the OPNs show insensitivity to

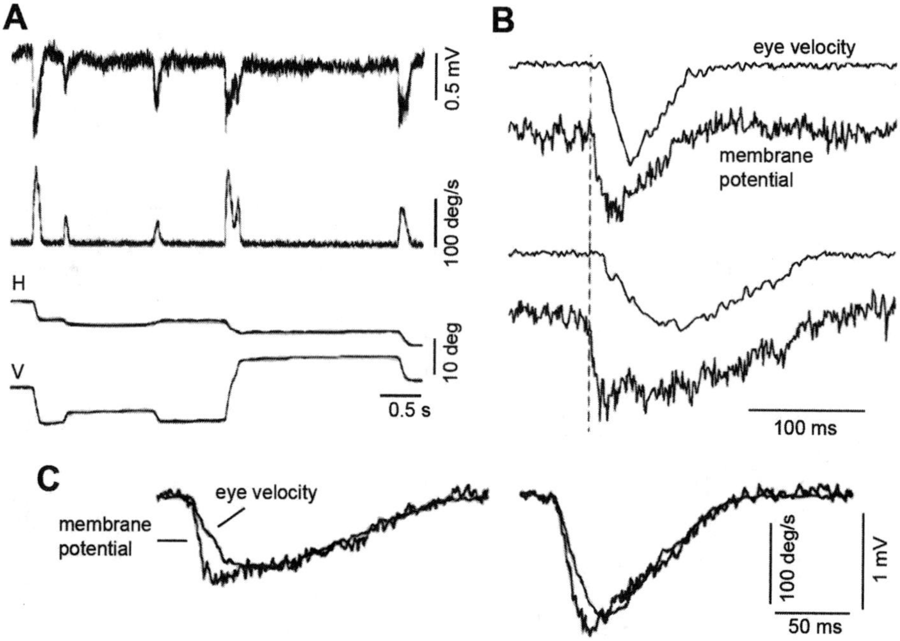

Figure 5.12
Intracellular recordings from omnipause neurons. **A**. Membrane potential, eye velocity, and eye position traces (H and V indicate horizontal and vertical eye position). The membrane potential is steady during fixation and then shows hyperpolarization with a pattern that strongly resembles eye velocity. Note that changes in membrane potential are independent of saccade direction. **B**. Eve velocity and membrane potentials aligned to change in potential for a short-duration and a long-duration saccade. **C**. Membrane potential and eye velocity aligned to onset of each signal. The initial change in membrane potential reflects a more abrupt change than eye velocity. After this initial phase, membrane potential precisely tracks eye velocity. (From Yoshida et al., 1999.)

this modulation during the reaction time period: they remain on while the eyes are holding still, then pause to allow the movement to take place.

Kaoru Yoshida, Hiroshi Shimazu, and colleagues (Yoshida et al., 1999) recorded intracellular membrane potentials of OPNs during spontaneous saccades in cats. (Stimuli were food pieces that were shown to the animal.) They noticed that the membrane potential was high during fixation but decreased with a pattern that was exquisitely similar to eye velocity (figure 5.12A). For example, when the saccade velocity was low, the membrane potential showed a small peak. When the saccade had a high, bimodal velocity profile, the membrane potential also had a large bimodal shape. The reduction in the membrane potential during saccades reflected hyperpolarization of the OPNs, and the authors were able to establish that this was due to inhibitory input to these cells.

The OPN membrane potentials showed a steep hyperpolarization around 16 ms before saccade onset (figure 5.12B). This hyperpolarization was due to temporal summation of inhibitory postsynaptic potentials that arrived at the OPNs. Beyond this initial steep hyperpolarization, the shape of the synaptic potential was strongly correlated with velocity profile during the saccade (figure 5.12C): the duration and peak change in membrane potential varied strongly with duration and peak saccade velocity. Yoshida et al. (1999) proposed that the pause in OPNs was caused by abrupt arrival of inhibition due to an intense input, likely from burst type neurons in the colliculus or the FEF. However, once the saccade was initiated, the further inhibition that OPNs received was likely from elsewhere, possibly from the burst generators and/or the cerebellum.

The threshold of activity needed to initiate a saccade may be set by the omnipause neurons.

If the threshold for allowing a saccade is set by the OPNs, then a critical question is whether the OPNs integrate input from the FEF and perhaps collicular neurons, as suggested by the results in figure 5.10E. In order to establish which FEF cells project to the OPNs, Mark Segraves (1992) stimulated the omnipause region while simultaneously recording from the FEF, thereby identifying the cortical neurons that projected onto the OPNs. Among the cortical neurons that were identified, the largest number encompassed burst-type saccade-related neurons. These FEF neurons produced little or no response to visual stimuli, but they were very active during a saccade (similar to the motor responsive neurons in figure 5.10D). A slightly smaller number encompassed fovea-related neurons, responding to rewarding visual stimuli on the fovea. Importantly, there were no neurons that projected to the OPNs from the FEF that had visual or buildup activity. That is, the projections to the gate-keeper OPNs from the FEF were either associated with prevention of motion (the fovea related neurons) or the promotion of motion (the burst-type neurons).

Like those in the FEF, the neurons in the superior colliculus that are active during fixation tend to project to the OPNs. Jean Buttner-Ennever and colleagues (1999) injected a tracer into the caudal and rostral regions of the intermediate layers of the superior colliculus and then looked for projections in the omnipause region. They found that from the rostral pole of colliculus, where the fovea-related neurons are located, there were strong projections that crossed the midline and terminated on omnipause neurons. (The projections are assumed, but not confirmed, to be excitatory.) From the nearby small saccade zone of the colliculus, the projections to the OPN region were significantly less dense. They found few projections to the OPN from the large saccade zone of the colliculus. When Neeraj Gandhi and Edward Keller (1997), using antidromic identification techniques, stimulated the OPNs and recorded from the superior colliculus, they found that around 70% of the cells that they identified in the colliculus were near the rostral pole. They then recorded from 30 collicular cells that they had antidromically identified and found that among them, only one was a burst neuron, whereas 11 were buildup neurons and 18 were fovea-related neurons. In contrast, 13 burst neurons, 16 buildup neurons, and only 4 fovea-related neurons did not project to the OPNs.

Neeraj Gandhi and David Sparks (Gandhi and Sparks, 2007) stimulated the OPNs as monkeys attempted to make head-free saccadic eye movements in order to place a visual stimulus onto their foveae. They found that the stimulation completely prevented saccade onset and significantly delayed start of the head movement. These results suggest that saccade initiation, and to a lesser extent head movement initiation, depends critically on the binary behavior of OPNs: these cells must halt their discharge in order for the movement to start. It is possible that these neurons are acting like a gateway allowing a movement to take place only when the converging input produces sufficient inhibition to overcome a threshold.

Missing from these data are tasks that involve decision-making; these tasks result in systematic variation of saccade latency. In such tasks, the long reaction times coincide with modulation of peak discharge in the bursting activity of both the FEF and collicular cells, suggesting that if there is a fixed threshold that enables movement initiation, that threshold is not present in these structures. If the threshold is set by the OPNs, we would expect their activity to show little or no modulation during the deliberation period. Thus, the limited data currently available raise the possibility that the OPNs act as the threshold, but they are not conclusive.

In summary, many of the fovea-related neurons of the FEF and the colliculus, as well as some of the burst-type saccade-related neurons of the FEF, project to the OPNs. The OPNs are highly active during fixation. Unlike fovea-related neurons of the FEF and the colliculus, OPNs show little or no change in their discharge during the reaction time period as the animal is deliberating the movement. OPNs pause just before all saccades (regardless of saccade direction) and then return to high levels of discharge around saccade offset.

Recordings from the OPNs are limited, but the available data suggest that the release from fixation is due to converging inputs accumulating and reaching threshold. The OPNs likely are excited by the fovea-related neurons of the FEF and the colliculus, and they are probably inhibited (indirectly) by the movement-related neurons of the FEF, and to a lesser extent, the colliculus. It is possible that the OPNs integrate these excitatory and inhibitory inputs and then pause when the sum exceeds a threshold, thereby allowing the movement that is encoded by the upstream neurons to take place. As the movement proceeds, OPNs receive inhibition that is approximately proportional to saccade velocity. This allows the OPNs to return to high discharge levels as the saccade ends and the eyes resume fixation.

5.7 The Premotor Cortex and the Decision to Reach

In the framework that we have been developing, during the period of deliberation, the utility of a potential act is integrated, as reflected in the activity of cortical neurons that prefer that movement; the activity of these neurons rises and eventually reaches sufficient levels to trigger that movement. The rate of rise and the magnitude of the peak activity appear to dictate reaction time and velocity of the ensuing movement. Although we have focused on saccades, we can ask whether this is a general concept that also applies to other kinds of movement.

During deliberation, saccade vigor increases as certainty increases regarding the reaching movement that will result in reward.

Paul Cisek and colleagues (Cisek et al., 2009) developed a novel task in which reward was acquired for making a reaching movement, but the probability of success depended on the information that changed on a moment by moment basis. In the tokens task (figure 5.14A), the trial began with a center location that had 15 tokens. Every 200 ms, one token at random went to one bin or the other. The animal's task was to pick the bin that would eventually have the most tokens, and then express its decision by reaching to the chosen bin.

As the hand arrived at the chosen bin (reach end, figure 5.14B), the remainder of the tokens sped up a little (at a rate of once every 150 ms) during the slow block, or sped up a lot (at a rate of once every 50 ms) during the fast block. At trial end, when all the tokens had been distributed, if the animal had chosen correctly and most of the tokens were in the chosen bin, it received a reward. Thus, the task involved a speed-accuracy tradeoff: the monkey could either wait to observe the fate of all the tokens, or decide earlier on the basis of less evidence and speed up the process a little (slow block) or a lot (fast block).

The session usually began with the slow block, and then after around 120 trials, switched to the fast block. In the slow block the animal was less likely to decide early: it usually waited more than a second before committing. This makes sense because in the slow block, after the reach ended, the monkey still had to wait a long time until all the tokens had been distributed. Thus, in the slow block, deciding early saved only a small amount of time. However, in the fast block, the animal tended to decide sooner (usually less than 1 s), which also makes sense because in this case the monkey saved a lot of time by deciding early. Time to decision was about 400 ms faster in the fast block, and as a result, in the fast block probability of success was lower, but the reward rate (drops/s) was higher.

During deliberation, the monkey made saccades to the two targets. Although this act was not rewarded, David Thura, Paul Cisek, and colleagues (Thura et al., 2014; Thura and Cisek, 2016) noted that saccade vigor nevertheless increased as the deliberation period unfolded (figure 5.13C). This may be because with passage of time, as more tokens were committed to each bin, the information content in each bin increased, thereby increasing the certainty of reward while reducing the time delay to that reward. Hence, the expected value of reward increased with the passage of time. Notably, during the fast block, saccade vigor appeared greater than in the slow block; this may have happened because in the fast block, time delay to the reward was shorter than in the slow block, thus increasing the subjective value of the reward.

Once the animal arrived at its decision, it reached for the chosen target, at which point the tokens accelerated. If the deliberation period was short, the reach had low vigor (figure 5.13C), but if the deliberation period was long, the reach was made with high vigor.

As in the case with saccades, this may have been because after a long deliberation period, there was increased certainty of reward as well as a shorter time delay to the reward, both leading to a higher expected value of reward. Also as in the case with saccades, reach vigor tended to be greater during the fast block; this again may be a reflection of the fact that in the fast block, the time delay for the reward following reach completion was shorter.

Thus, in the tokens task, visual information at each moment of time indicated the probability of reward for each of the two directions of reach. Saccade and reach vigor tended to be higher when there was greater certainty of reward (i.e., with increased time from trial start) and when the time to reward was shorter (i.e., in the fast block).

Activity in the premotor and primary motor cortex increases with evidence for the preferred reach. Reach vigor is greater when activity before movement onset is higher. Thura and Cisek (Thura and Cisek, 2014) recorded from the dorsal premotor (PMd) and primary motor cortex (M1) as the animal observed the tokens and made its decision. Many cells in these regions have a preferred direction of reach (Georgopoulos et al., 1982); these cells exhibit a buildup of activity during the delay period before their preferred reach (particularly in the premotor cortex and rostral M1) and then produce a burst of activity during the reach (particularly in the caudal M1) (Crammond and Kalaska, 1996).

During deliberation, some trials were easy because most of the tokens went to one side (figure 5.14A). Some trials were ambiguous because the tokens were evenly distributed. Finally, some trials were misleading in that the first few tokens went to one side, whereas the remainder went to the other side. These patterns provided the authors an opportunity to ask whether activity in PMd and M1 reflected the instantaneous evidence for the merits of reaching toward each of the two targets.

In the easy trials, if the tokens went toward the direction that the cell preferred, activity increased (figure 5.14B, left column), but remained roughly flat if the tokens went toward the other direction. In the ambiguous trials, the rise in the activity toward the preferred target was slower. Finally, in the misleading trials, the activity initially remained flat and then increased when the tokens reversed direction. Therefore, the duration of time that it took for the activity in the preferred direction to reach high levels was longer when the deliberation period was longer.

When the data were aligned to reach onset (figure 5.14B, right column), a striking pattern emerged: activity in PMd reached its peak approximately 250 ms before reach onset, and then it began to decline. This peak in PMd activity preceded the peak of activity in M1, suggesting that PMd was upstream to M1 in the decision-making process.

Reach vigor increased with duration of deliberation (figure 5.13C, right column), and indeed, the peak of activity was greater in PMd when the deliberation period was longer (misleading trials versus easy trials, figure 5.14B, right column). Thus, reaction time appeared to correspond to the duration of time it took for the activity to reach high

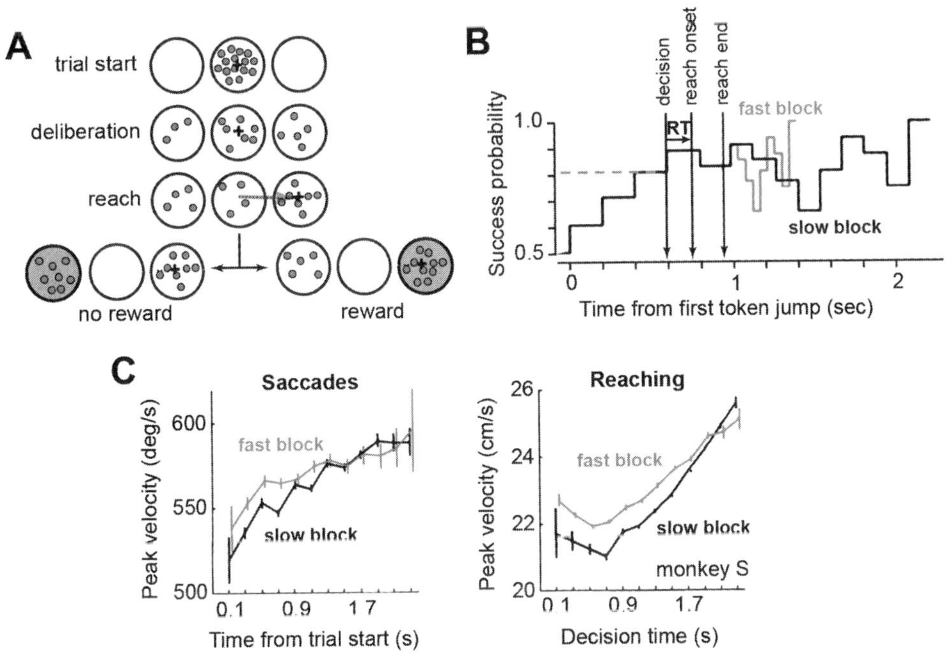

Figure 5.13
The tokens task. **A**. The trial began with 15 tokens at a center location. At 200 ms intervals, a token jumped randomly (uniform distribution) into one bin or the other. As the tokens were distributed, the subject decided which bin was likely to be the recipient of most tokens and expressed this decision by reaching for the chosen bin. Once the reach had ended, the tokens continued to be distributed, and the subject was rewarded if the chosen bin was indeed the one that acquired the most tokens. **B**. In both the fast and slow blocks, at trial start, the tokens were distributed at 200 ms intervals. Once the decision was made and the reach concluded, in the fast block, the tokens were distributed rapidly (once every 50 ms), but in the slow block, the distribution rate was lower (once every 150 ms). **C**. During deliberation, the animal made saccades to the two bins. Saccade velocity increased with deliberation (as did saccade amplitude) during both blocks, but the velocity was faster in the fast block. In trials in which the deliberation period was longer, the reach that expressed that decision had greater velocity. One of the two animals reached with greater velocity in the fast block. (From Thura et al., 2014.)

levels, and reach velocity appeared to be modulated by the magnitude of this activity in PMd and M1.

Activity during deliberation is inconsistent with accumulation of evidence, but suggestive of an urgency signal that pushes the system toward decision and movement.

The same pattern of tokens led to earlier decisions in the fast block than in the slow block. This pattern could have arisen because neuronal activity started similarly but rose more rapidly in the fast block. We would expect such a pattern from the drift diffusion models of decision-making in which evidence is accumulated for each potential action, but one is willing to sacrifice accuracy for speed.

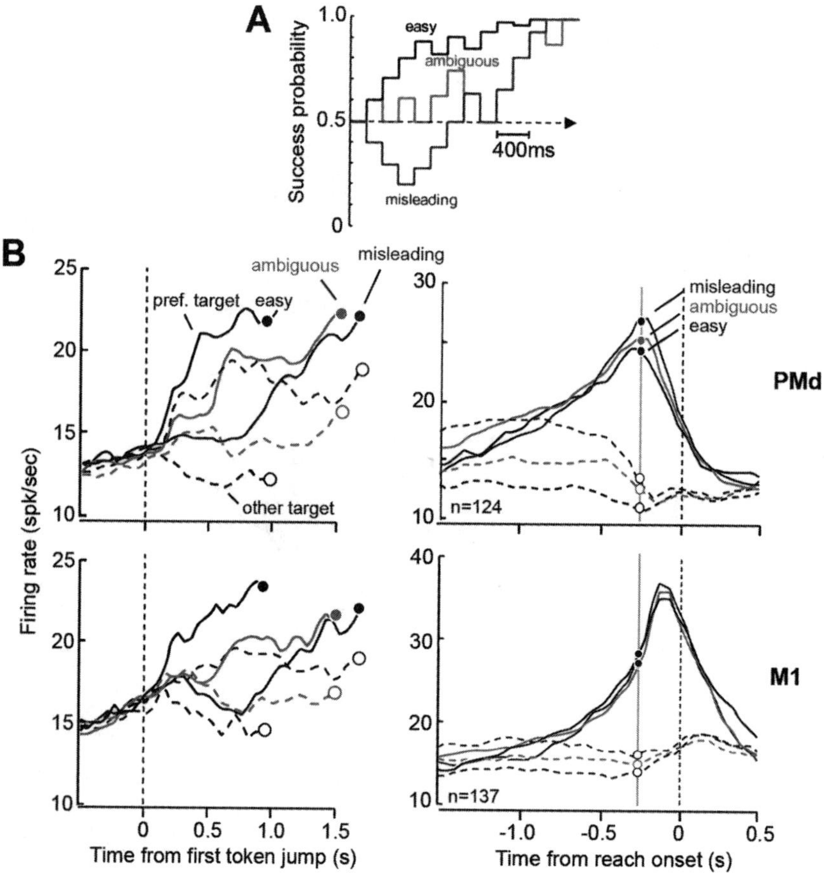

Figure 5.14
Decision-related activity in the dorsal premotor cortex (PMd) and primary motor cortex (M1) during deliberation. **A.** In the easy decision trials, most of the tokens went to one bin, while in ambiguous or misleading trials, the bin distribution was nearly uniform or reversing in direction, thereby making the decision more difficult. The deliberation period was shortest in the easy trial and longest in the misleading trials. **B.** Averaged activity of PMd and M1 cells that had expressed directional tuning during the deliberation period, aligned to the first token jump or to reach onset. Solid lines show activity when the decision was to reach in the preferred direction, and the dashed lines show activity when the decision was to reach in the other direction. (From Thura and Cisek, 2017.)

Contrary to this hypothesis, the authors discovered that in the fast block, neuronal activity was elevated even before the first token had jumped (figure 5.15A). The decision-making/movement-making system was primed in the fast block, thereby making the process of deliberation end sooner. In this scenario, neural priming is consistent with the idea of an urgency signal that boosts all potential actions.

A strong test of the urgency signal hypothesis is to consider activity in the neurons at the time when the evidence for both targets is equivalent (i.e., when both bins have the same

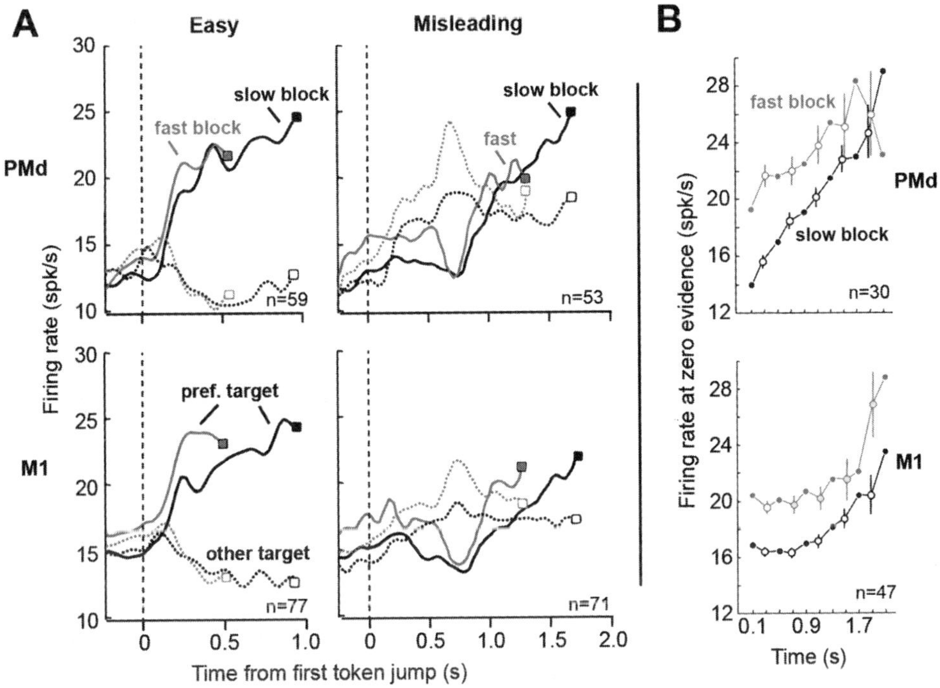

Figure 5.15
Context-dependent activity in dorsal premotor cortex (PMd) and the primary motor cortex (M1) during deliberation. **A**. Averaged activity levels during easy and misleading trials. **B**. Evolution of averaged activity levels (as a function of time from the first token jump), calculated for the situations when the evidence was equal for each bin. For odd-number jumps, the firing rate at 0 evidence was calculated by interpolation (filled circles). For even-number jumps, the vertical bar is the confidence interval. (From Thura and Cisek, 2016.)

number of tokens). Neural analysis revealed that under these conditions, the activity in both PMd and M1 was higher during the fast block than during the slow block (figure 5.15B).

These data do not support the hypothesis that early decisions come about because the evidence for them accumulates faster. Rather, it appeared that the context of the task elevated activity of all task-related neurons, regardless of their preferred direction, compelling the decision-making process to come to an end and produce a movement.

The results are particularly noteworthy when we compare them to the data from the FEF during the decision-making process for a saccade (figure 5.10). In both the FEF and PMd, activity did not reach a constant threshold to signal culmination of decision-making. Rather, in the easy reach trials and accurate saccade trials, activity peaked at a lower level than in the misleading reach trials and fast saccade trials. Thus, the decision to reach, like the decision to make a saccade, may not be due to activity in just one or two cortical areas arriving at a threshold, but perhaps convergence of these activities upon an (unknown) area that, like the omnipause neurons, integrates the various inputs and acts as a gate to allow the holding process to end and the movement process to commence.

In both the FEF and PMd, neural activity before the deliberation period was affected by the context of the task: in the accurate block of the saccade task, during fixation and before onset of deliberation, activity in the visual and motor responsive FEF neurons was low (figure 5.10). Thus, if the baseline is low, activity needs greater time during the deliberation period to reach high levels, and the decision-making process is slowed. Similarly, in the slow block of the reach task, at trial onset and before the start of deliberation, the activity in PMd and M1 was low (figure 5.15A). It took longer to decide on the winning direction and start the reach.

As deliberation started, in one context, the decision was made with lower peaks of activity (easy reach trials, accurate saccade trials), and in another context, the decision was made with higher peaks of activity (misleading reach trial, fast saccade trials). The higher peak of activity before movement onset often coincided with greater vigor in the ensuing movement.

Thus, the activities in both the FEF and PMd influenced not just the decision of what to do, but also the vigor with which to do it. When one is primed to make a decision, that decision is made sooner, but also the movement that signals that decision is made with greater vigor. This priming may be a reflection of an urgency signal, one that may be generated in the basal ganglia, a topic that we will consider in the next chapter.

5.8 Encoding Utility and Salience in the Parietal Cortex

Movements and decisions are affected by not only the utility of the stimulus, but also the low-level sensory properties of the stimulus, properties that make that stimulus salient, such as luminance, orientation, novelty, and motion. These properties attract our attention, and occasionally affect our movements. For example, in chapter 1, we saw that people made saccades to distractors that appeared suddenly or that had high contrast, particularly when they were under time pressure, even though such movements were counterproductive and not associated with reward. For many cells in the colliculus, the FEF, and the saccade region of the posterior parietal cortex (LIP), the initial response to the onset of a visual stimulus is a reflection of stimulus salience, not utility. With the passage of time after stimulus onset, the neural activity more closely reflects utility. Here, we will summarize the key experiments that have quantified encoding of utility and salience in the LIP.

Activity in LIP is modulated by magnitude and probability of reward.

The first experiment that established encoding of utility in the parietal cortex was performed by Michael Platt and Paul Glimcher (Platt and Glimcher, 1999). Each trial began with a fixation target (figure 5.16A). At 200–500 ms, two eccentric stimuli of different

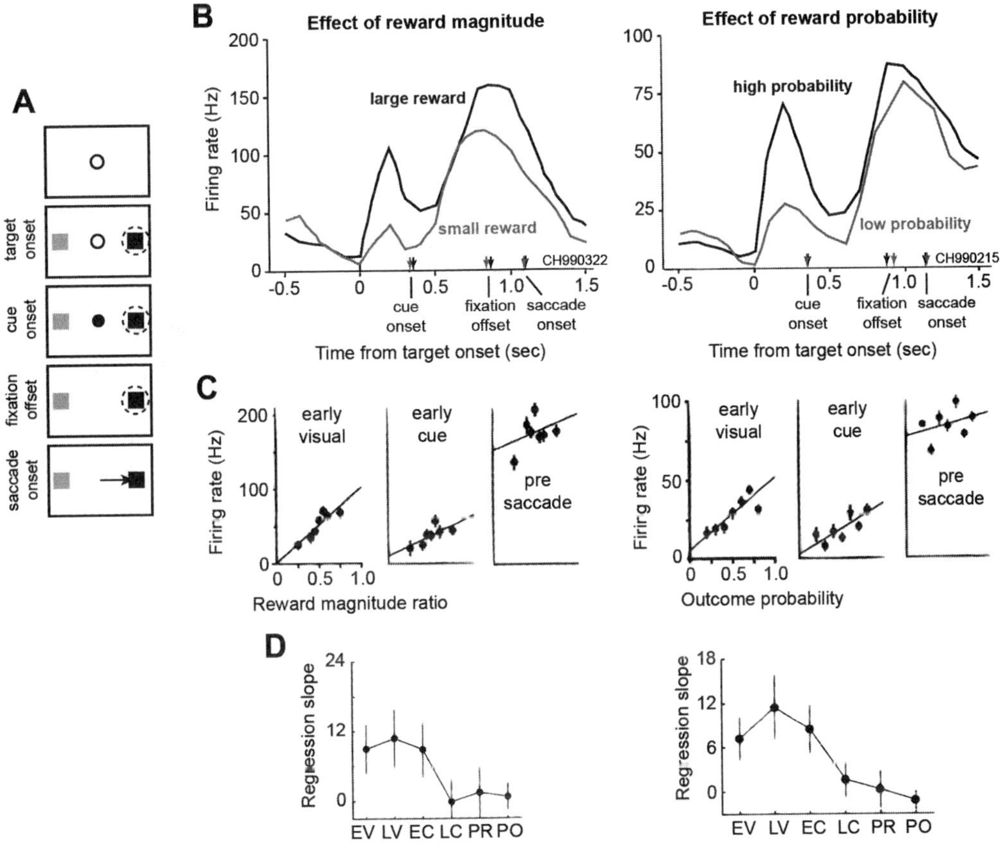

Figure 5.16
Activity of neurons in the lateral intraparietal sulcus (LIP) encodes economic value of stimulus. **A**. Experiment design. After fixation, two targets were presented, one of which was in the response field of the neuron (dashed circle). At cue onset, fixation stimulus changed color, identifying target of saccade. **B**. Response of two different neurons. In all trials that are displayed, the movement was to the target in the neuron's response field. Across blocks of trials, reward magnitude and selection probability were varied, while sum of reward magnitudes for the two targets was kept constant. For the data in the left column, reward magnitude was varied across blocks. For the data in the right column, the probability that the stimulus in the response field would be instructed was manipulated across blocks. **C**. Firing rate during different periods of time. Early visual: 0–200 ms after target onset. Early cue: 0–200 ms after cue onset. Presaccade: 0–200 ms before saccade onset. Reward magnitude ratio refers to reward for target in response field divided by sum of rewards for the two targets. **D**. For each neuron, firing rate during each period of time was fitted to a linear model of reward magnitude (or probability), saccade amplitude, peak velocity, and reaction time. The plot displays the regression slope for the reward magnitude or probability variable as a function of period of time. (EV, early visual period; LV, late visual period; EC, early cue period; LC, late cue period; PR, presaccade; PO, postsaccade) (From Platt and Glimcher, 1999.)

colors appeared, with one located in the response field of the neuron under study. After an additional 200–800 ms, the central fixation target changed color to instruct the goal target. Finally, after an additional 200–800 ms, the central fixation target was extinguished, and the monkey was rewarded if it shifted its gaze to the instructed target.

In experiment 1, using a block design, the authors varied the amount of reward (juice) the animal received from each of the two movements while keeping the sum constant. The probability of a target being instructed was 1/2. For example, in one block the reward for a saccade to the target in response field was 0.26 mL of juice, whereas the reward for the target outside of response field was 0.09 mL of juice. In trials in which the movement was into the response field, within 200 ms after target onset, the discharge rate was higher if the target was associated with greater reward (figure 5.16B, left column). They fitted the discharge rate of each neuron to gain ratio (reward for target in response field/sum of two targets), saccade velocity, amplitude, and reaction time (figure 5.16C, left column). The regression slope associated with gain ratio was significant during visual period and early cue period, but not in late cue and presaccade period (figure 5.16C, left column). They then varied the probability of a target being instructed across blocks of trials between 20% and 80%. This changed the expected value of the stimulus that was in the response field of the neuron. (Despite this, if the animal performed the correct movement, it received a reward in 100% of the trials.) During the visual and early cue periods, the regression slope of firing rate with respect to instruction probability was significantly positive. Therefore, the responses of LIP neurons near the time of cue and to the onset of visual stimuli varied linearly with the magnitude of the reward promised by the stimulus as well as the probability of that stimulus being instructed for action.

Activity in LIP is modulated by time delay to reward.

Kenway Louie and Paul Glimcher (2010) explored activity of LIP neurons in response to modulation of the stimulus value via time delay to reward. Animals prefer a reward if it is provided soon, suggesting that the delay to acquisition decreases the subjective value of that reward. Louie and Glimcher once again instructed the monkey to make a saccade to one of two stimuli, one associated with a small amount of juice, and the other associated with a larger amount. The larger amount was in the response field of the LIP neuron under study. The cue at fixation instructed which stimulus was the target of the saccade. If the saccade was performed correctly, the monkey received the promised juice after the prescribed time delay. The time delay was kept constant within a block of trials, allowing the animal to learn the parameters of the task. The authors found that in the 200 ms period following presentation of the targets, before the instruction cue was presented, LIP neurons responded with a greater firing rate when the stimulus promised the reward sooner (figure 5.17). Furthermore, the response at saccade onset also appeared to be larger when the movement was toward a stimulus that promised the reward sooner.

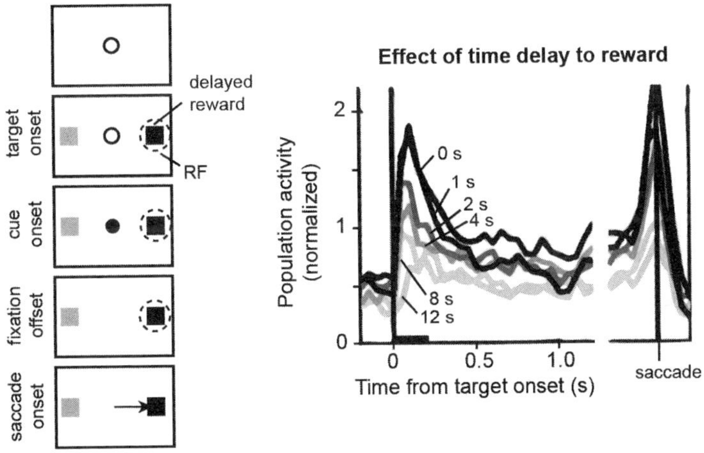

Figure 5.17
Activity of neurons in the lateral intraparietal sulcus (LIP) is lower when the stimulus in the response field (RF) promises reward after a longer time delay. The delayed reward target promised larger magnitude of reward, and was placed in the response field of the neuron under study. The amounts of reward were kept constant while the time delay for the larger reward was varied between blocks of trials. The time trace shows the normalized average firing rate of the population of neurons when the animal was instructed to make a saccade to the delayed target. (From Louie and Glimcher, 2010.)

In summary, the response of LIP neurons at stimulus onset increased with the magnitude of the reward promised by that stimulus and with the probability of that stimulus being instructed for action. However, the response of LIP neurons decreased with the time delay before acquisition of reward. All of these results are consistent with the idea that these neurons encode the subjective value or the utility of the stimulus in their response field.

Activity in LIP may be a reflection of both the utility of the stimulus and the amount of attention allocated to that stimulus.

A stimulus that is more rewarding is likely to gather more attention, as is a stimulus that is associated with more punishment or loss. Is the LIP activity reflecting utility of the action, or is it a reflection of attention to that spatial location?

Marvin Leathers and Carl Olson (2012) trained monkeys to learn the cues for each of four different outcomes: large reward, small reward, large penalty, and small penalty. The rewards were many or few drops of juice, and penalties were extended or brief periods of eccentric fixation. The task began with fixation of a central target. After fixation, two cues appeared, one of which was in the response field of the LIP neuron (figure 5.18). The monkey had learned the meanings of the cues and was associating them with the correct outcomes. The cues disappeared, and each was replaced with a dot in its location. Next, the fixation spot disappeared, instructing the animal to make its choice. The hypothesis

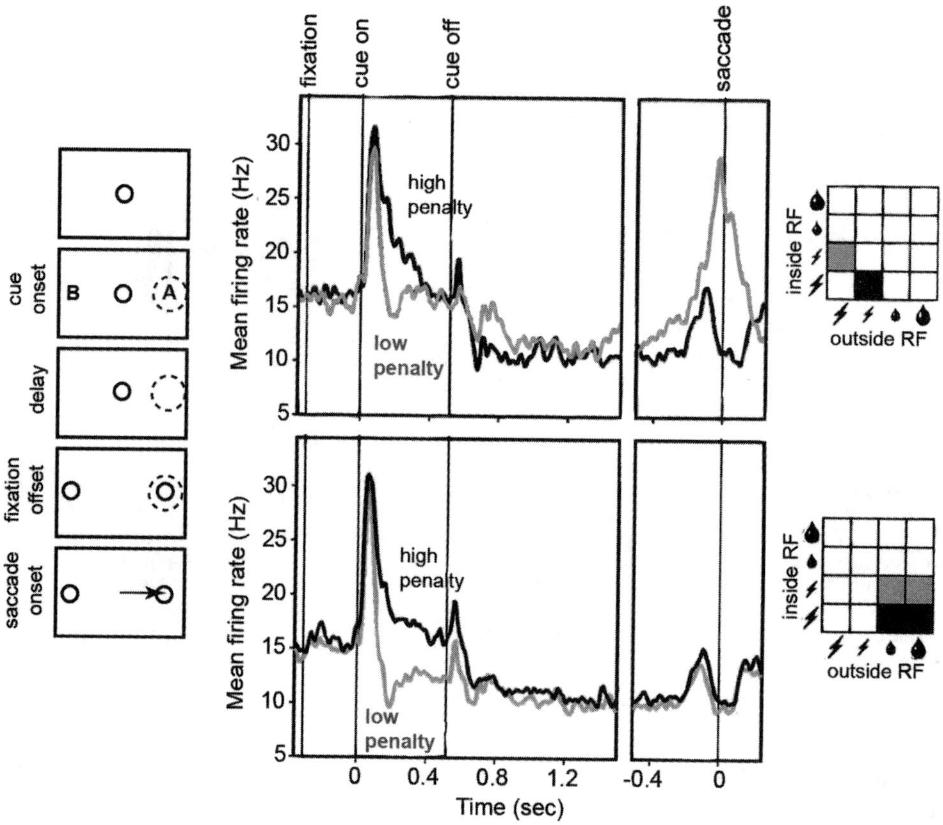

Figure 5.18
Activity of neurons in the lateral intraparietal sulcus (LIP) is greater in response to cues that are associated with greater penalty. The animals were trained to associate each of four cues with high reward, low reward, high penalty (eccentric fixation for a long period of time), or low penalty. After the presentation of two cues, the animal chose one cue by performing a saccade. The top plot shows average firing rate of n=67 LIP neurons under two sets of conditions: when the cue inside the response field (RF) indicated low penalty (gray line) and the other cue indicated high penalty, and when the cue inside the response field indicated high penalty (black line) and the other cue indicated low penalty. In both cases, the animal chose the low-penalty cue. Activity in the cue period was higher under the high-penalty conditions. The bottom plot shows average firing rate of LIP neurons under another two sets of conditions: when the cue inside the response field indicated high penalty (black line) and the other cue indicated low or high reward, and when the cue inside the response field indicated low penalty (gray line) and the other cue indicated low or high reward. In both cases, the animal chose the reward cue. Activity in the cue period was higher under the high-penalty conditions. (From Leathers and Olson, 2012.)

was that the large penalty was lower in value than the small penalty but was emotionally more potent. This acted as salience and was thought to influence response in the LIP.

In trials in which the choice was between large and small rewards, the animal very frequently chose the large reward. Consistent with observation of Platt and Glimcher (Platt and Glimcher, 1999), LIP population activity during the cue period was larger when the cue in its response field was associated with the larger reward. However, in trials in which the animal was offered a choice between a large penalty and a small penalty and chose the small penalty, the firing rates during the cue period were higher when the cue in the response field was for the larger penalty (figure 5.18, top subfigure). In trials in which the animal had to choose between a penalty (large or small) and a reward (large or small), and the penalty cue was inside the response field, firing rates during the cue period were again higher when the cue in the response field was for the larger penalty than for the smaller one (figure 5.18, bottom subfigure). Therefore, the response was higher when the cue in the response field promised the larger reward (compared to the smaller reward), but it was also higher when the cue promised a greater penalty (compared to the smaller penalty). This argues against the idea that during the visual response period (cue-onset to cue-offset), LIP neurons are encoding only the utility of the cue. Rather, it suggests that during the cue period, much of the LIP response is due to attention that is being focused on the stimulus that is salient and if chosen (by mistake), will be punished by a large penalty.

Interestingly, Leathers and Olson (2012) also found that while saccade reaction time was shortest when the choice was toward the high reward cue, it was longest when the choice was toward the cue that promised a penalty. Although they did not report saccade velocities, it seems possible that movement vigor was an accurate reflection of the utility of the cue, even though the LIP activity in the visual response period to that cue did not differ between penalty and reward.

For constant reward, reaction times are faster toward novel stimuli.

The interaction between utility and salience was studied by Nicholas Foley, Jacqueline Gottlieb, and colleagues (Foley et al., 2014). They performed an experiment in which stimulus utility (reward value) and stimulus salience (novelty) were independently controlled. To manipulate utility, wire-frame shaped images were used as cues to indicate whether a particular movement would be rewarded or not (figure 5.19A). To manipulate salience in each session, two of the stimuli were familiar, having been used in many previous sessions, while the other two were novel. In all sessions, the familiar stimuli were consistently associated with receiving a reward or receiving no reward. A number of stimuli that had never been seen before were used in each session and also stably associated with a receiving a reward or not receiving a reward. Each novel stimulus was presented an average of 32 trials during each session, thereby allowing the animal to learn the reward value of that stimulus. In each trial, one place holder and one stimulus were presented.

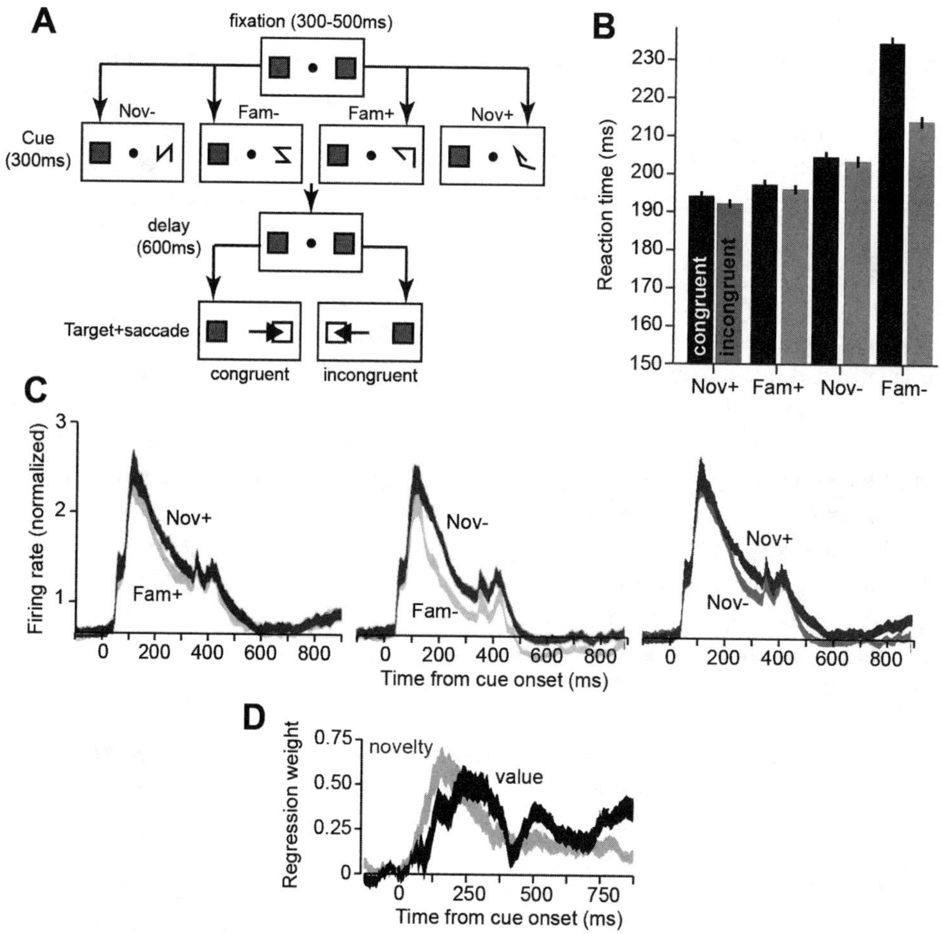

Figure 5.19

Lateral intraparietal sulcus (LIP) neural activity in response to novelty and value of the stimulus. **A**. After a fixation period during which two place-holder stimuli were presented, one place-holder was replaced with a cue (wire-frame stimulus). The cue could be novel (Nov) or familiar (Fam), rewarding (+) or non-rewarding (-). After a delay, subject made a saccade to a location that was congruent or incongruent with the location of the cue. **B**. Saccade reaction times were faster for Nov+ than for Fam+ cues. **C**. Firing rate of LIP neurons with response fields centered on the cue. Novel cues triggered greater activation than familiar cues. Rewarded cues triggered greater activation than non-rewarded cues. **D**. Activity for individual cells was fitted to a linear model of stimulus novelty and value. The graph displays the regression weight of these two factors. (From Foley et al., 2014.)

The stimulus could be novel or familiar and could be rewarded or not rewarded. After a delay, the stimulus was removed and in its place another place holder was displayed. Then, one of the two place holders was removed, instructing the subject to make a saccade. In the congruent trials, the saccade was made to a location where previously a familiar or novel stimulus was located.

The authors found that reaction times were longest when the stimulus was Fam- (familiar, no reward). However, reaction times were significantly lower if saccade was away from Fam-, rather than toward it (even though no reward was attained in either case). Novelty affected reaction times: latency in Nov+ trials (novel, reward) was slightly shorter than in Fam+ trials (familiar, reward), although the animal was less certain that it would get reward from the novel stimulus. Therefore, salience as manipulated through novelty affected reaction times. When the stimulus was novel, reaction times to the location where the stimulus used to be located were shorter.

Salience is encoded early in the activity of the LIP cells, but then subsides as value-related activity becomes more prominent.
The authors recorded from LIP neurons with response fields that included the stimulus. When the stimulus appeared in the response field, LIP neurons had a transient response to the visual onset, and then sustained post-cue responses that were stronger for reward cues than for no-reward cues (Fam+ versus Fam-, figure 5.19C). Importantly, responses were stronger for novel relative to familiar stimuli (Nov+ vs. Fam+, Nov- vs. Fam-). The authors fitted the trial-by-trial response of each cell to a two-variable (novelty and value) linear model of time. They found that the novelty factor initially played a larger role in accounting for the discharge of the cells, but then its influence subsided and value become the prominent factor (figure 5.19D).

In summary, we can attend to important stimuli, but choose not to move toward them. Here, we considered to what extent LIP activity reflects our locus of attention instead of only the utility of moving toward that location. When a cue appears in the response field of an LIP neuron and that cue is associated with a movement that is associated with reward, the activity of LIP neurons in response to the cue onset increases linearly with the magnitude and probability of reward. However, this activity is not merely a reflection of the utility of the movement. Rather, the activity in the period that follows cue onset is also affected by salience of the stimulus: the activity is higher when the cue is associated with greater penalty or loss. Therefore, in the period after the cue appears, the degree with which it gathers our attention is a key factor that affects LIP activity. Indeed, LIP neuronal activity is higher when the stimulus is rewarding and novel than when it is rewarding and familiar. It appears that stimulus salience is encoded early in the activity of the LIP cells but then subsides as the utility-related activity becomes more prominent.

5.9 Good-Based versus Action-Based Models of Decision-Making

Many of the decisions that we have considered thus far have the feature that the choice shares fundamental spatial properties with the action that expresses that choice. For example, in the unique-stimulus detection task (figure 5.5, figure 5.10) or the fly detection task (figure 5.9), the perceptual decision of which stimulus is the goal is expressed by making a saccade toward it (figure 5.5). If we are recording from a region of brain that has neurons with a certain movement field, and the stimulus appears in that movement field, we observe that during deliberation, activity in the neuron rises and approaches a threshold, at which point the animal makes a movement toward the stimulus. Because the subject necessarily reports the outcome of the deliberation via a movement that has spatial properties aligned with the stimulus, one can argue that the neural activity is a reflection of preparation for that specific movement. That is, the close spatial relationship between the decision and the movement that reports that decision makes it difficult to know whether the neural activity is a reflection of preparation for movement or the deliberation that concluded in the choice of that stimulus.

To make this distinction, consider the period when you are deciding between two possible choices, but have not been instructed on how you will report your choice. Only after you made your choice you are given instructions regarding how to report it. For example, suppose you are given options A and B (say, two different types of pizza), and you deliberate between the two options. After you indicate that you have made your choice, further instructions are provided: if you chose A, you will press the blue key, but if you chose B, you will press the green key. The brain regions that have movement fields associated with actions toward the blue and green keys will not be activated during the deliberation period and play no role in the deliberation process. However, you will still make eye movements toward the symbols that represent options A and B. Is there a place in the brain where one can find evidence for the utility of each option but is not sensitive to movement parameters such as the spatial location of the option?

One can make a decision without making a movement. In the colliculus, such decision-making engages only a small minority of movement-related neurons.

The *good-based model* of decision-making asserts that the brain maintains an abstract representation of the value of each option. In contrast, the *action-based model* asserts that the value of the stimulus is represented in terms of the movement that is required to acquire that stimulus. In the good-based model, the computation of value depends on action cost but is represented in a nonspatial way. The good-based model separates the decision-making process from the action generation process by evaluating the abstract representation of the stimuli first and then committing that selection to the action generating machinery. Here, we will examine experiments in which the animal deliberates between options and makes a choice, but may not have instructions regarding how to express that choice. We will start

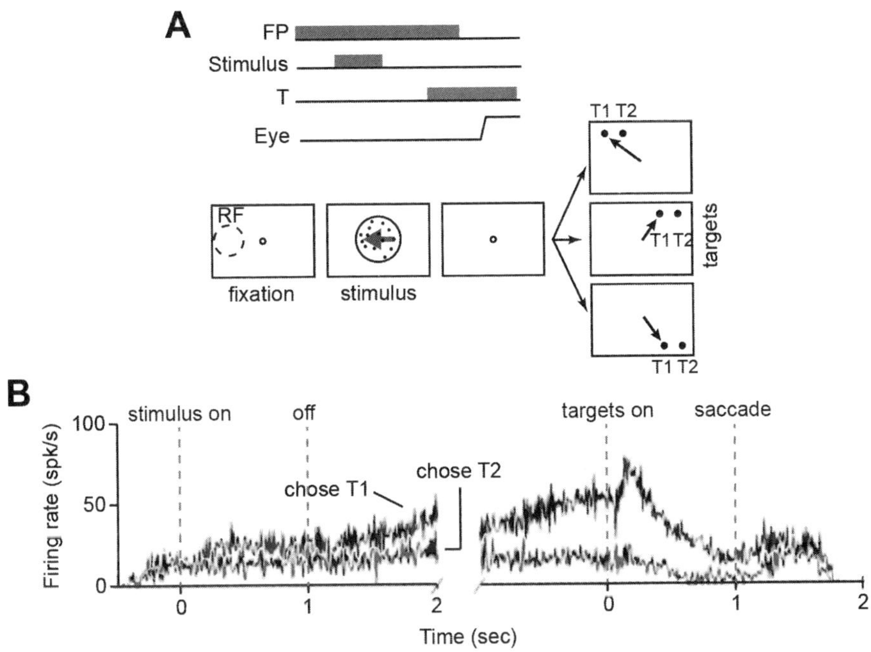

Figure 5.20
Activity in the superior colliculus during a task in which the movement that reports the choice is not known until after the deliberation period. A. Task design. After identification of the response field of a collicular neuron (indicated by RF), a motion stimulus moved toward or away from that response field. After removal of the stimulus, two targets were presented at random locations. In the case shown here, the motion was to the left, and the correct answer was to make a saccade to target T1, the leftmost target on the screen. B. Example of a choice-predictive neuron. The cell responded more strongly in trials in which the animal ultimately chose target T1, although the required saccade could not be planned until the target presentation. (From Horwitz et al., 2004.)

with the colliculus and its response when the deliberation involves accumulating information to make a decision without instructions for a particular movement.

Gregory Horwitz, Aaron Batista, and Bill Newsome (2004) designed a task in which monkeys were instructed to maintain fixation. After 300 ms had passed, a motion stimulus appeared at the center of gaze, moving in one direction or another (figure 5.20A). The stimulus was then removed and after a period of fixation, two targets appeared somewhere on the screen. If the motion stimulus had been moving to the left, the animal made a saccade to the leftmost target (in this case, target T1). Notably, target T1 was not always to the left of the center of fixation, but always to the left of an arbitrary target T2. Because the targets were randomly positioned, target locations were unknown to the subject during deliberation. In this way, during the period in which the animal made the perceptual decision ("leftward motion"), it did not know the action that it would need to make to express that decision. Would the cells that encoded movements to the left be engaged in the deliberation period?

Horwitz and colleagues recorded from cells in the superior colliculus, identifying the region of space for which the cells had a response fields (shown by the RF region in figure 5.20A). They then presented the motion stimulus in such a way that the motion was either toward or away from the cell's response field. If the motion was toward the response field, then the correct target was labeled T1. Otherwise, the correct target was labeled T2. They found that in about 90% of the movement-related cells that they encountered, the responses during the stimulus presentation period did not differ between the two directions. Therefore, as expected, when the information during the deliberation period did not indicate a movement, most movement-related cells were not engaged.

However, contrary to their expectations, they found a small number of collicular cells that did show activity that was a predictor of the upcoming choice. An example of one of these choice-predictive cells is shown in figure 5.20B. When the choice was to saccade to T1, activity during the stimulus presentation period and after the delay period was higher in this cell than when the choice was to saccade to T2. That is, when the motion was toward the response field of this cell, its activity was a predictor of the future choice that the animal would make, even though the upcoming movement that reported that choice was sometimes away from its response field. Thus, from the activity of these cells, the authors could predict the perceptual judgment of the subject: the activity was related to the perceptual decision-making and not the movement that would report that decision.

It would appear that during deliberation, if one does not have information on the nature of the movement, a vast majority of collicular cells are not selectively active. However, a small fraction does become active differentially on the basis of whether the motion is toward or away from their response field. Thus, even in this abstract task, a small part of the motor system is engaged and represents the information that is being used to make the decision.

Neurons in orbitofrontal cortex maintain an abstract representation of the value of each option without encoding the movement needed to choose that option.

As we move away from the sensorimotor regions of the brain, neurons no longer have response fields that are sensitive to the location of the stimulus with respect to the fovea or response fields that are sensitive to movement direction. Among these regions, a particularly crucial location is the orbitofrontal cortex (OFC), where damage can produce disorders in patterns of decision-making, including abnormal risk-seeking, excessive gambling, increased impulsivity, and personality disorders. One finds evidence for the good-based model of decision-making in the activity of the cells in the OFC.

Camillo Padoa-Schioppa and John Assad (2006) trained monkeys to choose between two juices offered in variable amounts. The amounts were specified by the number of squares on a screen with the color of square referring to juice type. In a trial (figure 5.21A), the animal was offered a choice between three drops of juice A and one drop of juice B (1B:3A). After a fixation period, the animal received a go signal, and then made a saccade toward A or B, indicating its choice. They varied amounts of A and B and recorded the

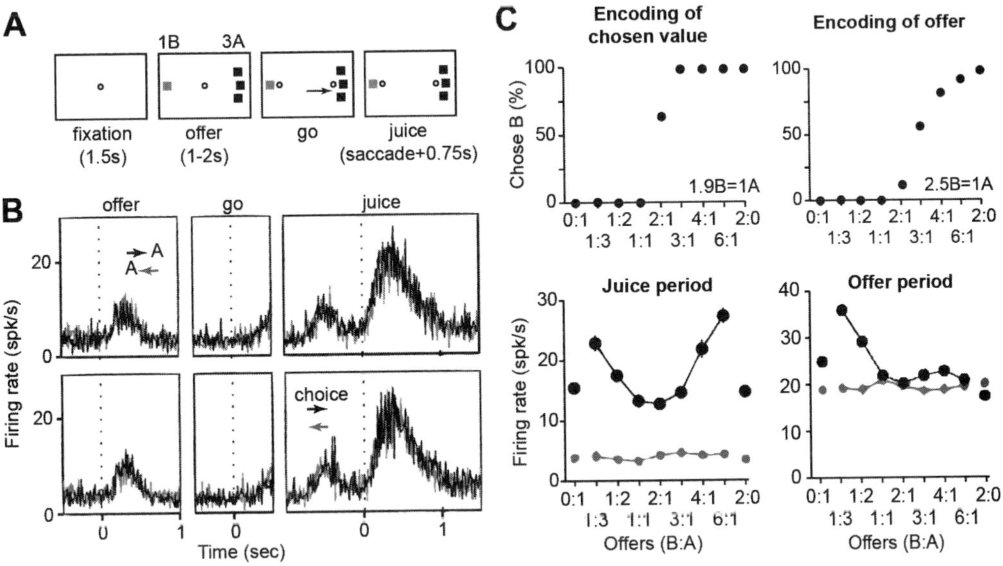

Figure 5.21
Activity in the orbitofrontal cortex during decision-making. **A**. Monkeys chose one of two offers: 1 drop of juice B or 3 drops of juice A. To express its choice, the monkey made a saccade toward one or the other option after the removal of the fixation point (go cue). Juice was delivered 750 ms following the saccade. **B**. Activity of one neuron in the offer period, go period, and juice-delivery period. Activity was similar regardless of whether option A was to the left or right of fixation, or whether the animal expressed its choice by making a saccade to the left or right. **C**. The choices made in separate sessions, and activity of two different neurons, one in each session. The top plots show the choices made. The x-axis reflects the offers (number of drops of B versus number of drops of A). The lower plots show the activity of two neurons. For one neuron, activity in the juice period correlated with the value of the chosen item. For another neuron, activity in the offer period correlated with value of one of the offers (value of A). Gray dots are baseline activity. (From Padoa-Schioppa and Assad, 2006.)

choices that the monkey made (figure 5.21C, top plots). In one session, the monkey's behavior indicated that 1 drop of A was equivalent to 1.9 drops of B, while in another session, 1 drop of A was equivalent to 2.5 drops of B.

They recorded from hundreds of OFC cells and found that neurons modulated their response with respect to three variables: offer value (the value of only one of the two juices), the chosen value (the value chosen by the monkey in the trial), and taste (a binary variable identifying the chosen juice). Importantly, the neurons did not have a spatial or movement-related response field. That is, their activity did not reflect the spatial position of the stimulus during the offer period or the direction of the saccade during the go period. This is illustrated for one neuron in figure 5.21B. Activity in this neuron increased transiently after the onset of the offer period and then increased again when the choice was made and juice was delivered. (Juice was delivered 750 ms after saccade.) However, the activity did not reflect the location of the stimulus in the offer period or direction of saccade in the go period. Indeed, less than 5% of the OFC neurons were modulated by the spatial configuration of the offers.

However, activities of about 50% of the neurons were modulated at various times with offered value, chosen value, or taste. For example, the cell in left column of figure 5.21C had a U-shaped response during the period after juice delivery. The discharge was correlated with the value of the chosen option (for this session, behavior indicated that A=1.9 B). For example, when the offer was between 4 drops of B and 1 drop of A and the animal chose B, activity was high. However, the same neuron was also very active when the offer was between 1 drop of B and 3 drops of A and the animal chose A. This neuron's response during the taste period appeared to reflect value of the chosen option.

In contrast, the activity of another cell varied with the value of A during the offer period (right column of figure 4.19C). When the offer included large amounts of juice A (1 drop of B versus 3 drops of A) and the animal chose A, the cell responded strongly. In contrast, when the offer included small amounts of A (4 drops of B versus 1 drop of A) and the animal chose B, the cells responded near baseline. This neuron's response during the offer period appeared to reflect the offer value of stimulus A.

In summary, there are occasionally scenarios in which one deliberates and comes to a decision without having to perform a movement that is associated with the expression of that decision. In this case, during deliberation, a vast majority of movement-related neurons in the colliculus do not show activity that indicates preparation for a particular movement. The decision-making in this case is partly due to activity of neurons in OFC, maintaining an abstract representation of the value of each option. Whereas OFC damage causes decision-making deficits such as impulsivity, gambling, and eating disorders, parietal cortex damage causes neglect. Padoa-Schioppa and Assad (2006) contrasted the good-based model from the action-based model by writing: "From a computational perspective, a modular design separating the mental operation of choosing and moving is more parsimonious."

5.10 Deciding When to Leave

Although utility is often associated with moving toward a stimulus, one of the key aspects of computing utility is with regard to the decision of when to abandon the current reward site: how long should one stay and consume a diminishing reward before moving away? In chapter 2, we considered this question in the framework of optimal foraging. The decision regarding when to leave can be made if one compares the local capture rate during harvest to the global capture rate: you should abandon the current reward site and leave when the local capture rate falls below the global rate. What is the neural basis of control of the harvest period? That is, what brain regions are responsible for determining when the current reward site should be abandoned?

Neurons in dorsal anterior cingulate encode the value of ending the harvest and leaving the patch.

One of the first experiments that explored the neural basis of decision-making during foraging was by Benjamin Hayden, John Pearson, and Michael Platt (Hayden et al., 2011).

They developed a foraging-like task (figure 5.22A) in which monkeys chose one of two targets: one that indicated that they wanted to remain in the patch, thereby receiving an amount of juice that became smaller with each repeated decision to stay, and one that indicated that they wished to travel to the next patch, thereby incurring a time delay (and a reward magnitude reset) before the next choice opportunity materialized. The authors recorded from neurons in the dorsal anterior cingulate cortex (dACC). They chose ACC because when people are resisting a craving in the context of inhibiting a maladaptive behavior (like the urge to smoke a cigarette or the urge to finish the whole container of ice cream), there is usually increased activity in ACC and other prefrontal cortex regions of the brain (Tang et al., 2015). Increased activity in ACC seems to be important for self-control.

Monkeys chose one of two actions: (1) stay and get rewarded progressively less or (2) leave and incur a travel cost. During each trial, a center fixation stimulus appeared with two options (figure 5.22A). One option was a dot, meaning the choice was to stay. The second option was a bar, with the length proportional to delay (travel time), meaning the choice was to travel. After a 500 ms fixation period (deliberation period), the animal made a saccade to one of the options. If the choice was to stay, it got juice following a 400 ms delay period. If this was the first time the choice was to stay in this patch, the amount of juice was 0.3 mL. With each repeated decision to stay, the amount of juice decreased by 0.02 mL. Therefore, after n decisions to stay, the total reward received at that patch was as follows:

$$r(n) = 0.3n - 0.02\left(\frac{1}{2}n(n-1)\right) \tag{5.2}$$

To formulate a harvest function, we need to consider that there is an energetic cost associated with staying that grows with the number of times the animal chooses to stay. During trial i, if the subject chose to stay n_i number of times, then the amount of reward it accumulated minus the energetic cost of staying is this:

$$f(n_i) = r(n_i) - kn_i \tag{5.3}$$

If the choice was to leave, the animal incurred a travel time t_i and an energetic cost kt_i. After each choice, there was also an intertrial period of 1 s. Based on experimental data, the time cost of staying was roughly 2 s. Optimal foraging suggests that the cost to maximize is the global capture rate, defined as the total amount of juice received, minus total energy spent, divided by total time:

$$\bar{J} = \frac{\sum_i f(n_i) - kt_i}{\sum_i 2n_i + t_i + 1} \tag{5.4}$$

Assuming that the choice in any particular patch does not affect resources and behavior at other patches, we can write equation 5.4 as follows:

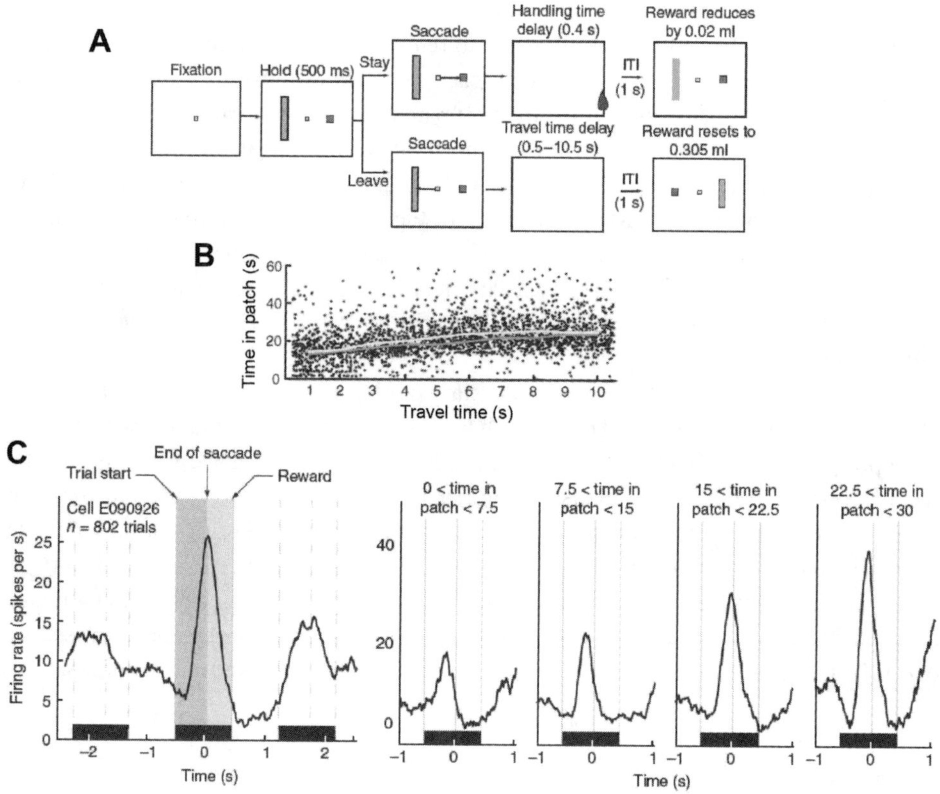

Figure 5.22
Activity in the dorsal anterior cingulate cortex during a foraging task. **A.** Monkeys chose one of two actions: (1) stay and get reward with diminishing return or (2) leave and incur a travel cost. They signaled their decision by making a saccade toward a dot (stay in patch) or a bar (leave the patch). The bar length indicated length of time to the next opportunity for reward. **B.** Animals generally opted to stay in the patch for a longer period of time when the travel time to the next patch was longer. **C.** Average activity of a sample neuron. Time 0 s indicates end of saccade. The neuron's response was enhanced around time of saccade and then fell to near baseline level. The size of the burst increased with the time already spent in patch. (From Hayden et al., 2011.)

$$\bar{J} = \frac{f(n_i) - kt_i + A}{2n_i + t_i + 1 + B} \tag{5.5}$$

The animal controls the variable n_i, the number of times it chooses to stay and harvest. The maximum $\bar{J}$ is achieved when the derivative of the above expression with respect to n_i is equal to 0. This occurs when the following condition is met:

$$\left.\frac{df(n_i)}{dn_i}\right|_{n_i^*} = 2\bar{J}\Big|_{n_i^*} \tag{5.6}$$

Solving the above expression for n_i, we find that the optimum number of times to stay increases with the duration of time t_i that is required to travel to the next patch.

$$n_i^* = \frac{1}{2}(t_i^2 + 200k(t_i - 1) + 64t_i - 400A + B^2 + B(64 - 200k + 2t_i) + 63)^{1/2} \\ -\frac{1}{2}(1 + B + t_i) \tag{5.7}$$

Indeed, the monkeys behaved roughly as expected (figure 5.22B, white-shaded line), staying longer when the option of leaving required a longer travel time.

The authors recorded from neurons in dACC and found that during the fixation period (i.e., the deliberation period), a majority of neurons produced a burst (figure 5.22C). These bursts reached their peak discharge around the time of the saccade. (The saccade indicated the choice that the animal had made.) Unlike the bursts in the FEF or collicular cells, the bursts in the dACC were unrelated to direction of the saccade. Rather, the magnitude of the burst was associated with how long the animal had stayed in the current patch. In the first stay trial, the burst magnitude was small. As number of stay trials increased, the burst magnitude became larger, until it seemed to reach a threshold, predicting that this would be the last trial that the animal would stay. (In the next trial, the animal switched, and burst magnitude returned back to being small.) In many cells, the burst magnitude had a significant positive correlation with time already spent in the patch. In a few cells, the burst magnitude had a negative correlation. In the various trials, among the cells in which burst magnitude increased with time in patch, the neural activity always rose to approximately the same level, at which point the animals switched.

In summary, in many scenarios, deciding to stay longer produces diminishing returns. To investigate the neural basis of deciding when to leave, Hayden et al. (2011) measured activity in dACC. Unfortunately, the authors did not report vigor measurements. Rather, they focused on decision-making: whether to stay and harvest reward or incur a travel time delay and wait for the opportunity to renew harvest at another patch. Activity in dACC, a region important for self-control, encoded a decision-variable that indicated the value of leaving the patch (or alternatively, the negative value of staying). Travel time between patches largely affected the rate at which this decision variable rose to a threshold: when the travel time was long, the animal chose to stay for more trials, and the rate of rise of the burst was slower as a function of stay period. There are other potential interpretations: dACC activity could be tracking the difficulty of making the decision rather than the value associated with staying (Shenhav et al., 2014). Notably, dACC activity did not uniquely represent any single variable in the marginal value theorem equation, suggesting that the decision-making process is due to interactions between multiple regions.

Limitations

Although damage to V1 results in loss of the ability to consciously perceive visual stimuli, it does not eliminate the ability to make movements toward those stimuli. If the unconscious movements are made mainly through the retina-collicular pathway and not through engagement of the cortex, then we might expect that their vigor of these reflexive movements will not be influenced by reward and effort. This prediction has yet to be tested.

Damage to the FEF appears to give the colliculus greater autonomy in reflexively reacting to visual stimuli. This makes it likely that utility computation is purely a cortical phenomenon, but that also remains to be tested.

The concept of a decision variable rising to a threshold is a useful framework to consider patterns of decision-making, but neuronal activity in the FEF and superior colliculus do not support the concept of a fixed threshold. For example, movements that have a shorter reaction time typically are preceded by activities that not only rise faster, but also reach a higher level. One possibility, as suggested by Heitz and Schall (2012), is that the threshold for initiating a movement is set not via the activity of cortical or collicular cells, but by cells in a region of the brainstem (omnipause neurons) that integrate the input that they receive from various sources, including the FEF and the colliculus. However, missing from the omnipause data are experiments that involve decision-making. If the threshold is set by the omnipause neurons, we would expect that their activity to show little or no modulation during the deliberation period and not be influenced by the utility of the stimuli.

Summary

We do not need the cerebral cortex to move, but we need the cortex to compute the utilities of our options. That evaluation allows us to suppress reflexive behaviors that attract movements toward the most salient stimuli and instead generate movements that maximize reward. Here, we considered neuronal activity in various cortical regions that make utility computations: the frontal eye field, premotor cortex, parietal cortex, orbitofrontal cortex, and anterior cingulate. In the FEF, neurons encode both the utility of staying still and the utility of moving. During reaction time, as we deliberate, the movement-related neurons in the FEF and PMd increase their activity, and the movement starts when this activity reaches high levels. Urgency raises the baseline activity of these neurons, encouraging earlier decisions. Reward increases the rate of rise and peak discharge of the neurons; a more vigorous movement soon follows. In the parietal cortex, neurons encode utility of the stimulus, but they are also modulated by novelty and salience. In the orbitofrontal cortex, neurons do not have movement response fields. Rather, these neurons encode stimulus utility without concern for the spatial properties of the movement that needs to be performed to acquire the promised reward. Finally, the decision to stop harvesting and leave the reward site (i.e., explore other opportunities) appears to depend on neurons in the anterior cingulate. The decision to move the eyes, however, is not trusted to any one particular cortical region. Rather, inputs from various cortical regions converge onto cells in the brainstem omnipause neurons, which are responsible for maintaining gaze.

The superior colliculus can generate a saccade, and the cortex can provide it with excitatory inputs that indicate where to move the eyes, but the cortical excitation is typically not enough to make a movement. Rather, the colliculus also needs the basal ganglia to reduce their inhibition. Control of this inhibition depends on the risks and rewards that the basal ganglia assign to the upcoming movement, defining an implicit motivation to move. That motivation depends on dopamine, the star of our next chapter.

6

Basal Ganglia and the Motivation to Move

> The birds around me hopped and played:
> Their thoughts I cannot measure,
> But the least motion which they made,
> It seemed a thrill of pleasure.
>
> William Wordsworth

Patient A2 was a 62-year-old man with advanced Parkinson's disease (PD). He suffered from severe rigidity and was confined to a wheelchair, unable to walk unassisted, and lived on the first floor of his home. A hurricane approached the city and his wife left to get some supplies from the drugstore. His neurologist described what happened next (Schwab and Zieper, 1965): "As a result of the storm the harbor overflowed 10 feet into the street. Sitting in his wheelchair, he suddenly saw the door blown in and a wall of water entered the house. Exactly how he did it is not clear, but he managed to get out of his wheelchair and climbed the steps to safety on the second floor where he was found several hours later by his wife, the waters having subsided. She found him seated in a chair as helpless as he was before." Thus, under extraordinary circumstances the patient showed spontaneous improvement, implying that his symptoms were not due to an intrinsic inability to produce movements. Rather, the movements could not be expressed unless there was urgency.

We need not place the patient's life in danger to test this idea. We can increase urgency by manipulating the consequences of failing to make the movement: when asked to reach as fast as possible to pick up a ball, Parkinsonian patients reach much slower than normal if the ball is stationary, but at near normal speed if the ball is moving (Majsak et al., 1998), and even faster if the ball is about to fall to the floor (Ballanger et al., 2006).

The mechanism with which the PD brain produces these feats remains a mystery. Yet, the observations do hint that latent in the PD brain is the ability to make fairly normal movements. However, without increased urgency, the movements are unavailable for expression. Why?

About a decade ago, Pietro Mazzoni, Anna Hristova, and John Krakauer (2007) approached this question by asking PD patients to reach to a target. Until their work, the prominent theory was one of movement scaling. For example, in their review of bradykinesia (slowness of movement), Berardelli et al. (2001) wrote: "Bradykinesia seems to result primarily from the underscaling of movement commands in internally generated movements. This may well reflect the role of the basal ganglia in selecting and reinforcing appropriate patterns of cortical activity during movement preparation and performance." However, this view did not explain why urgency unmasked the latent ability to move.

To explore this question, Mazzoni and colleagues removed visual feedback of the hand at reach onset and told the patients that the trial was successful only if the speed of the reach was within the requested range. Crucially, the subjects had to repeat the trial if the speed was outside the requested range. They made the surprising observation that for a given reach speed, the endpoint accuracy of the movement in PD was similar to controls. That is, the patients were capable of producing movements of normal speed and accuracy. However, they required many more attempts in order to produce a reach that was as fast as the requested speed. This implied that under typical circumstances, the patients reached slowly not because of a strategy to improve accuracy or a fundamental scaling of movements, but because of another cause. The authors hypothesized that this cause was a greater than normal *implicit motor motivation* that the PD brain required to produce the large motor commands required for a fast movement. That is, the PD brain appeared to be shackled with a greater motor cost.

What might be this implicit motor motivation, and how might it be related to dopamine, the neurotransmitter that is lost as the disease destroys the cells that produce it in the basal ganglia? Is motor motivation regulated by dopamine?

Here, we will consider the roles of the basal ganglia and dopamine in controlling movements. Much of our focus will be on saccades and the neural circuits that control them. Whereas the cortical structures that compute the utility of a movement excite the superior colliculus, the basal ganglia's output to the superior colliculus is conveyed via inhibition. Regulation of this inhibition occurs via cells in the output nucleus of the basal ganglia, the substantia nigra reticulata (SNr). SNr cells appear to carry two kinds of information: (1) the reward at stake for making the movement and (2) the risk and costs associated with that movement. When the reward at stake is larger, SNr cells reduce their activity more, removing more of the inhibition that they impose on the superior colliculus, allowing the saccade to take place with greater vigor. When there are risks with performing the movement, SNr cells burst, increasing the inhibition sent to the superior colliculus and ultimately discouraging the movement.

These two types of SNr cells receive their information directly and indirectly from the striatum, the input stage of the basal ganglia, where cells also encode costs and rewards of movement. The striatal cells (particularly in the caudate) are themselves recipient of two kinds of information: excitatory input via glutamate (from the cortex), and modulatory input via dopamine. Dopamine regulates how the striatal cells respond to cortical input and, therefore, influences SNr output.

At stimulus onset, dopamine in the striatum is largely a reflection of the reward value of that stimulus and not the effort required to acquire that reward. However, during the movement, dopamine encodes the effort requirement of the task, and it is dispersed in greater amounts when more effort is required to acquire the reward. In turn, this higher concentration of dopamine makes it easier for excitatory cortical inputs to drive caudate activity, facilitating the removal of the inhibition that SNr cells impose on the colliculus and resulting in a more vigorous movement.

In a sense, dopamine in the striatum acts like the Roman god Janus, with one face noting the subjective value of reward when the stimulus appears and another face supporting the effort needed to attain that reward when the movement begins.

6.1 Anatomy of the Basal Ganglia

The largest nucleus of the basal ganglia is the striatum, which is composed of the putamen, the caudate, and the ventral striatum. (The ventral striatum is collectively the nucleus accumbens and the olfactory tubercle, but here the accumbens is the relevant region.) The striatum serves as the main input region of the basal ganglia, receiving projections from all cortical areas (figure 6.1). The cortical inputs to the striatum are glutamatergic, producing excitation of striatal neurons. The principal neurons of the striatum are GABAergic medium spiny neurons which (in rodents) constitute more than 90% of all striatal neurons. These neurons have axons that project locally as well as collaterals that project to other basal ganglia structures. The medium spiny neurons are typically divided into two groups of roughly equal number. The direct pathway constitutes one group of striatal neurons that

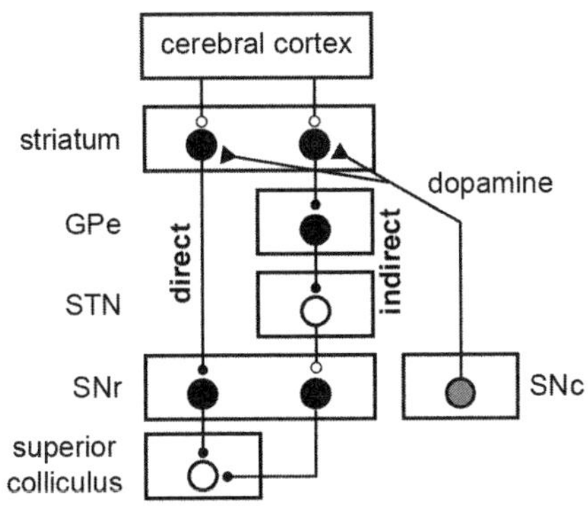

Figure 6.1
Simplified schematic of the anatomical organization of the basal ganglia. Dark filled circles indicate inhibitory neurons; unfilled circles are excitatory neurons. Dopamine neurons are shown as having triangular synapses originating from the SNc. Striatum includes the putamen, the caudate, and ventral striatum, which is also known as nucleus accumbens. GPe is globus pallidus external region. STN is subthalamic nucleus. SNr is substantial nigra reticulata region. SNc is substantial nigra compacta region, which houses dopamine-producing cells.

project to the output nuclei of the basal ganglia, the substantia nigra and the internal segment of the globus pallidus (GPi). The indirect pathway constitutes another group of striatal neurons that project to the external segment of the globus pallidus (GPe).

Striatal neurons that participate in the direct pathway can be identified by their expression of D1 dopamine receptors, and neurons that participate in the indirect pathway can be identified by their expression of D2 dopamine receptors. Although the direct and indirect pathways are not truly separate (some neurons projecting to GPi and substantia nigra also send projections to GPe, and some neurons express both D1 and D2 dopamine receptors), this classification remains a useful heuristic.

All neurons in GPe, GPi, subthalamic nucleus (STN), and SNr are autonomous pace markers, capable of firing on their own without synaptic input. Thus, the direct pathway, which constitutes GABAergic neurons from the striatum, can decrease the activity of SNr neurons, whereas the indirect pathway can increase the activity of SNr through disinhibition.

Dopamine facilitates transition of D1 receptor striatal neurons to up-state, and hinders that transition for D2 receptor neurons.

The medium spiny neurons of the indirect pathway are more excitable than their counterparts in the direct pathway, as evidenced by how they respond to current injection to their soma: indirect pathway neurons in the striatum respond more strongly to a given current injection than their direct pathway counterparts (Gertler et al., 2008). At rest, the membrane potential in both groups of neurons is far from threshold (around −90 mV). This resting level is called the down-state. In response to converging excitatory input, the medium spiny neurons transition to an up-state, which can last hundreds of milliseconds, leading to production of spikes.

A major modulator of the state of the striatum is dopamine. Arrival of dopamine affects the glutamatergic synapses: its effect on D2 receptors is to impede the transition to the up-state, whereas the effect on D1 receptors is to enhance this transition (Gerfen and Surmeier, 2011). Arrival of dopamine activates D2 receptors and thus reduces depolarizing currents, increases hyperpolarizing currents, and diminishes the presynaptic release of glutamate. On D1 receptors, dopamine has largely the opposite effect; depolarizing currents are increased.

As a result, arrival of dopamine impedes the transition of the indirect-pathway medium spiny neurons to their up-state, thereby discouraging the production of spikes. In contrast, the arrival of dopamine facilitates transition of the direct pathway striatal neurons to their up-state, hence encouraging the production of spikes.

6.2 Substantia Nigra Reticulata

Substantia nigra is one of the output nuclei of the basal ganglia. It is divided into two parts, the reticulata (SNr), and the compacta (SNc). SNr cells project to the thalamus and the superior colliculus. SNc contains dopamine cells, which project to many regions of the brain, but most strongly to the striatum.

SNr cells are active during fixation and respond to stimuli in a broad region of the contralateral visual space.

One of the first examinations of the SNr during an oculomotor task was by Okihide Hikosaka and Robert Wurtz (Hikosaka and Wurtz, 1983). They trained monkeys to touch a bar in order to start a trial, resulting in presentation of a fixation point. In one block of trials, the goal for the animal was to detect dimming of the light at the fixation spot (figure 6.2A). When the animal detected this change, it released the bar and received a reward. In another block of trials, a target appeared after fixation, and the animal made a saccade to it. The goal here was to detect dimming of the target light (and not the fixation light). Therefore, animals were not rewarded for making saccades, but rather for directing their attention to a spatial location and then detecting the change in stimulus intensity. The block design allowed the animals to anticipate whether the important information was at the fixation point or at the target.

The authors recorded from the SNr and found that most cells had a high discharge rate (around 100 Hz) when the animal was fixating (figure 6.2B). When the target stimulus was turned on, these cells usually reduced their discharge after a delay of around 130 ms, pausing briefly. This response was for the target on the side of fixation contralateral to the recorded nucleus, but the modulation was smaller if the target was on the ipsilateral side. Importantly, the response field was quite large, extending over 40° of visual angle spanning both the ipsilateral and contralateral visual fields (figure 6.2B).

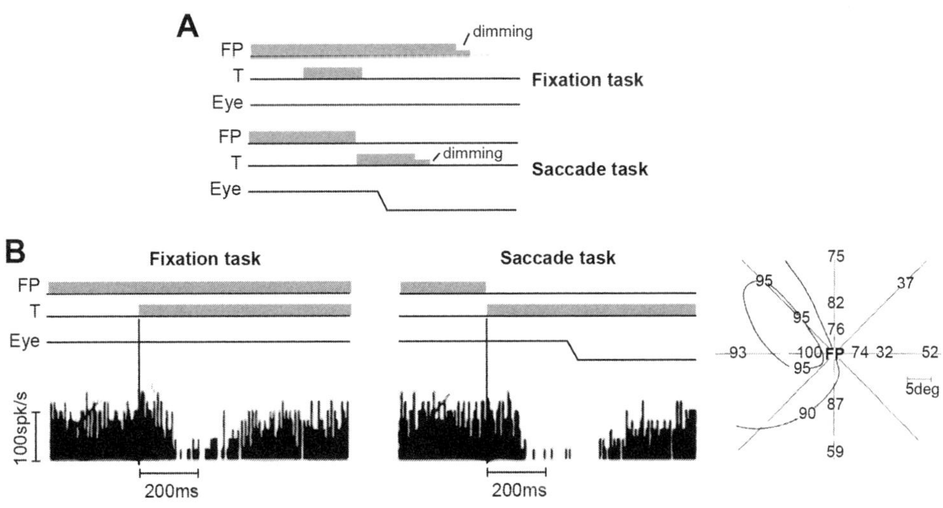

Figure 6.2
Responses of two substantia nigra reticulata (SNr) cells to visual stimuli. **A**. Task design. The monkeys learned to detect dimming of the stimulus (either at fixation or at the target) and release a handle in order to receive reward. **B**. Response of a neuron in the right SNr in the fixation task (left) and in the saccade task (middle). In the fixation task (left), the target acts as a distractor. Target is presented left of fixation in both cases. The neuron shows a pause in discharge around time of saccade. The figure on the right shows percent decrease in discharge after presentation of the visual stimulus in the saccade task (FP, fixation point). Note that the response field is quite large. (From Hikosaka and Wurtz, 1983.)

Hikosaka and Wurtz (Hikosaka and Wurtz, 1983) found that SNr cells responded more strongly (deeper reduction in discharge) when the target stimulus was not a distractor, but rather indicated the location where valuable information would be available (dimming of the light, which if detected would lead to reward). As a result, while the animals fixated on the central location, many SNr neurons fired strongly, producing sustained inhibition upon the superior colliculus and thus preventing saccades. When a stimulus appeared that contained valuable information, some of these cells reduced their discharge at a latency of around 130 ms, making it possible for the colliculus to produce a saccade.

The rather large response field of SNr cells was confirmed by experiments performed by Ari Handel and Paul Glimcher (1999). They considered an overlap saccade task in which a target light came on while the fixation spot was still present. Removal of the fixation spot was the go-cue that instructed the animal to make a saccade to the target. They found that a majority of the SNr cells either reduced their response to presentation of the target and maintained their reduced discharge up until saccade time, or paused primarily around time of saccade. Importantly, while the response of the cells was greater (larger modulation) for stimuli on the contralateral side, the encoding of space with respect to fixation (modulation of discharge) was not Gaussian, but approximately linear, growing larger as the target distance increased from fixation.

Therefore, unlike cells in the superior colliculus and the frontal eye field, SNr cells did not have a "preferred target location". Rather, SNr cells modulated their discharge roughly linearly as a function of distance of the stimulus from the fovea. This implies that output of the basal ganglia to the colliculus is unlikely to specify a movement. Rather, this output is likely a modulator of the movement selected by other regions that also project onto the colliculus.

Some SNr cells burst in response to contralateral stimuli, others pause.

Handel and Glimcher (1999) also discovered a number of cells that Hikosaka and Wurtz (1983) had missed: these SNr cells increased their discharge in response to onset of the visual stimulus in the contralateral field, and maintained this increased activity until after the saccade. Thus, there were at least two types of task-related cells in SNr: a group that responded with reduced discharge after target onset, and another that responded with increased discharge. Both the type 1 and type 2 SNr cells responded to visual stimuli in the contralateral field, and for both this response broadly encoded the visual space.

However, the presence of potentially two distinct types of SNr cells, ones that increased their discharge in response to a visual stimulus in the contralateral field, and ones that decreased their discharge, introduced a puzzle. What is the functional difference between these cells?

One possibility was that the SNr cells that showed increased activity projected to the contralateral superior colliculus. Another possibility was that one cell type was part of the direct pathway in the basal ganglia, while the other cell type was part of the indirect pathway.

Masaharu Yasuda and Ohikide Hikosaka (2017) examined this question by identifying which colliculus an individual SNr cell projected to. Across three monkeys, using antidromic stimulation they found that the SNr cells that projected to the colliculus did so almost exclusively to the ipsilateral side (at a ratio of 10 to 1, ipsilateral to contralateral). Furthermore, among the ipsilateral projecting SNr cells, some paused in response to the visual stimulus (cell type 1, figure 6.3), while others produced a burst (cell type 2, figure 6.3). Therefore, it was not the case that bursting and pausing SNr cells projected to different sides of the colliculus.

The question of whether these two types of cells belonged to the direct and indirect pathways remains unanswered. However, on the basis of results from experiments concerning

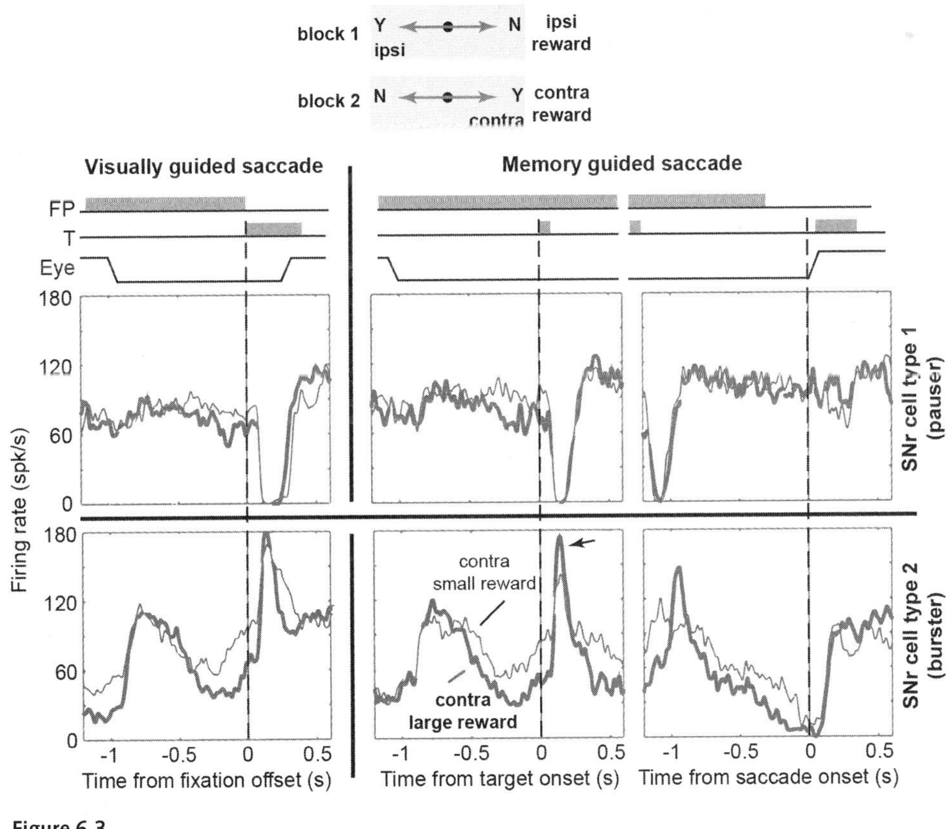

Figure 6.3
Substantia nigra reticulata (SNr) cells that project to the ipsilateral superior colliculus exhibit two different kinds of responses to visual stimuli. The data here are for stimuli that appeared in the contralateral visual field (saccades to the contralateral side). Cell type 1 showed a pause in response to the visual stimulus, while cell type 2 showed a burst. The thick lines indicate trials in which the contralateral saccade was rewarded by a large amount of juice. Note that in the memory-guided task, cell type 2 showed a larger response following the visual target. (From Yasuda and Hikosaka, 2017.)

the activity of caudate neurons, it is likely that the SNr cells that paused in response to the rewarding stimulus were part of the direct pathway, whereas the SNr cells that burst were part of the indirect pathway.

Importantly, the response of these two types of cells also differed around saccade onset. To illustrate this, Yasuda and Hikosaka (2017) considered two types of tasks: visually guided and memory-guided (figure 6.3). In the visually guided task, a block consisted of trials that required saccades to the contralateral side and trials that required saccades to the ipsilateral side. Furthermore, in one block, the contralateral saccades were rewarded, whereas in the next block the ipsilateral saccades were rewarded. This pattern was repeated in the memory-guided task.

The data in figure 6.3 are for the trials in which the target was located on the contralateral side and the monkey made correct saccades to that side. In the visually guided saccade task, SNr cell type 1 produced a pause in response to the target, while SNr cell type 2 produced a burst. In the memory-guided task, when the animal had to withhold making a saccade to the target, the responses to target onset were again a pause for cell type 1 and a burst for cell type 2. However, around saccade onset in the memory-guided task, cell type 1 showed little or no modulation, whereas cell type 2 showed a pause. Both of these cells projected to the ipsilateral colliculus. As a result, when the target was presented, the pause produced by cell 1 would encourage a saccade toward it, whereas the burst produced by cell 2 would discourage that saccade. Presence of this discouragement is critical in the memory-guided task because in that task, the animal should wait and refrain from making a saccade to the target.

Different SNr cells may encode the reward and the risk of performing a movement.
We still do not fully understand the function of the various types of SNr cells. One possibility is that the cells that paused in response to the contralateral target were reflecting the reward value of the movement toward that target, whereas cells that produced a burst were encoding some aspect of the risk or penalty associated with that movement.

In figure 6.3, the reward value encoded in cell 1 is difficult to see because this cell essentially stops firing in response to the visual stimulus under both high-reward and low-reward conditions. (We will come back to this issue by examining activity in other SNr cells that also reduce their activity in response to the visual stimuli.) However, the risk or penalty might be gleaned from the discharge of cell type 2 during the memory guided task (figure 6.3). In the memory-guided task, in trials in which a reward was associated with the target on the contralateral side, cell type 2 showed a larger burst after target onset when the reward value was high than when it was low (arrow, figure 6.3). In contrast, in the visually guided task, cell type 2 showed similar responses to the visual stimulus under both the high-reward and low-reward conditions.

Makoto Sato and Okihide Hikosaka (2002) speculated that the SNr cells that paused in response to the visual cue were part of the direct pathway and received input from the caudate, whereas the SNr cells that produced a burst were part of the indirect pathway, receiving input from the subthalamic nucleus (figure 6.1). Both of these types of SNr neurons likely converge via inhibition onto cells in the ipsilateral superior colliculus, thereby affecting the probability of performing an action as well as the vigor of that action. However, SNr cells in the direct pathway are likely to encourage performing the action by pausing in response to the visual stimulus, whereas SNr cells in the indirect pathway are likely to discourage performing that action by bursting. In this framework, the discharge of the type 2 cells would be particularly critical during the memory-guided saccades, when the animal needs to inhibit the desire to saccade to the target.

Deactivation of SNr makes it hard to maintain fixation and withhold saccades to the contralateral target.

The special role of the SNr in preventing unwanted saccades was illustrated by an experiment performed by Hikosaka and Wurtz (1985) in which they injected muscimol into this nucleus. Muscimol is a GABA agonist. Therefore, its action is to reduce activity of all SNr cells, which in turn reduces the inhibition that SNr imposes onto the superior colliculus. The authors found that as they deactivated the SNr, visually guided saccades were barely affected: saccades to the contralateral side showed a slight decrease in reaction time, with no change in peak velocity. However, memory-guided saccades were profoundly affected: SNr inactivation resulted in premature saccades to the contralateral target during the time in which the animal should have waited for the removal of the fixation cue. Indeed, the animal had trouble maintaining fixation; it made repeated saccades to the contralateral side even before the target was shown. In sharp contrast, these asymmetries to the contralateral side were muted for visually guided saccades.

These observations in the memory-guided saccade task suggested that the role of the basal ganglia was particularly important in discouraging the colliculus to respond to a visual stimulus. With deactivation of the SNr, movements to the opposite visual field were executed sooner and were more vigorous. However, the main effect was an inability to withhold movements to the contralateral side.

Reward modulates the pauser type SNr cells.

We speculated that the SNr cells that paused in response to the visual stimulus were part of the direct pathway, encouraging the animal to move toward the stimulus, whereas the SNr cells that produced burst in response to the visual stimulus were part of the indirect pathway, signaling the cost of that movement. Perhaps the direct pathway carries a signal associated with the reward that may be gained by performing the action, whereas the indirect pathways carries a signal associated with the risk or penalty of performing that action. There is some evidence for this conjecture.

Let us first consider the information that may be present in the direct pathway, potentially affecting the SNr cells that pause in response to the visual stimulus. Sato and

Hikosaka (2002) quantified the effect of reward on the activity of SNr cells. They trained monkeys to perform the memory-guided saccade task in which targets were presented randomly in one of two locations. Initially, regardless of the target direction, the monkeys were rewarded if they made the correct saccade (AllDir task). They then trained the animals during blocks of trials in which saccades to one direction were rewarded, while those to the other direction were not (1Dir task). After 40 successful trials, the previously rewarded target was no longer rewarded, while the previously unrewarded target was now rewarded. Importantly, the amount of reward in the 1Dir trial was twice that of the AllDir trial (so that the average rate of reward remained the same in the 1Dir and AllDir blocks).

The authors found that before the target was presented, many cells began modulating their discharge in anticipation of the target appearance; bursters increased their rates while pausers decreased them, as shown in figure 6.4A and figure 6.4B. Notably, in the

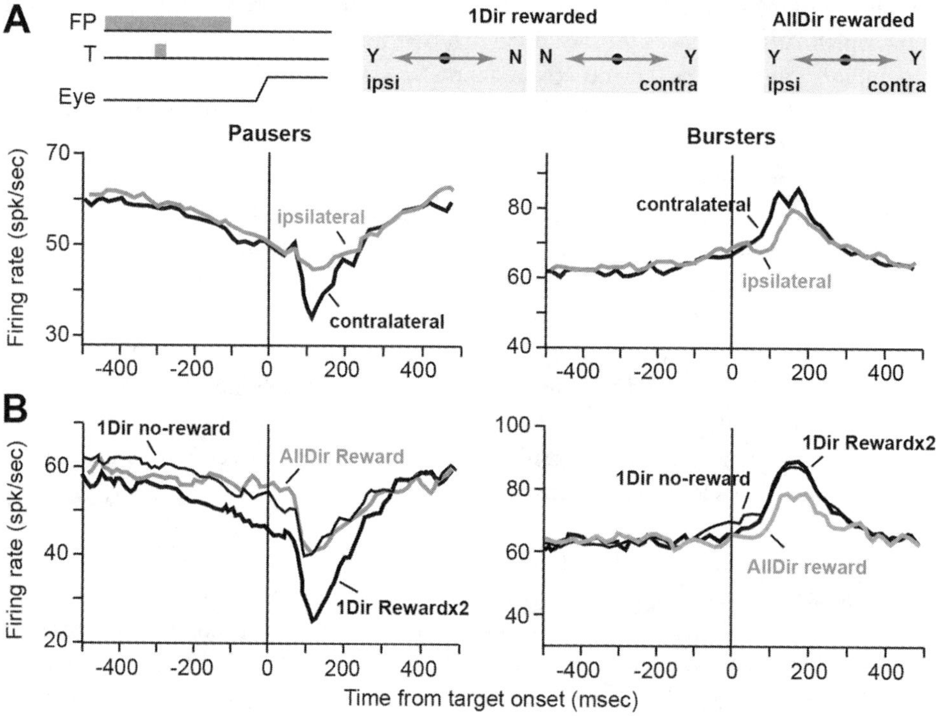

Figure 6.4
Reward-dependent response of two groups of substantia nigra reticulata (SNr) neurons during a memory-guided saccade task. **A**. Average response of pause-type and burst-type SNr neurons to target presentation in the contralateral and ipsilateral visual fields. **B**. Response to reward. In the 1Dir task, one of the two targets is consistently paired with reward, while the other target is not rewarded. In the AllDir task, both target directions are paired with reward. The pauser cells tend to be strongly modulated by possibility of reward and pause more strongly after target presentation. (From Sato and Hikosaka, 2002.)

1Dir task, if the contralateral target was paired with reward, then the activity of pausing type SNr neurons was lower even at the very the start of the trial. Thus, the SNr was priming the colliculus to direct the eyes toward the rewarded side.

Around 110–130 ms after the presentation of the target, the pausers exhibited a sharp reduction in their discharge, whereas the bursters exhibited a sharp increase. Generally, the level of modulation of the response was stronger when the target was on the contralateral side of the fixation. Notably, the effect of the reward was noticeable primarily in the pausers: in the 1Dir task, if the target was associated with reward, the response was a stronger reduction in discharge than when the same target was not associated with a reward. In contrast, the reward seemed to have no effect on the bursters in the 1Dir task.

However, when both directions were rewarded, the response of the pausers was similar to the response when the target was not associated with a reward. Therefore, when rewarding targets were presented among non-rewarding targets (1Dir task), the response to the rewarding target was a stronger modulation of discharge among the pauser cells of the SNr (putative direct pathway recipients), but not the bursting cells (putative indirect pathway recipients).

Reward-dependent activity of the pauser type SNr cells plays a role in control of saccade vigor.

Does the reward dependent modulation of SNr cells that pause in response to a rewarding visual stimulus contribute to saccade vigor? Yasuda and Hikosaka (2017) examined the response of these cells in the visually guided task. When the saccade was toward the contralateral side, and that side was highly rewarded, many SNr cells began reducing their discharge in anticipation of target presentation in the contralateral side (figure 6.5). These SNr cells essentially stopped firing around the time that the saccade was performed. In contrast, when the target on the contralateral side was associated with a smaller reward, before target presentation the cells showed little or no reduction in activity, and then a weaker modulation of response following target presentation.

Thus, at least some of the cells that inhibited the colliculus reduced the level of that inhibition when anticipating a more rewarding target on the contralateral side; by doing so, these inhibitory cells primed the colliculus to direct a movement if the target appeared there and then stopped inhibiting the colliculus altogether around the time of the saccade. These results suggest that the reward-dependent activity of the pausing SNr cells could help bias the decision-making toward the contralateral, more rewarding side and could help control saccade vigor toward that side.

We can also consider activity of SNr cells when the stimulus type predicts reward but not its direction. In this scenario, the more rewarding stimulus can appear at either side, and therefore the SNr activity should not differentiate between directions, but rather reflect the type of stimulus.

Yasuda, Yamamoto, and Hikosaka (2012) examined activity of SNr cells in a task in which the value of the stimulus changed from one block to the next. In this flexible-value

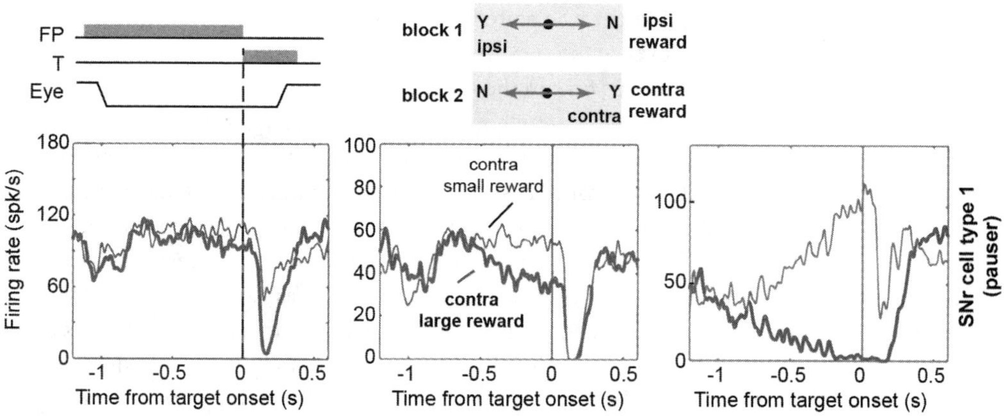

Figure 6.5
Reward-dependent modulation of activity in three cells of the substantia nigra reticulata (SNr) near the time of saccade in a visually guided task. The target appeared on the contralateral side. The activity of the cells in the period before presentation of the target and around the time of saccade was modulated by the reward value of the target in the contralateral side. This reward-dependent modulation of response appears sufficient to regulate reaction time and velocity of the saccade toward the rewarding direction. (Y, rewarded; N, not rewarded.) (From Yasuda and Hikosaka, 2017.)

association task (figure 6.6A), each forced-choice trial began with a fixation followed by the presentation of one of two randomly selected images presented at the neuron's preferred side, which was usually contralateral. Removal of the fixation spot instructed the animal to make a saccade to the target. The amount of reward was small for image 1 and big for image 2 for the duration of a block of 40 trials. In the subsequent block, the rewards associated with the images changed (image 1 was now associated with big reward, image 2 with a small reward). During ¼ of the trials, both images were presented and the monkey made a choice (free-choice trials). Behavior during choice trials demonstrated that the animal quickly switched to the high value image on the basis of its reward history (figure 6.6B).

The authors found that in the forced-choice trials, the activity did not differ across the SNr cells before image onset. This makes sense because the target usually appeared on the contralateral side, but its reward value was unknown. After the presentation of the image on the contralateral side, activity of the SNr cells as a population declined. Importantly, the extent of the decline differed between low-value and high-value images both during the overlap period and around the time of the saccade (figure 6.6C). As a result, the sum total of inhibition upon the colliculus from SNr was lower when the stimulus had greater reward value, and the saccade to the high value image occurred after a shorter reaction time and had greater velocity.

Finally, let us consider the SNr cells that we had conjectured were part of the indirect pathway. These cells produced a burst in response to the visual stimulus during the memory-guided saccade task (figure 6.2), thus withholding the saccade until the go

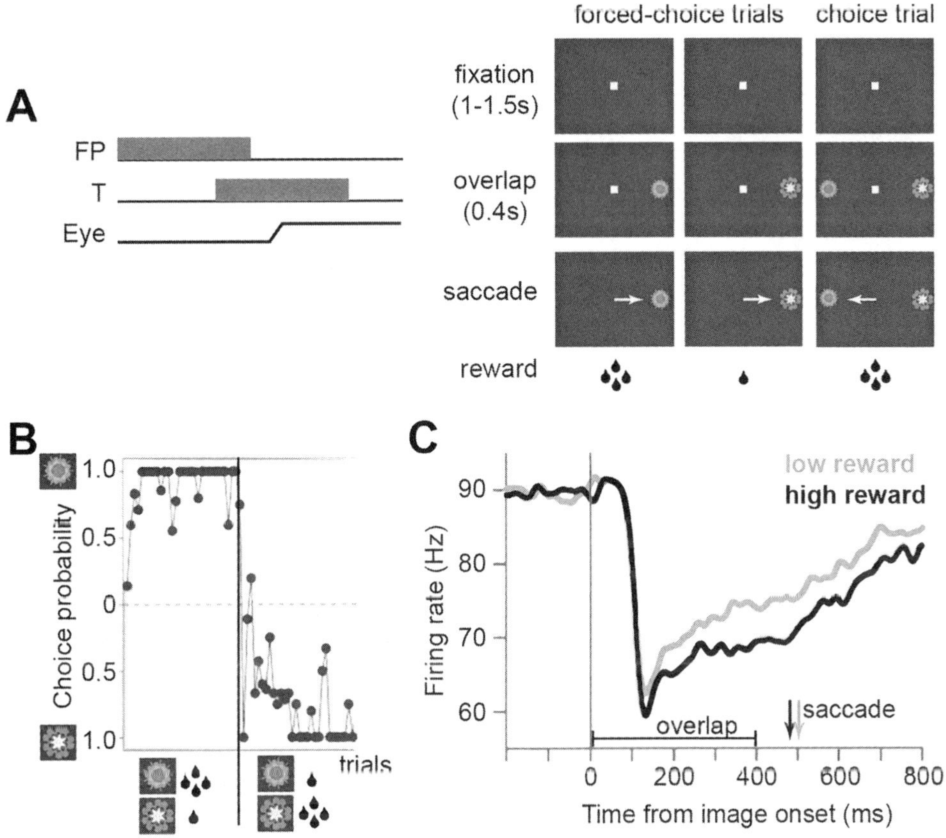

Figure 6.6
Encoding of stimulus value in a flexible-value task by cells of the substantia nigra reticulata (SNr) before and during movements toward the stimulus. **A**. In forced-choice trials, each trial began with a fixation. Next, one of two randomly selected images (overlap period of 0.4 s) was presented at the neuron's preferred location (usually contralateral). Removal of the fixation spot instructed the animal to make a saccade to the target. The amount of reward was low for image 1 and high for image 2 for the duration of a block of 40 trials. In the subsequent block, the reward-image association changed (image 1 was now associated with high reward, image 2 with low reward). In ¼ of the free-choice trials (labeled "choice trial"), both images were presented and the monkey made a choice. **B**. Selection in the free-choice trials. The image-value association changed from block 1 to block 2, which was then followed by a change in the choices that the animal made. **C**. Averaged activity across n=37 neurons in the SNr of one monkey in forced-choice trials for stimuli appearing in the response field. There is greater reduction in discharge if the stimulus promises greater reward. The average saccade times under the high- and low-reward conditions are shown by the arrows. (From Yasuda et al., 2012.)

signal. What kind of information is transmitted by neurons in the indirect pathway? No data is currently available regarding SNr activity during saccades that carry risk, effort, or other forms of cost. However, in the next section we will consider activity in the caudate, and there we will find evidence suggesting that caudate neurons belonging to the indirect pathway encode aspects of risk and effort, whereas those belonging to the direct pathway encode aspects of reward and gain.

In summary, these results present a number of ideas regarding encoding of stimulus value and control of saccade vigor by one of the output nuclei of the basal ganglia, the SNr. First, during periods of holding still, while one is fixating a stimulus, the tonic activity of the SNr cells imposes constant inhibition upon the colliculus, thereby discouraging saccades. Reduction of this inhibition (via injection of muscimol in the SNr) impairs the ability to maintain fixation; the animal produces intrusive saccades to the contralateral visual field, and is particularly impaired in tasks in which it needs to withhold movements in response to the onset of a contralateral visual stimulus.

Second, unlike cells in the superior colliculus and the frontal eye field, cells in the SNr have rather large response fields that cover much of the visual space contralateral to fixation. This would imply that whereas collicular and frontal eye field cells can differentiate between targets at 10° and 20° and specify one as the goal, SNr cells probably cannot. That is, it seems unlikely that SNr output can determine by itself which movement one should produce. Rather, SNr can bias the selection process and then modulate the vigor of the ensuing movement.

Third, in response to a target in the contralateral direction, many SNr cells show a reduction in discharge, whereas other SNr cells show an increase. The pausers may be part of the direct pathway, encoding the reward value of the stimulus, whereas the bursters may be part of the indirect pathway, encoding the effort or cost associated with acquisition of that stimulus. As a result, the total inhibition that the SNr provides to the superior colliculus appears to be modulated by the value of the stimulus. If there is greater reward at stake for a movement toward the contralateral side, inhibition is reduced before the target is presented, thus biasing the decision-making process to select the contralateral direction, and then inhibition is reduced again during the ensuing movement, invigorating the saccade. Therefore, SNr output to the colliculus appears sufficient to influence both the process of decision-making (via its output before and during the period when the options are being evaluated) and movement vigor (via its output during the period of the saccade).

6.3 Caudate

The inhibition that SNr provides to the colliculus is a major factor in influencing where we will direct our saccade and the vigor with which we will move our eyes during that saccade. This raises the question of how the SNr gets its information about the utility of the stimuli in the visual field. A major input to the SNr is from the striatum, in particular the caudate nucleus via the direct pathway. This input originates from the medium spiny neurons in the caudate, which are inhibitory. Activity of caudate neurons appears to encode stimulus value, thus influencing decision-making as well as vigor.

Caudate neurons have large contralateral response fields.

One of the first investigations of the saccade-related activity in the caudate was by Hikosaka and colleagues (Hikosaka et al., 1989). In the memory-guided task, activity in some cells tended to build as the animal waited for instruction to make the saccade; around time of saccade or just before it, the activity of these cells reached a maximum. Like the cells in the SNr, the cells in the caudate had large response fields, generally for movements toward the contralateral side, without obvious preference for amplitude. Notably, the first hints that these cells regulated only rewarded movements, and not all movement, was suggested by the behavior of these cells under two conditions. The caudate cells were largely silent during the prelude to spontaneous saccades in the dark, but they were active for similar saccades toward stimuli that promised reward.

Lauwereyns, Hikosaka, and colleagues (Lauwereyns et al., 2002) recorded from putative medium spiny neurons in the caudate nucleus while monkeys performed a cued saccade task. The task began with a cue that indicated which of two locations was the correct site for a rewarded saccade. One second after the cue was removed, two targets appeared, and the monkey was rewarded for choosing the target that corresponded to the previously shown cue (figure 6.7). A typical task-related caudate neuron increased its discharge after the onset of the cue, and produced a greater response when the cue was to the contralateral side. Similarly, the neurons fired before onset of the saccade (after presentation of the

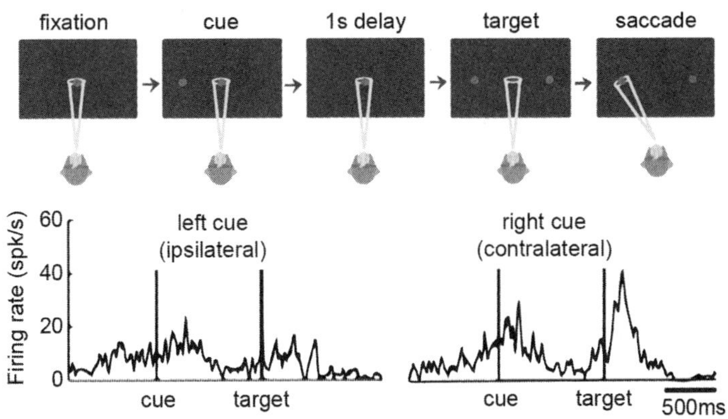

Figure 6.7
Saccade-related caudate neurons tend to have a visual response field on the contralateral side. In this experiment, reward was provided for all correctly performed saccades. Data are from a single neuron. Response builds in anticipation of the cue in the contralateral field and rebuilds to a burst after presentation of target in the contralateral field. (From Lauwereyns et al., 2002.)

targets) and fired more if the saccade was toward the contralateral direction. That is, the response field of the saccade-related neurons in the caudate, as in the SNr, was located in the contralateral visual field.

Some caudate neurons prefer high reward value stimuli, while others prefer low value stimuli.

Notably, the effect of reward on caudate cells suggested presence of two separate groups of neurons: some that showed their largest response for high-reward stimuli, and some that showed their largest response for low-reward stimuli. This diversity of encoding was illustrated in an experiment performed by Hyoung Kim and Hikosaka (2013). They trained monkeys in the flexible value task, (figure 6.6A and figure 6.6B) in which after fixation, a target (one of two fractal images) appeared to the contralateral side. After a 400 ms overlap period, the fixation spot was removed; this event cued the animal to make a saccade to the target (figure 6.8A). During a given block, one of the fractal images was paired with a large amount of juice, whereas the other was paired with a small amount. In the next block, the reward association switched.

The authors recorded from cells in various regions of the caudate (head, body, and tail, figure 6.8B). They found that most of the task-related cells were in the head and body regions, and modulated activity after the onset of the target in the contralateral field. Notably, in the head and body regions of the caudate, a group of cells showed their largest response when the target was paired with high reward, whereas another group of cells showed their largest response when the target was paired with low reward. For example, in the head of the caudate, 40% of the cells produced a large burst after the presentation of the high-reward target, and 18% of the cells produced a large burst after the presentation of the low-reward target. (These neurons are labeled high value and low value preference neurons [figure 6.8D].) These cells maintained their reward-modulated discharge until the fixation dot was turned off and the saccade took place. Therefore, many caudate cells, particularly in the head and body regions, differentiated the value of the target in their response field in this flexible value task. Importantly, some caudate cells preferred the high-value stimulus, whereas others preferred the low-value one.

An additional point highlighted by Kim and Hikosaka (2013) was that not all regions of the caudate responded similarly to the rewarding value of the visual stimulus. Whereas in the head and body regions there were cells that were affected by the value of the stimulus in the flexible value task, most cells in the tail of the caudate responded to the stimulus but did not differentiate on the basis of its reward value (figure 6.8C, tail of the caudate). The authors discovered that after they trained the monkeys for many days so that they learned a long-term, stable association between each image and its reward value, a value-dependent modulation of response developed in the tail region of the caudate. Notably, in this stable value task, some neurons in the tail region preferred high-value stimuli, whereas other neurons preferred low-value stimuli.

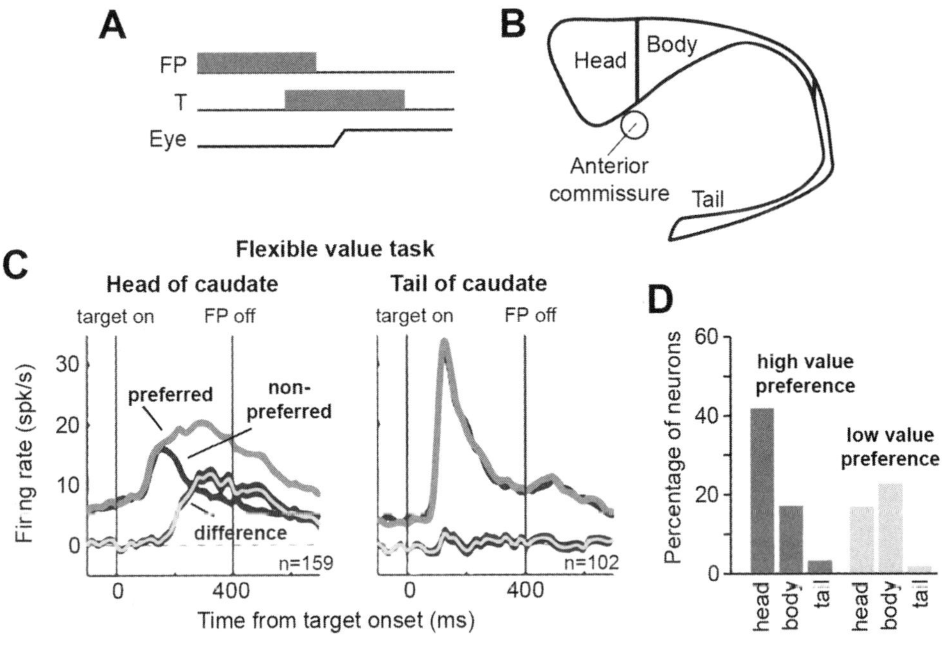

Figure 6.8
Diversity in encoding of reward value of a saccade target in the caudate. **A**. In the flexible-value task, a target (one of two fractal images) appeared to the left or right of fixation. After an overlap period (400 ms), the fixation point disappeared and the monkey made a saccade to the target to receive reward. In a given block, one of the images was paired with a high amount of juice, while the other image was paired with a low amount. In the subsequent block the reward pairing was switched. **B**. Recordings were made from the head, body, and tail regions of the caudate nucleus. **C**. Response of neurons in the head and tail regions. Cells in each region are grouped based on whether they responded more strongly to the high-reward or low-reward image as it appeared in their response field (usually contralateral direction). **D**. Preferences for percentages of task-related neurons in each region. (From Kim and Hikosaka, 2013.)

In summary, although different regions of the caudate specialized in encoding value of the object in terms of whether that value was flexible or stable, within each region, some cells preferred the low-value objects, whereas other cells preferred the high-value ones.

Positive reward value preferring caudate neurons may project via the direct pathway. This diversity in how the caudate cells encoded reward value provided an important clue: perhaps the positive-value caudate cells were part of the direct pathway, whereas the negative-value caudate cells were part of the indirect pathway. In this framework, the positive-value caudate cells might encode a measure of reward, whereas the negative-value caudate cells might encode a measure of cost or effort. The positive-value caudate cells would engage the pause type SNr cells, whereas the negative-value caudate cells would indirectly engage the burst type SNr cells.

It does seem that the direct pathway carries information regarding the reward value of the stimulus, whereas the indirect pathway carries information regarding its cost or effort requirements. However, this idea remains largely untested in primates. Nevertheless, there is indirect evidence in support of this hypothesis. Kim, Amit, and Hikosaka (2017) argued that if the caudate cells that preferred the negative-value stimuli principally projected via the indirect pathway, then cells in the external segment of the globus pallidus (GPe) should preferentially exhibit a reduction in discharge in response to low-value stimuli. That is, given that negative-value preferring caudate cells show a burst in response to low-reward stimuli, and given that caudate cells inhibit the GPe, then a majority of task-related GPe cells should show a reduction in discharge after the presentation of a low-reward stimulus.

To test this hypothesis, Kim et al. (2017) trained monkeys on various stimuli that promised low or high volumes of juice. Over many training sessions (figure 6.9A), the monkeys learned the value of a few fractal visual stimuli. (This is the stable value task, as compared to flexible value, in which the amount of juice assigned to each fractal changed within a single session.) During each training trial, the animals were presented with one stimulus at a random location, and after making a saccade to it, they received an amount of juice. Some stimuli were paired with a large volume of juice, while others were paired with either a low volume of juice or none at all. Then, during subsequent testing trials (figure 6.9B), while the animals fixated on the central target, the stimuli were presented one at a time in the neuron's preferred location (usually contralateral visual field). After presentation of a few stimuli, the monkeys received their rewards. Importantly, the animals maintained fixation during the testing trials while the stimuli were passively presented in the contralateral field. (The reward was unrelated to the visual stimuli and was related only to maintaining fixation.)

Kim et al. (2017) recorded from about 200 task-related GPe neurons during the test trials. While the animal fixated, the cells on average had a tonic discharge of around 70 spikes per second. After the presentation of a stimulus in the contralateral field, if the stimulus was associated with a low-value reward, some cells showed a reduction in discharge while others showed an increase. Notably, the response of GPe neurons as a population (averaged across all cells) was a reduction in discharge for low-value stimuli and a slight increase in discharge for high-value stimuli (figure 6.9C).

This result in GPe is consistent with the hypothesis that caudate neurons that prefer negative-value stimuli likely transmit their output to the GPe, ultimately affecting the SNr through the indirect pathway, whereas positive-value preferring caudate neurons likely transmit their output to the SNr, affecting it through the direct pathway.

Positive-value preferring caudate neurons may contribute to regulation of vigor.

Is there a relationship between activity in the caudate neurons and the control of vigor? To explore this question, Lauwereyns et al. (2002) focused on the activity of the caudate neurons that preferred the high-value targets. They designed a visually guided saccade

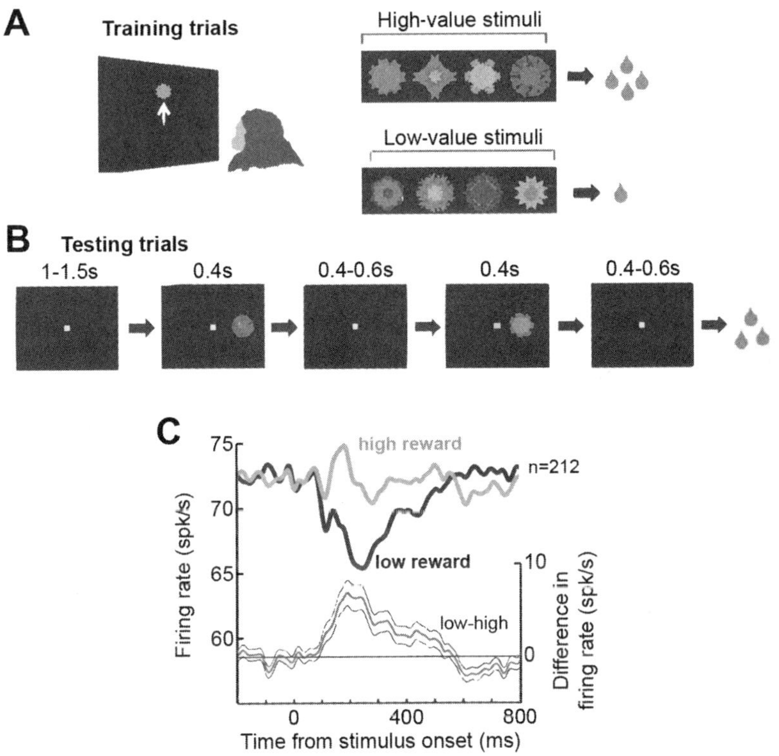

Figure 6.9
Activity across a group of global pallidus (GPe) neurons in response to learned value of visual stimuli. **A**. During the training trials, animals were provided with a visual stimulus (fractal image) and those who followed a saccade to that stimulus were provided with reward. Some stimuli were paired with a high-value reward, while others were paired with a low-value reward. Training on the same stimuli continued for weeks. The associated reward value was stably maintained during training. **B**. During testing, the animals fixated on a center position while the various stimuli were passively presented to the left or right. The animals maintained fixation. C. Averaged activity of n=212 task relevant GPe neurons during the testing trials, and the within-neuron difference in activity in response to low- and high-value stimuli. On average, the population showed a decline in activity following a low-value stimulus. (From Kim et al., 2017.)

experiment in which the animal was rewarded for only one of the two possible target locations (1Dir task). In this experiment (figure 6.10A), after a 1.5 s fixation period, a target appeared to the left or right, and the animal made a saccade to it. However, in one block (20–40 trials) only the saccades to the left target were rewarded, whereas in the subsequent block, only the saccades to the right target were rewarded.

As expected, saccades to the rewarded side exhibited shorter reaction times, and greater peak velocity (figure 6.10B). Notably, in the block in which the rewarded target was contralateral to the caudate, the positive-value preferring cells began firing soon after fixation and long before presentation of the target. Once the target was presented, the cells

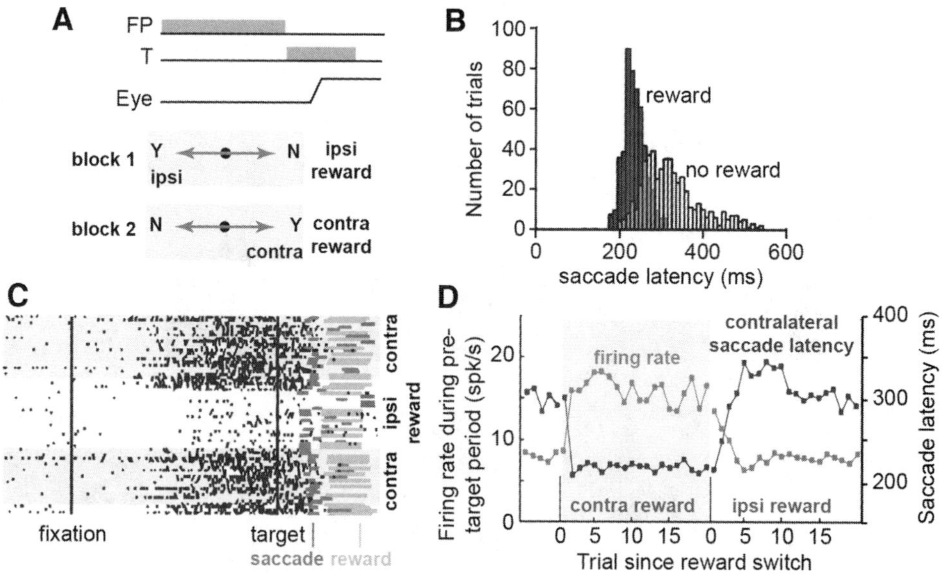

Figure 6.10
Correlates of saccade vigor in the caudate nucleus. **A**. In this task, reward was provided for saccades made to only one of two directions. The rewarded direction remained constant within a block of 20–40 trials, and then switched to the other side. **B**. Reaction-time for saccades toward the rewarded and unrewarded directions. **C**. Discharge of a single cell during consecutive blocks. In each block, saccades were made to both contralateral and ipsilateral directions. However, in the first block (top) only the contralateral direction was rewarded. In this block, the cell's discharge was high before the target is presented, and the discharge remained high up until time of saccade. In contrast, in the second block, the ipsilateral direction was rewarded. In this block, the cell's discharge was low before the target is presented, and the discharge remained low up until time of saccade. Therefore, this is an example of a high-value preference neuron. **D**. Data for saccades toward the contralateral side only. The data are average of n=25 neurons with preference for high-value target in their response field. As the reward switched from ipsilateral to contralateral, the animal learned quickly, and produced a rapid reduction in reaction time. This coincided with an increase in pretarget discharge in the high value preferring caudate cells. (From Lauwereyns et al., 2002.)

maintained their high firing rate until the saccade was made (figure 6.10C). In contrast, in the block in which the contralateral target was no longer rewarded, the cells showed a much weaker discharge during all periods. That is, expectation of reward for an action toward the contralateral side produced a strong increase in the discharge of cells during the period both before the target was presented (anticipation period) and after it was presented (movement preparation and execution period).

Was this increased discharge linked to increased vigor? During the experiment, the direction of reward was kept constant in a block of 20–40 trials and then switched to the other side without warning. Therefore, the animal's expectation of reward changed from trial to trial on the basis of its experience of reward. The animal's behavior suggested that it learned rapidly when a previously unrewarding saccade was now rewarded (positive-reward prediction error), but learned more slowly when a previously rewarding saccade was no longer rewarded (negative-reward prediction error). That is, the animal took longer to "unlearn" the initial positive association. To illustrate this, consider the data shown in figure 6.10D, which presents saccades made during two blocks only to the contralateral side (ipsilateral saccades were also made in these blocks, but they are not shown). After the reward switched from the ipsilateral to the contralateral side, the contralateral saccade that was previously unrewarded was now suddenly rewarded (trial 1, contralateral reward condition, figure 6.10D). After experiencing this positive-reward prediction error, during the next trial, saccade-reaction time toward the contralateral side substantially decreased; the monkey needed only one trial to learn that in this block, the contralateral saccade was rewarding. Remarkably, the average firing rate of the positive-value preferring caudate neurons during the pretarget period also showed a sharp increase in that trial. In contrast, after the reward switched from the contralateral to the ipsilateral side, the reaction times of contralateral saccades (i.e., saccades that were no longer rewarded) increased slowly, requiring about four trials to reach steady state. During the pretarget period of each of these trials, the average activity of the caudate neurons showed a gradual trial-by-trial decrease, needing about four trials to reach baseline.

Therefore, when the reward value of the contralateral target changed, there was a coincident change in saccade vigor. The trial-by-trial pattern of change in vigor covaried with dynamics of change in discharge of this positive-value preferring caudate neuron. As the authors wrote, the pretarget activity seemed to "act as a spatial response bias, which prioritized eye movement toward the neuron's preferred position if that position was associated with high reward value." This biasing would result in increased inhibition onto the SNr, which would in turn reduce inhibition onto the superior colliculus. As a result, movements toward the unrewarded side would be hindered by an inhibition that prevented rise of collicular activity to threshold, whereas movements toward the rewarded side would be helped by the reduction in inhibition, which in turn would encourage the rise of activity to threshold.

Inhibition of the direct pathway striatal neurons reduces velocity of the movement.
The results in figure 6.10D are correlative, suggesting a relationship between activity in the positive-value–encoding caudate neurons and saccade vigor. Is there a way to establish whether this relationship is a causal one? To test for this, one needs to manipulate the direct pathway projecting striatal neurons. This technique is currently possible in rodents; Babita Panigrahi, Joshua Dudman, and colleagues (2015) performed such an experiment. They trained mice to use their forelimb to push a joystick and produce a target displacement to receive water as a reward. Within a session, the amount of displacement that was required gradually increased and, along with it, so did the effort required to push the joystick. As the displacement increased, so did the peak velocity of the movement. In randomly selected trials, the researchers employed optogenetic stimulation to silence the

neurons in the dorsal striatum that projected via the direct pathway without affecting the neurons of the indirect pathway. They found that this stimulation (in three mice) reduced reach velocity. Therefore, there appears to be a causal relationship between activity in the direct pathway striatal neurons and vigor: reductions in the activity of these neurons result in less vigorous movements.

Encoding of reward in the caudate is relative to other potential rewards, and not in terms of the absolute value of that reward.

Did the reward value that the caudate neurons encode in the pre-cue period reflect the absolute value of the action or the relative value with respect to other potential actions? To answer this question, consider a scenario in which there are four potential movements that can be performed. Is the value of reward for the preferred movement encoded with respect to the absolute amount of the reward, or is it encoded with respect to other possible rewards (possibly the sum of all rewards)?

An experiment performed by Yoriko Takikawa, Reiko Kawagoe, and Hikosaka (Takikawa et al., 2002) provided insights into this question. They trained a monkey to make memory-guided saccades to targets that appeared in one of four locations (figure 6.11). In block 1, only saccades to the rightmost target were rewarded (i.e., the animal made saccades to all four targets, but three of the four targets were not rewarded). In block 2, only saccades to the upper target was rewarded. After four blocks of this 1Dir scheme, all

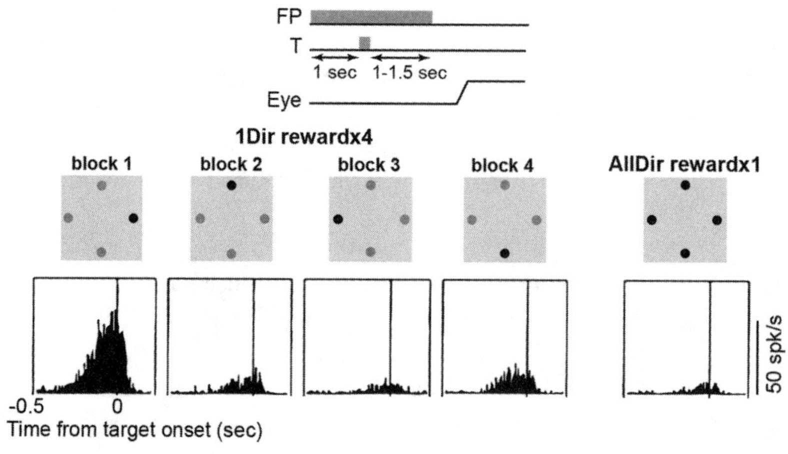

Figure 6.11
In the caudate, reward encoding among the positive-value coding neurons may be relative to all other potential rewards. Monkeys performed a memory-guided saccade task in which either one target or all targets were associated with reward (1Dir scheme versus. AllDir scheme). The plot shows the activity of a single cell in the left caudate nucleus. Before the target was shown, as the animal gazed at the fixation point, the cell began firing more if the target in its response field (saccade to the right) was rewarded. Notably, this pretarget activity was much higher if the rightward target was the only one among the four that was associated with reward (1Dir scheme), than when all four targets were associated with reward (AllDir scheme). (From Takikawa et al., 2002.)

targets were rewarded as they were in the AllDir scheme. Crucially, the amount of reward per trial was constant in both schemes, which means that in each successful 1Dir trial, the animals received 4 times the amount of reward than they did in a successful AllDir trial.

During the pretarget period, as the animal gazed at the fixation spot, it could not predict which target would be presented. Regardless, this positive-value preferring cell discharged more during the pretarget period if the rewarded target in that block was located in the contralateral visual field. For example, for the left caudate cell shown in figure 6.11, activity rose in the pretarget period as the animal waited for the target to appear. Interestingly, when the same target was associated with a reward during the AllDir scheme, in which the other targets were rewarded as well, the discharge during the pretarget period was barely noticeable. Although the amount of reward in the 1Dir scheme was only 4 times that in the AllDir scheme, the discharge during the 1Dir trials appeared to be 10 or more times higher than in the AllDir trials. The authors performed a control experiment to show that this was unlikely to be because of the order in which the schemes were performed.

Taken together, the results suggested that the activity that preceded execution of a movement (in the positive-value coding neurons) was not reflecting the absolute amount of reward at stake, but rather the relative amount with respect to all other potential rewards. One possibility is that this relative encoding is the reward available for the preferred movement with respect to the total reward available for all movements.

Activity in the striatal neurons of the indirect pathway makes the animal less likely to choose the risky option.

What information is carried by neurons in the indirect pathway? Consider the SNr neuron that produced a burst in response to the presentation of the visual stimulus in the memory-guided task, as shown in figure 6.2. This burst would discourage production of a saccade toward the stimulus; the animal would be allowed to sustain fixation and wait, hence preserving the possibility of acquiring the reward. In a sense, the discharge in this SNr neuron indicates the risk or cost associated with performing the saccade: when reward increases, there is more at stake, and the neuron produces a larger burst in response to the visual stimulus (arrow, figure 6.3). As time passes beyond presentation of the target, the discharge in the neuron falls, reaching a minimum around saccade time. The saccade takes place when indication of risk or cost has reached a minimum. If we assume that this SNr cell is part of the indirect pathway, is there any indication that cells that are part of the indirect pathway encode risk or cost associated with performing a movement?

Kelly Zalocusky, Karl Deisseroth, and colleagues (2016) performed an experiment that sheds light on this question. They trained rats in a task in which in each trial, the animals chose between pressing two levers: one was associated with a 100% probability of a small food reward, and the other lever provided a 25% chance of a very large reward and 75% chance of an extremely small reward (figure 6.12A). That is, the expected value of the two

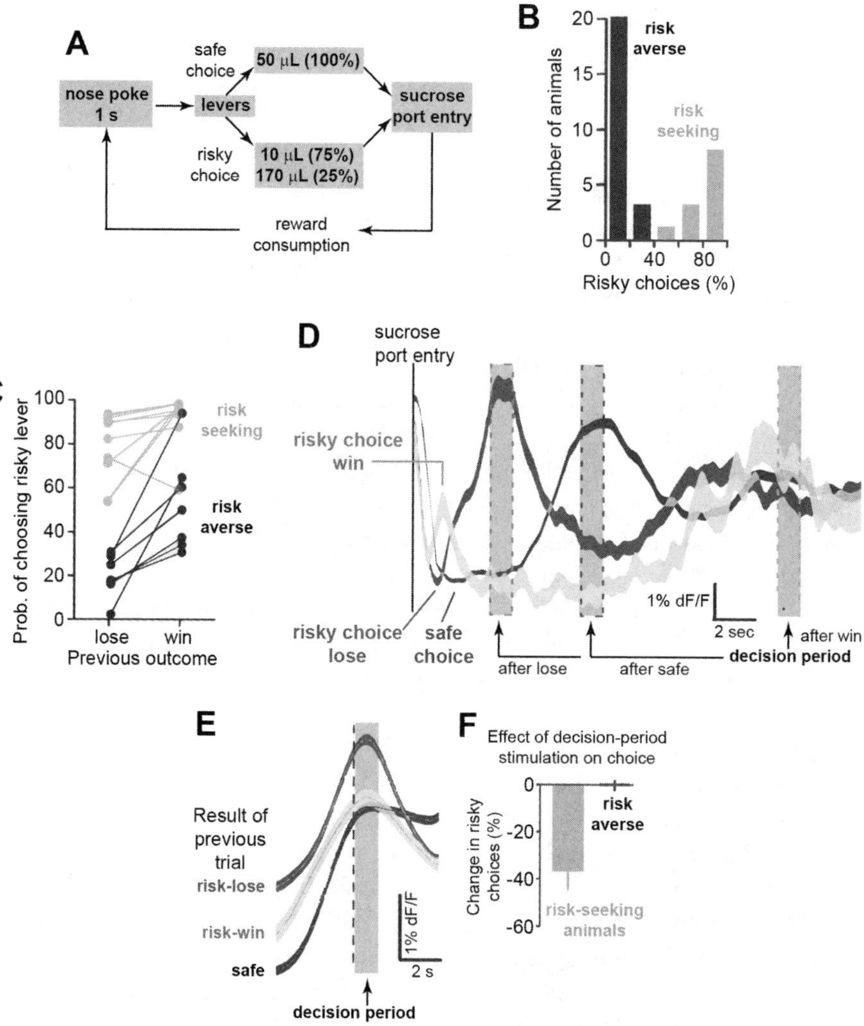

Figure 6.12
Activity in D2 receptor expression neurons in ventral striatum acts as a bias that encourages the animal not to choose the risky option. **A**. Experiment setup. **B**. Histogram of risk preference. Risk averse was defined as <50% risky choices, risk seeking was defined as >50% risky choices. **C**. Both groups of animals were more likely to choose the risky lever after a win than after a loss. However, this effect was larger in the risk-averse rats. **D**. Time course of optical imaging data from D2 receptor expressing neurons in NAc. Data are aligned to sucrose port entry. **E**. Data are aligned to onset of decision period (nose poke), and sorted based on result of the previous trial. **F**. Effect of stimulation of the D2 receptor expressing neurons during the decision-making period in the two groups of animals. (From Zalocusky et al., 2016.)

levers was the same, and so there was no advantage in picking one over the other. Regardless, there were differences in the responses among the rats. Some exhibited risk-seeking behavior, while others were risk averse (figure 6.12B). About a quarter of the rats were risk averse: in less than 40% of trials, they chose the risky lever. The majority of the rats were risk seeking; they picked the risky lever in a majority of the trials. In both groups, the choice of seeking the risky lever depended on what had transpired in the previous trial (figure 6.12C). After a loss, the risk-averse rats were much less likely to try the risky lever again than were their risk-seeking counterparts. In contrast, the performance after a win (getting a large reward from the risky lever) was roughly equal in the two groups. Therefore, the choice in a given trial was largely predicted by whether in the previous trial the rat had chosen the risky lever and received the negligible reward. A bad experience with the risky lever resulted in switching to the safe lever. A good experience was often followed by picking the risky lever again.

To determine the neural basis of this decision-making process, the authors gave the rats a drug (pramipexole [PPX]) that acted as a D2 dopamine receptor agonist. (This drug is also administered in the treatment of Parkinson's disease.) The authors found that PPX increased risk-seeking behavior, whereas a D1 dopamine receptor agonist did not have this effect. They then used optical methods to record from D2 dopamine expressing cells (cells that project in the indirect pathway) located in the nucleus accumbens (NAc), a region in the ventral striatum. They noted that as the animal entered the sucrose port and experienced the result of its choice, activity in the D2 dopamine receptor neurons dissociated whether it had won or lost (figure 6.12D). There was a burst when the animal had taken a risk and lost, but no burst when the animal had taken a risk and won or when it had taken the safe choice and received the safe reward.

After a loss in the previous trial, as the animal entered the nose poke port and deliberated on what to do, activity in the D2 dopamine receptor neurons showed a large burst. In contrast, if the animal made a safe choice in the previous trial, the activity in these neurons during the decision period was somewhat weaker. The activity was weakest following a win. These results are further illustrated in figure 6.12E, where the data are aligned to the decision period and sorted on the basis of whether the animal had won, lost, or chosen the safe lever in the previous trial. During the decision period, activity of the D2 receptor neurons in NAc cells was higher if the rat had experienced a loss in the previous trial than if it had experienced a gain or safe outcome.

To test whether the activity during the decision period was causally influencing behavior, the authors optogenetically stimulated the D2 receptor neurons and found that the stimulation had no effect on risk-averse rats. However, in risk-seeking rats, stimulation dramatically decreased the probability of choosing the risky lever (figure 6.12F). Indeed, animals that showed greater activity in D2 receptor expressing neurons during the decision period after a loss also tended to avoid the risky lever more, thereby exhibiting a risk-aversive tendency.

These results illustrate that when there is increased activity in the D2 receptor expressing neurons of the ventral striatum (neurons which are likely to project via the indirect pathway), the animal is biased away from choosing the risky, more costly option.

In summary, the reward and cost sensitivity that SNr cells display is likely because of the information that is conveyed to them from the caudate via the direct and indirect pathways, respectively. Like the SNr, saccade-related cells in the caudate have large response fields that are more sensitive to stimuli on the contralateral side. And, like the SNr cells, caudate cells show diversity in how they encode the reward value of the visual stimulus; some caudate cells show a preference for stimuli that promise a large reward, whereas others show a preference for stimuli that promise little or no reward. It is possible that high-value preferring cells in the caudate are part of the direct pathway, whereas the low-value preferring cells are part of the indirect pathway. The encoding of reward value appears to be relative to other potential rewards and does not seem to reflect the absolute value of reward. While activity in the direct pathway caudate neurons encodes a measure of reward or gain associated with performing a movement or choosing an option, activity in the indirect pathway encodes a measure of risk or cost.

These results suggest that there is a link between activity of striatal cells and control of vigor: as the value associated with a stimulus changes, the change in vigor often covaries with the change in the discharge of the positive-value preferring caudate neurons. Inhibition of the direct pathway projecting neurons in the dorsal striatum reduces the vigor of the movement. Thus, the reward-dependent modulation of discharge in the striatum is partly responsible for the reward-dependent modulation in the output nuclei of the basal ganglia, which in turn affect the vigor of the ensuing saccade.

6.4 Dopamine and the Pleasure of Moving

If the reward-dependent modulation of SNr activity is driven by reward-dependent activity in the caudate, how do the caudate neurons acquire their knowledge about utility of the stimulus? The principal source of input to the caudate is from the cerebral cortex, particularly the frontal eye field and the posterior parietal cortex. The excitation that these regions provide activates caudate cells, but the sensitivity of the caudate cells to this excitation is modulated by the availability of dopamine in the striatum. Dopamine alters the gain of the cell's response to their input, controlling the amount of change in discharge for a given amount of excitation. A critical factor in reward-dependent response of the caudate cells, and therefore control of vigor, is dopamine.

Arrival of dopamine modulates the glutamatergic synapses of the medium spiny cells of the caudate nucleus. The effect of dopamine on D2 receptors is to impede the transition to an up-state, a state in which they are more responsive to excitatory inputs, whereas the effect on D1 receptors is to enhance this transition (Gerfen and Surmeier, 2011).

Some subjects do not show reward-dependent modulation of vigor. They may lack reward-dependent sensitivity in their dopamine release.

Over the course of decades that Okihide Hikosaka had trained monkeys to perform various tasks, he came across one monkey that was peculiar. During the 1Dir trials, it performed the task fine, but its saccades were not modulated by reward. This led him to hypothesize that if dopamine was critical for generating vigorous saccades toward reward, then this animal that did not display reward-depend modulation of vigor may also lack reward sensitivity in the response of its dopamine and caudate neurons.

Reiko Kawagoe, Yoriko Takikawa, and Okihide Hikosaka (Kawagoe et al., 2004) trained monkeys on the 1Dir and AllDir memory-guided tasks (figure 6.13). After fixation, a target briefly appeared in one of four locations. The animal maintained fixation for 1.0–1.5 s before the fixation point disappeared, and then it made a saccade to the remembered target location. Among the four monkeys that the authors trained, three showed clear reward-related modulation of vigor; their saccades toward the rewarding target in the 1Dir task exhibited reduced reaction time and increased peak velocity. (Data from one of these monkeys, monkey M, are shown in figure 6.13.) Curiously, one of the monkeys did not show this reward-related modulation of vigor (monkey G).

For the three reward-sensitive monkeys, in the 1Dir task, SNc dopamine neurons showed a burst at the onset of the fixation point. If the location of the target was associated with reward, then another burst was seen at target onset. If the location of the target was associated with no reward, the cells showed a slight decline in discharge after target presentation. Notably, the dopamine neurons did not respond when the reward was delivered because presentation of the stimulus fully predicted the reward. Unlike in the 1Dir trials, in the AllDir trials, the dopamine neurons did not respond to the target. Rather, in the AllDir trials, the dopamine neurons responded to fixation point onset, whose timing was unpredictable but whose occurrence in the AllDir task (but not the 1Dir task) fully predicted the reward.

In all monkeys, dopamine cells responded to the presentation of the fixation point. However, unlike in the reward-sensitive monkeys, in monkey G, dopamine cells did not respond differentially to presentation of the rewarding target in the 1Dir task; target presentation did not result in reward-dependent modulation of dopamine.

The authors next examined how cells in the head and body region of the caudate responded to reward. Figure 6.13 shows the population averaged activity of the task-related cells that were recorded. (Most of the recorded cells would be classified as positive-value encoding.) In the 1Dir task, neural activity differed between the reward-sensitive monkeys and the reward-insensitive monkey. In the three reward-sensitive monkeys, during fixation (before target presentation), activity built up at a faster rate if the contralateral side was paired with a reward. Then after target onset, the caudate cells showed a strong response if the target on the contralateral side was associated with a reward.

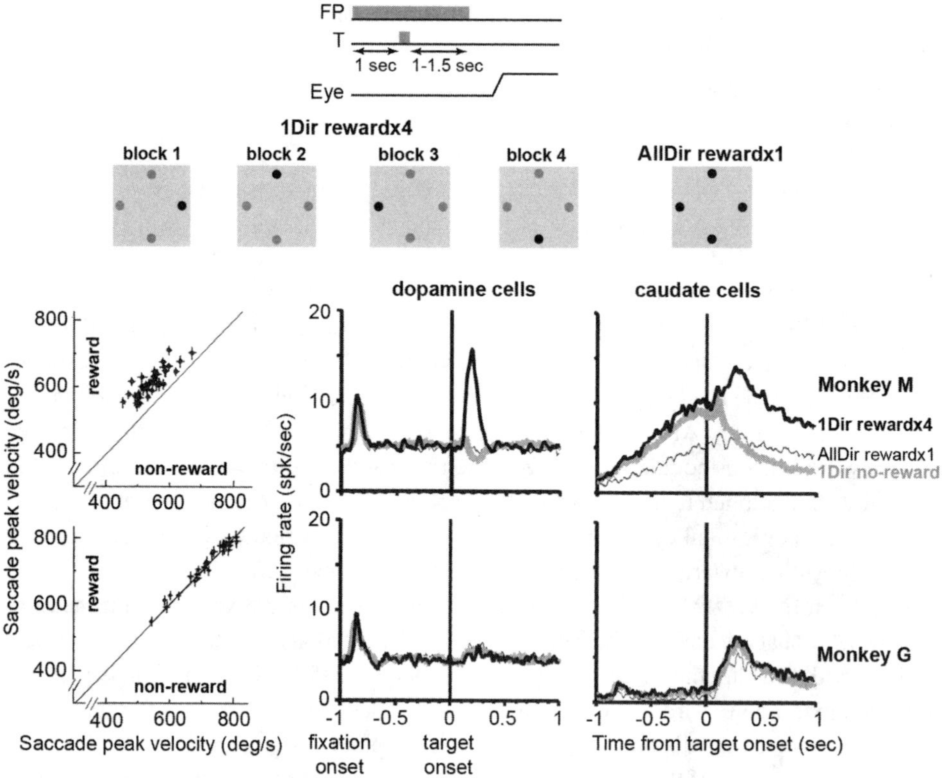

Figure 6.13
In a monkey that does not show reward-dependent change in vigor, dopamine and caudate response does not differ between reward and non-reward trials. Monkeys were trained in the 1Dir and AllDir tasks. Both monkeys learned the tasks. In the 1Dir task, monkey M showed reward-dependent modulation of saccade vigor: saccades toward the rewarded target exhibited greater peak velocity than saccades to non-rewarded targets. However, monkey G did not exhibit reward-dependent modulation of saccade vigor. In both monkeys, after fixation onset, SNc dopamine cells exhibited a burst. (SNc, substantia nigra reticulate.) In the reward-sensitive animal (monkey M), dopamine neurons produced a burst at target onset if that target was associated with reward in the 1Dir task, but not in the AllDir task. In the reward-insensitive animal (monkey G), dopamine neurons did not react to target onset, regardless of whether it was associated with reward. In the reward-sensitive animal, caudate neurons with response fields near the rewarding target showed a rise in their activity as the animal waited for the cue. This rise was missing in the reward-insensitive animal. After target presentation, potential for reward in the 1Dir task modulated response of the caudate neurons in the reward-sensitive monkey, but not in the reward-insensitive animal. (From Kawagoe et al., 2004.)

In contrast, in monkey G, the caudate response before target onset was flat, not modulated by that fact that the target on the contralateral side was rewarding. At target onset, activity did not differ with the reward value of the target. Having observed this monkey's behavior outside of this task, the authors speculated that its symptoms resembled those of schizophrenia.

These results also demonstrated that in this population of sampled neurons, the dopamine cells were not spatially selective: they responded to the reward value of the target regardless of whether it appeared in the ipsilateral or the contralateral side. In contrast, caudate neurons were selective, firing more for contralateral targets. The spatial selectivity of caudate neurons was likely due to the inputs that they received from the frontal eye field (FEF) and lateral intraparietal cortex (LIP).

In summary, all animals made the saccades to the correct location, but saccade vigor in one subject was insensitive to reward. In that subject, before target onset, there was no rise of activity in the positive-value coding caudate neurons to bias behavior toward the contralateral, rewarding stimulus. After the rewarding target was presented, dopamine release did not indicate that this target was different from other targets, and there was no reward-dependent activity in the positive-value preferring neurons in the caudate. Thus, a lack of dopamine-based differentiation of stimulus reward value coincided with a host of abnormalities in the reward-dependent activity in the striatum, which concluded with a lack of reward-dependent modulation of saccade vigor.

Reward-dependent changes in saccade vigor are eliminated after D1 dopamine receptor antagonist in the caudate.

Presence of dopamine in the striatum is thought to alter the properties of caudate neurons, making it easier for cells with D1 receptors to transition to an up-state upon arrival of excitatory inputs. One way to test whether dopamine is required for reward-dependent modulation of vigor is via manipulation of the D1 dopamine receptors in the caudate.

Kae Nakamura and Okihide Hikosaka (Nakamura and Hikosaka, 2006) trained monkeys in the 1Dir version of the visually guided saccade task (figure 6.14). Each trial began with a fixation. Once the animal fixated on the central spot for about 1 s, the fixation spot disappeared and immediately a target appeared on the left or right. In a block design (20–28 trials per block), large (0.4 mL) or very small (0.0 mL or 0.05 mL) amounts of reward (juice) were paired with each target and delivered after a correct saccade. Before injection of a D1-antagonist drug, reaction times for saccades were smaller by around 70 ms for the large-reward target. When they injected the D1-antagonist into the caudate, they found that saccades to the contralateral direction did not change under the small-reward conditions. However, when the contralateral target was paired with a large reward, injection of the D1-antagonist increased reaction times. They then performed an experiment in which all saccades to targets were rewarded equally, regardless of target direction

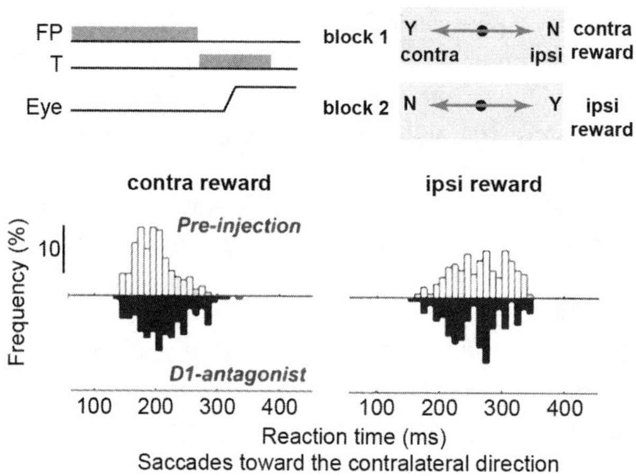

Figure 6.14
Injection of a dopamine Dl antagonist into the right caudate reduced the effect of reward on vigor of leftward saccades. The experiment relied on visually guided saccades. In blocks of 20–28 trials, the target for which the animal was rewarded switched from left to right and back. Before injection, in the block in which contralateral saccades were highly rewarded, reaction time was on average about 70 ms shorter than ipsilateral saccades (which were either not rewarded or rewarded by a very small amount). After dopamine Dl antagonist injection, reaction time for ipsilateral saccades did not change. However, reaction time for contralateral saccades increased. (From Nakamura and Hikosaka, 2006.)

(AllDir). The authors found that the presence of the Dl-antagonist again increased the reaction time for contralateral saccades but had no effect on ipsilateral saccades.

This established that reward-dependent changes in saccade vigor were partly due to the effect that dopamine had on the Dl receptor neurons in the caudate (i.e., cells that are part of the direct pathway to the SNr).

Dopamine discharge reflects reward prediction error.

Discharge of dopamine neurons does not reflect the reward value of the stimulus or the animal's past rate of reward (Mohebi et al., 2019). Rather, dopamine discharge is a reflection of the difference between the value of the reward that one expects to receive and the value that one has received, with greater discharge often reflecting a more positive-reward prediction error (i.e., the reward was higher than expected). Indeed, many dopamine neurons do respond in this way, responding strongly to positive-value events. However, there are also many other dopamine neurons that respond strongly to negative-value events instead.

Masayuki Matsumoto and Okihide Hikosaka (2009) trained monkeys to associate reward (juice) to some visual stimuli, and an air puff to other visual stimuli. In the reward block of trials, after a random interval, the trial began with a start cue that was displayed on the screen for 1 s, followed by one of three stimuli that predicted reward at 100%, 50%, or 0% probability. In the air puff block of trials, the trial began as before, but a different set of three stimuli predicted an air puff at 100%, 50%, or 0%. Although the animals did not have any particular task to do in order to earn or avoid the air puff, after seeing the reward-predicting stimulus, they licked the juice spout at a rate that increased with probability of reward, and after seeing the air-puff-predicting stimulus, they blinked their eyes at a rate that increased with probability of an air puff. Therefore, when the stimulus predicted either a high probability of reward or a high probability of an air puff, the animal behaved with high vigor: a high licking rate followed the stimulus that predicted the high reward, and a high blinking rate followed the stimulus that predicted the high likelihood of an air puff.

In the reward block, after the onset of the stimulus, many dopamine neurons responded at around 150 ms with a magnitude of discharge that was highest for the 100% probability stimulus and lowest for the 0% probability stimulus (figure 6.15, left column). A higher probability of reward generally coincided with a higher rate in the discharge of the dopamine cells. However, among these neurons, during the air-puff block of trials, some responded by reducing their activity, while others responded by increasing their activity (figure 6.15, right column). That is, for some dopamine neurons, the discharge rate was higher for larger reward values and lower for negative, aversive stimuli. However, for other dopamine neurons, both large positive and large negative (aversive) stimuli led to higher discharge.

This suggests that the activity of some dopamine neurons does reflect the utility of the stimulus (increases with positive value and decreases with negative value), whereas the activity of other dopamine neurons reflects the salience of the stimulus (increases with the absolute value of the stimulus). Increases in the activity of either type of neuron may be followed by more vigorous bodily actions. Unfortunately, the authors did not present data for the period around time of the movement. As we note below, other experimental results suggest that during the movement, greater dopamine release often coincides with greater vigor.

Dopamine release before movement onset increases vigor.

The link between the release of dopamine and the vigor of the ensuing movement was investigated by Joaquim Alves da Silva, Rui Costa, and colleagues (da Silva et al., 2018). The researchers placed mice in an open field with no external cues or rewards and recorded from dopamine neurons in the SNc. The mice were often standing still, and occasionally walking about (figure 6.16A). Most of the dopamine neurons increased their firing rate around 300 ms before bouts of walking (52%, figure 6.16B). However, a minority decreased their firing rate soon after the movement began (28%, figure 6.16C). Across all of the dopamine neurons that they recorded, on average there was a transient increase before movement initiation (trace labeled "all", figure 6.16C).

The dopamine neurons that increased their discharge rates did so without an apparent preference for direction of movement, suggesting that the transient activity was not action

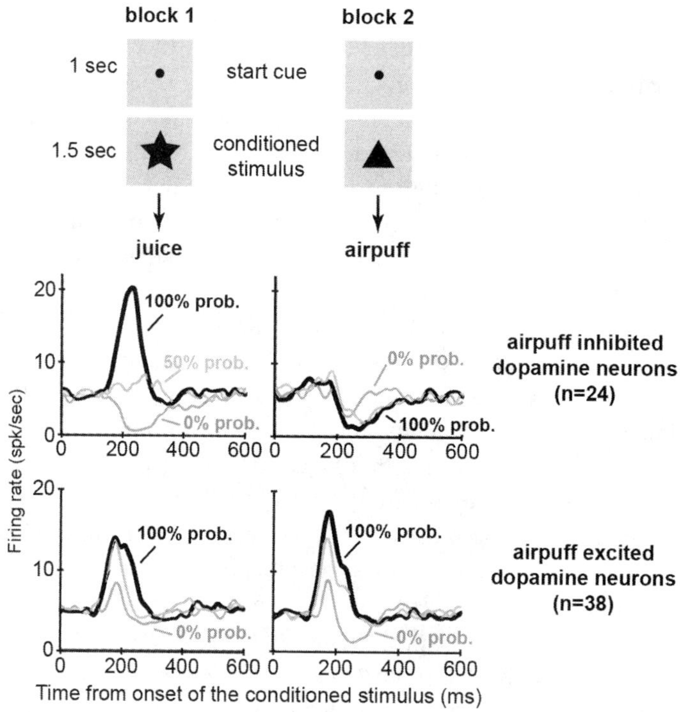

Figure 6.15
Some dopamine cells respond positively to stimuli that have a negative utility, while others respond negatively. In this experiment, in a block of trials, 3 stimuli that indicated a reward were presented, while in a subsequent block, 3 stimuli that indicated punishment were presented. The reward was juice; the punishment was an air puff. Each stimulus was associated with a probability of juice or air puff. Of the 103 dopamine neurons in substantia nigra reticulate (SNc) and ventral tegmental area that responded to the conditioned stimuli, all responded positively to stimuli that predicted reward. However, a fraction of those neurons were inhibited by stimuli that predicted an air puff (24/103 air puff inhibited dopamine neurons), whereas a larger fraction were excited by stimuli indicating air puffs (38/103 air puff excited dopamine neurons). (From Matsumoto and Hikosaka, 2009.)

specific. Notably, in a trial-by-trial analysis, about 40% of all the dopamine neurons had significantly higher activity before high-vigor movements than before low-vigor movements (figure 6.16D, left). Despite this, some dopamine neurons increased their activity before movement onset, but this change was insensitive to changes in vigor of the ensuing movement (figure 6.16D, right). Taken together, it appeared that many SNc dopamine neurons transiently increased their discharge around 300 ms before the onset of a voluntary movement.

Next, the authors waited for periods of immobility (immobile for at least 900 ms), and then optogenetically activated the dopamine neurons for 500 ms in 50% of trials. They found that this produced overt bouts of walking that lasted a few seconds. Movement initiation after stimulation was more vigorous than self-initiated movement without stimulation

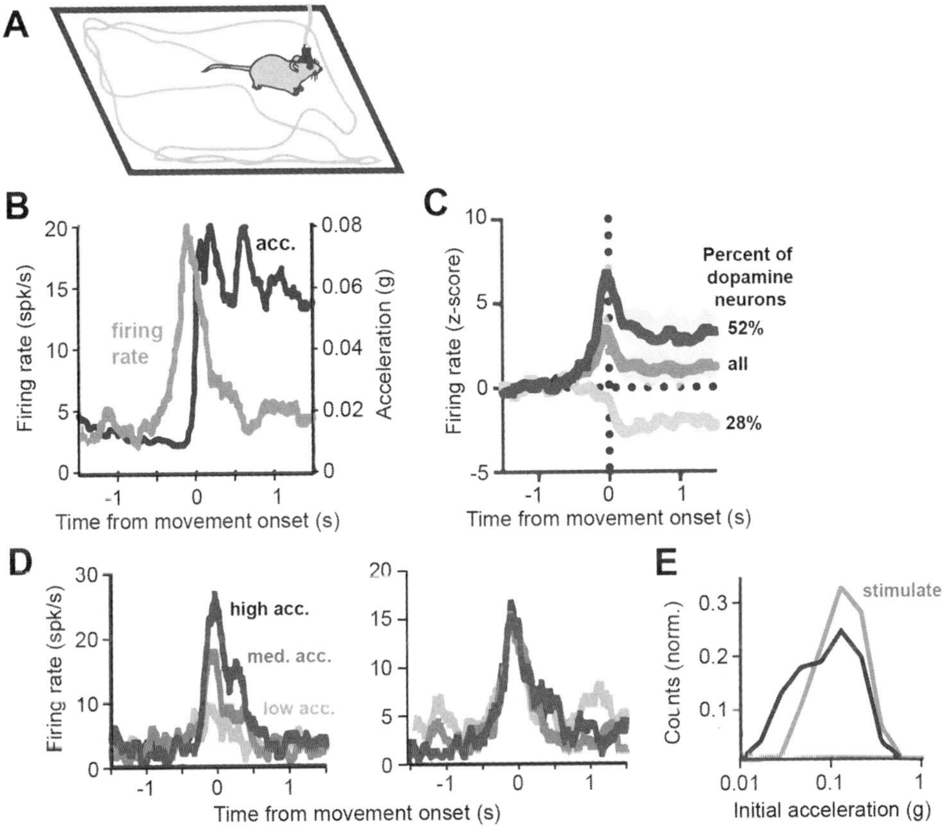

Figure 6.16
Activity in some SNc dopamine neurons before onset of a self-generated movement correlates with vigor of that movement. (SNc, substantia nigra reticulate) **A**. Mice were placed in an empty, open field. **B**. Example of a single SNc dopamine neuron. Acceleration refers to motion of the head and body. **C**. Activity of n=25 dopamine neurons. A portion of cells (52%) produced a burst before onset of the movement, whereas a portion (28%) reduced their discharge around movement onset. "All" refers to average across all cells. **D**. Example of a vigor-related neuron (left) and a neuron unrelated to vigor (right). **E**. Acceleration values for initiation of self-generated movements and for movements that followed activation of dopamine neurons. Movements were more vigorous after stimulation. (From da Silva et al., 2018.)

(figure 6.16E). When the researchers inhibited the dopamine neurons during the periods in which the animal was standing still, they found that movement initiation was impaired. In contrast, when they inhibited the neurons during periods of movement, they observed no significant change in the ensuing acceleration.

These results established that even without explicit rewards, just before onset of a movement, there is transient discharge in the dopamine neurons. Artificial activation of dopamine neurons invigorates the ensuing movement.

Reward prediction error drives dopamine release and invigorates the ensuing saccade. One aspect of dopamine release appears to follow a simple rule: in general, when the acquired reward is unexpectedly large, many neurons burst, but if the same reward is expected and received, these neurons no longer respond (Schultz et al., 1997; Bayer and Glimcher, 2005). We saw an example of this in the dopamine response to the target in the AllDir task (Monkey M, figure 6.13). In this respect, many dopamine neurons encode the difference between the predicted stimulus value and the actually acquired value, termed reward prediction error (RPE). The transient encoding of RPE provides an interesting prediction with regard to vigor: if dopamine release in the milliseconds before movement onset contributes to control of vigor, then movements that follow a positive RPE (+RPE) should exhibit high vigor, and those that follow a negative RPE (-RPE) should exhibit low vigor. That is, vigor modulation should depend on reward prediction error, not reward itself.

The hypothesis that RPE (and not reward per se) drives vigor has been difficult to test because in a typical experiment, the RPE event occurs after a movement has been completed and the reward acquired, not before the onset of a movement. Ehsan Sedaghat-Nejad in our laboratory (2019) designed an experiment that overcame this limitation. In our experiment, we relied on the idea that the viewing of images carries some of the hallmarks of reward: given the option of choosing from various image categories, people prefer face images and are willing to spend a greater amount of effort in exchange for gazing at those images (Aharon et al., 2001; Yoon et al., 2018). Furthermore, viewing of face images activates the brain's reward system (O'Doherty et al., 2003). We therefore probabilistically controlled image content to induce RPE events. We asked whether induction of an RPE event in the milliseconds before a movement influenced the vigor of that movement.

Subjects made saccades to view an image (figure 6.17A), and upon initiation of the saccade, we randomly altered the position and content of that image. The position change forced the subjects to follow their initial saccade with a secondary saccade. Our concern was vigor of this secondary saccade, which usually took place at a latency of less than 150 ms with respect to completion of the primary saccade. In some trials, the value of image A (primary image) was higher than that of image B (secondary image), while in other trials, the value of B was higher than that of A. As a result, in some trials, subjects expected to view a low-value image, but upon completion of their primary saccade, were presented with the opportunity to gaze at a high-value image. This resulted in conditions in which during the milliseconds before the onset of the secondary saccade (as A was replaced by B), there was a +RPE (B>A) or a −RPE (B<A) event.

To produce a +RPE event, the trial began with a noise image (figure 6.17A, left column). As subjects initiated their primary saccade, we probabilistically erased the noise image and replaced it with a face image at a new location (NF trials, first column figure 6.17A). Similarly, in order to produce a −RPE event, the trial began with presentation of a face image, which, after saccade onset, was probabilistically replaced with a noise image

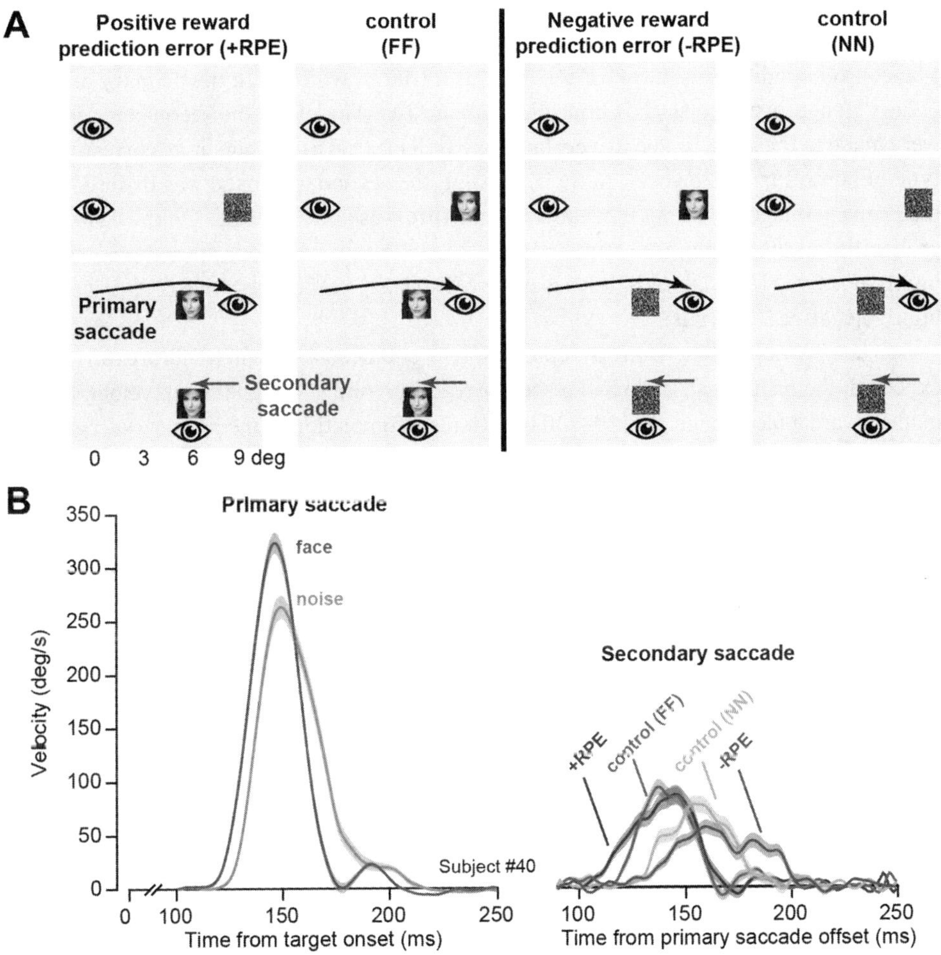

Figure 6.17
A reward prediction error event before onset of a saccade modulates vigor of the ensuing saccade. **A**. Each trial began with a fixation dot near the center. Following a random fixation interval, we presented a primary image at 9° to the right along the horizontal axis. As the primary saccade took place, we erased the primary image and replaced it with a secondary image. In +RPE trials, a noise primary image was replaced with a face secondary image. A face-face trial served as control for the +RPE trial. In −RPE trials, a face primary image was replaced with a noise secondary image. A noise-noise trial served as control for the −RPE trial. **B**. Saccade velocity for a representative subject. Primary saccade exhibited a shorter reaction-time and a higher velocity in response to a face image. Data for the secondary saccade are plotted with respect to termination of the primary saccade in the same trial. The secondary saccades exhibited shortest reaction-times in +RPE trials, and longest reaction-times in −RPE trials. (From Sedaghat-Nejad et al., 2019.)

(FN trials, third column figure 6.17A). The control condition for the +RPE event was a trial in which both the primary and secondary images were faces (FF control, figure 6.17A). The control condition for the −RPE event was a trial in which both the primary and secondary images were noise (NN control, figure 6.17A). Therefore, the secondary saccades were made to the same image in a certain type of RPE trial as the one in its corresponding type of control trial. However, in the RPE trial, the secondary image was different from the primary image, resulting in what we conjecture was an RPE event. On the basis of the probability of the events, we estimated that the trials produced four magnitudes of RPE: highly positive (NF trials), slightly positive (FF trials), slightly negative (NN trials), and highly negative (FN trials).

Saccade velocity data from a representative subject is shown in figure 6.17B. As expected, the primary saccade had a shorter reaction time and higher peak velocity when made toward a face image. At 100–130 ms after the completion of the primary saccade, the subject generated a secondary saccade. We measured the reaction time of the secondary saccade as latency with respect to the end of the primary saccade. The reaction time and peak velocity of the secondary saccade were affected by not just the image at the destination (i.e., the secondary image), but more importantly, by the sign of the RPE event. The reaction time of the secondary saccade appeared shortest after the +RPE event and longest after the −RPE event. Indeed, the properties of the secondary saccade appeared to follow a striking pattern: shortest reaction time and highest velocity for the most positive RPE event (NF), longest reaction time and lowest velocity for the most negative RPE event (NF), and in between for the mildly positive (FF) and mildly negative (NN) RPE events.

These results suggest that in humans, the control of saccade vigor is influenced by the RPE event that precedes that movement. If we view RPE events as modulators of dopamine release, then this would imply that in humans, saccade vigor is modulated by release of dopamine in the moments before the onset of the movement.

Discharge of dopamine cells in response to stimulus presentation is largely a reflection of its reward value, with relatively weak encoding of effort.

The effort required to obtain a reward devalues that reward. If dopamine release precisely signals utility prediction error (and not reward prediction error alone), then increasing effort requirements should reduce dopamine release. Surprisingly, this prediction is currently not supported by the available data.

Benjamin Pasquereau and Robert Turner (2013) trained monkeys to reach to a target under varying reward and effort conditions (figure 6.18A). At the start of the trial, the cue indicated whether the reach would be rewarded with 1, 2, or 3 drops of juice, and whether the reach had to be made against 0 N, 1.8 N, or 3.2 N of force. Movement vigor was clearly affected by reward and effort contents of the trial: as reward size increased, reach reaction time decreased (figure 6.18B), and acceleration and velocity of the reach increased.

However, contrary to expectation, reach reaction time tended to be lowest under the high effort conditions (figure 6.18B). (Note that in this experiment, we cannot compare vigor during reaching under conditions of varying effort levels because the changes in force requirements of the reach necessarily alter movement speed. As a result, it becomes hard to determine whether changes in velocity are due to changes in utility of the reach or simply to the greater resistance offered by the manipulandum. However, the fact that reaction time was reduced in the high-force condition suggests that for these monkeys, vigor was higher in the high-effort condition.)

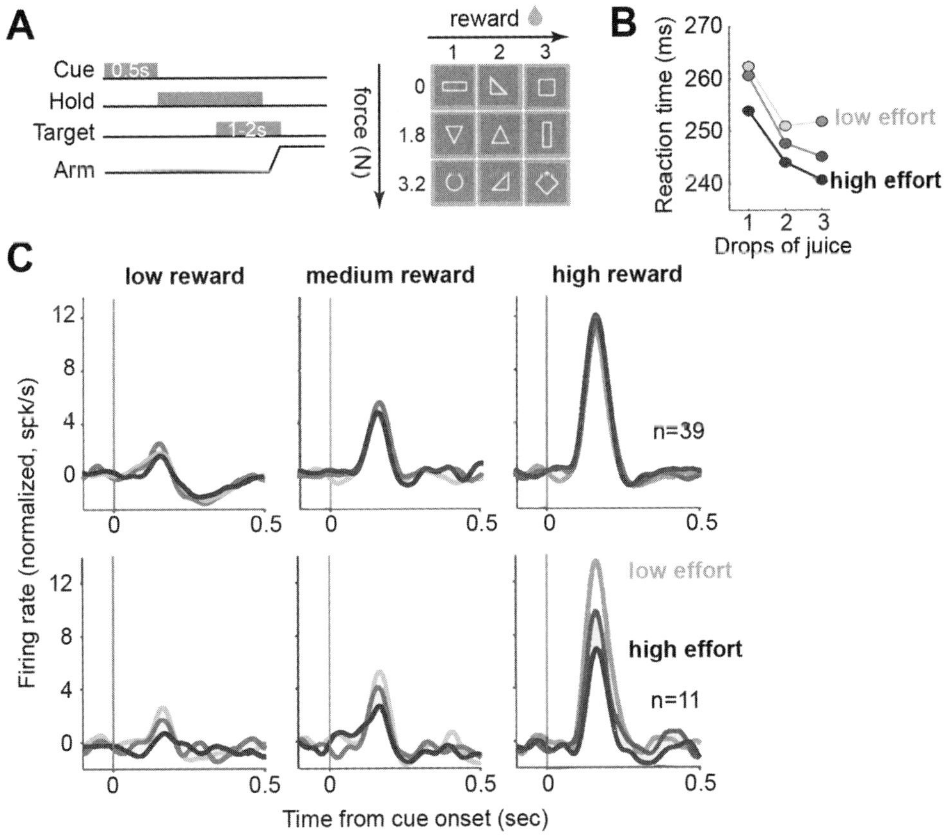

Figure 6.18
Cue-related SNc dopamine activity is mostly driven by reward value, not effort. (SNc, substantia nigra reticulate.) **A**. The task required the monkey to make a reaching movement against varying amounts of force in order to receive a reward. The cue at trial onset indicated the amounts of reward and effort. **B**. Reach reaction time for one of the monkeys. **C**. Dopamine response with respect to cue onset. In many task-related neurons (n=39), the response to cue was associated with reward value, not effort (top row). In a smaller number of neuron (n=11), response was modulated by both reward and effort. (From Pasquereau and Turner, 2013.)

The authored recorded from SNc contralateral to the behaving arm and found 39 cells that, at cue onset, responded with an encoding for reward. These cells produced a burst that varied positively with the reward value; however, these cells were largely unaffected by the varying effort. The authors also found 12 cells that, at cue onset, encoded both reward and effort. These cells produced a burst with a magnitude that increased with greater reward and decreased with greater effort. Therefore, in response to the cue, most SNc dopamine cells appeared to encode the reward value of the cue, but a small fraction of them encoded both reward and effort.

As the monkey waited, the go cue appeared (disappearance of the hold stimulus), instructing the animal to start its reach. This led to another burst of dopamine, which exhibited a latency of around 170 ms. Importantly, this movement-related response of the dopamine cells as a population was not modulated by reward or effort properties of the ongoing movement (Pasquereau and Turner, 2015; Pasquereau et al., 2019).

As the monkeys worked during the recording session, they consumed the reward and eventually became satiated, resulting in loss of motivation. This was evident in their behavior via changes in reaction time: as the trials increased, reaction time and movement duration also increased. Interestingly, Pasquereau and Turner (2013) noted that as the trials wore on, there was a reduction in discharge of some dopamine cells. In 11 cells, the mean firing rate in response to the cue declined across the recording session. (Mixed among the cells were those that encoded for both reward and effort and those that encoded only reward.).

Together, these results suggested that the cue response of SNc dopamine cells was largely a reflection of reward prediction error, with relatively weak encoding of effort. There was further dopamine release before reach onset, but in this data set, the magnitude of that release did not correspond to the reward or effort value of the task.

Dopamine neuron firing rates may not be sensitive to effort.

The results shown in figure 6.18 cast some doubt on the contention that dopamine is a factor in modulating vigor. However, other results in primates and rodents do appear to make this link. Chiara Varazzani, Sebastien Bouret and colleagues (Varazzani et al., 2015) presented monkeys with a cue that indicated how hard they had to grip a force transducer (three levels of force), and how much reward was at stake (one, two, or four drops of juice) (figure 6.19A). After the presentation of the go signal, the animals squeezed a bar until the force reached threshold, at which point the fixation point turned blue. They then maintained this force for 600 ms to obtain the promised reward (figure 6.19B). Whereas the amount of reward varied by fourfold (4 drops of juice versus 1 drop), effort levels varied by a smaller amount: effort level 3 required roughly 70% more force than effort level 1 (figure 6.19B).

The authors recorded from SNc and observed that in response to the cue, dopamine discharge was strongly modulated by reward; the dopamine neurons fired more when more reward was indicated (figure 6.19C), a finding that confirmed the results of

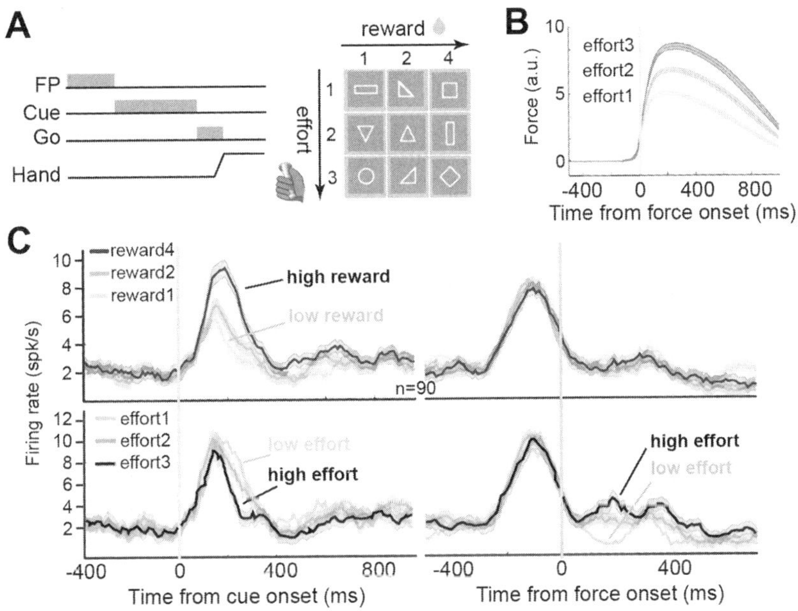

Figure 6.19
Activity of a population of SNc dopamine neurons in response to reward and effort. (SNc, substantia nigra reticulate) **A**. The cue provided information about reward (juice drops) and effort requirements (squeeze force) in each trial. **B**. Average force produced during each trial. **C**. Average responses of dopamine neurons to cue onset and force onset. (From Varazzani et al., 2015.)

Pasquereau and Turner (2013). After cue onset, dopamine neurons were only weakly modulated by effort: the neurons as a population fired at a higher rate for low-effort cues than for high-effort ones. Therefore, dopamine appeared to loosely reflect the utility of the stimulus, though the encoding of effort was considerably weaker than the encoding of the reward.

Just before the monkey began force production, there was another burst in the dopamine neurons. The magnitude of this dopamine burst was unrelated to effort or reward, another finding that confirmed the results of Pasquereau and Turner (2013). As the hand started producing the required force, there was additional modulation of dopamine activity; discharge was somewhat higher when the animal squeezed with greater force (bottom right subfigure, figure 6.19).

The authors also recorded from locus coeruleus (LC) as the animals performed this task. The LC contains neurons that release noradrenaline (NA). NA release is required for the actual production of force, because the body needs to recruit muscles and engage the autonomic nervous system. In response to the cue onset, LC neurons showed reward modulation (greater activity for greater reward), but not effort modulation. However, just before the onset of the force production, there was another burst of activity in the NA neurons, and the magnitude of the burst was associated with effort: more discharge preceded production of greater effort. Furthermore, during force production, the discharge of NA neurons was much higher for production of greater force.

The release of NA before onset of force production appeared consistent with the need to engage the cardiovascular and skeletal system to support production of greater force. In contrast, release of dopamine during the force production also appeared consistent with the need to assist "drive", making it easier for D1-receptor neurons in the caudate to transition to their up-state.

In summary, in this study, dopamine discharge in response to the cue reflected reward value robustly, but effort requirement poorly. Before force production began, there was another dopamine burst, but its magnitude did not vary with reward or effort. However, during force production, dopamine release was weakly modulated by the magnitude of force: a slightly greater amount of dopamine was released during the production of greater force. The results demonstrated that in the post-cue period, dopamine discharge appeared to reflect reward-prediction error associated with the stimulus, whereas during force production, it appeared to weakly support the need to generate greater drive.

Dopamine levels in the ventral striatum are higher during production of effort.
At this point we have arrived at a puzzle: there is conflicting evidence that after cue onset or before movement onset, the spiking activity of dopamine neurons reflects the required effort of the task (compare figure 6.16 with 6.18 and 6.19). Furthermore, there is no evidence that tonic spiking activity of dopamine neurons reflects the history of the animal's reward rate, something that would be critical for history-dependent modulation of vigor (as in marginal value theorem). Instead, what we have seen is evidence for encoding of reward-prediction error in the discharge of the dopamine neurons, a variable that is essential for learning the reward value of the stimulus. But what is missing is the mechanism with which modulation of past reward rate, or expectation of effort, affects vigor.

To examine this puzzle, we need to step away from work that has recorded spiking activity of dopamine neurons and instead consider work that has measured dopamine levels in the downstream regions, such as the striatum. Remarkably, our expectation that discharge correlates with concentration will turn out to be wrong (Mohebi et al., 2019).

Emilie Syed, Mark Walton, and colleagues (Syed et al., 2016) trained rats to perform a nose poke for 0.5 s and then presented the rats one of three auditory cues. The cues instructed the rats to go left, to go right, or simply stay still and wait (figure 6.20A). If the instruction was to go left or right, the animal went to the appropriate lever and pressed it two times, and then went to the food port to receive its reward. The stay instruction required the animal to stay in the nose-poke region for 1.7–1.9 seconds during a sound, and then upon termination of the sound, go to the food port and get reward. In all cases, the amount of reward was the same. However, in two out of the three scenarios, the animal had to work for reward (press lever), whereas in the other scenario, it had to wait.

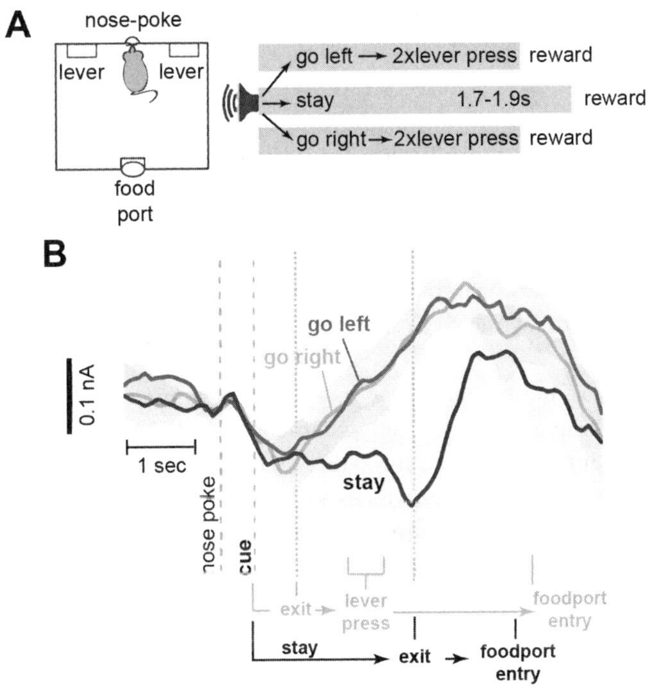

Figure 6.20
Dopamine concentration in the ventral striatum is modulated by the requirement to work for reward, rather than just wait for reward. **A**. Experimental setup. An auditory cue instructed the animals to go left and press the bar, go right and press the bar, or simply stay in the nose-poke location. The gray region indicates the period for which the auditory cue stayed on. **B**. Average concentration of dopamine (n=7 rats) in the nucleus accumbens as estimated via voltammetry, aligned to entry into the nose-poke location. (From Syed et al., 2016.)

Thus, if dopamine is related to movement invigoration, then perhaps there will be greater amounts of dopamine available during the period of working for the reward than during the period of simply waiting for the same reward.

The authors recorded dopamine levels in the nucleus accumbens by using cyclic voltammetry, a technique that has a temporal resolution of about 10 Hz. They observed that if the cue indicated left or right, during the period of lever press there was an increase in dopamine levels, which continued until the rat entered the food port to receive its reward (figure 6.20B). In contrast, if the subject had to wait for the same reward, there was no change in dopamine levels during the period of waiting. Rather, dopamine levels showed an increase only after the wait period ended and the animal began walking to the food port.

To test whether the magnitude of dopamine concentrations was related to the expected reward amount, the authors trained another group of animals with four cues: large and small reward for a go cue, and large and small reward for a stay cue. They found that reward magnitude affected vigor: the latency of leaving the nose-poke region was shorter for large rewards as compared to small rewards, and the same pattern held true in the case of stay trials. During the period of lever press, dopamine levels were much higher for high-reward cues than they were for low-reward ones. Similarly, during the stay trials, when the animal was simply waiting, dopamine concentrations were higher for the high-reward cues than for the low-reward ones. Regardless of whether the instruction was go or stay, the amount of reward associated with the large label and the amount associated with small label remained the same. Nevertheless, across instruction sets, when the reward amount was the same, dopamine levels were higher for the go cue than for the stay cue. (For high-reward cues, the dopamine levels were greater when the rats were instructed to go than when they were instructed to stay. The same pattern held true for the low-reward situations.)

Another example of movement-related increase in dopamine response is in an experiment that was performed by Claire Stelly, Matthew Wanat, and colleagues (Stelly et al., 2019). The authors initially trained rats so that they pressed a lever to obtain food. Next, the researchers placed the rats in a different chamber where regardless of what they did, they received a mild foot-shock. In this chamber, the trial began with a 30 s period of a tone, after which there was a 60 s period of tone and foot shocks and then a final 90 s period without foot shocks or tone. Thus, lever pressing had no consequences: the shock would be given regardless.

Some rats pressed the lever during the cue and shock periods anyway and were classified as high-pressing, whereas other rats pressed the lever much less. In this inescapable foot-shock session, when the cue came on, there was a large decrease in the dopamine levels recorded from the ventral striatum. Notably, the rats that pressed the lever had higher levels of dopamine. Thus, the presentation of a cue that predicted an aversive condition resulted in reduced dopamine levels, but the animals that were more active during that period experienced a smaller reduction.

Thus, unlike dopamine firing rates, dopamine levels appeared sensitive to the effort that the animal was expending. Dopamine levels in the ventral striatum were higher when the animal pressed a lever to acquire a reward then when the animal simply waited to acquire the same reward. Dopamine levels were higher when the animal expected a greater reward, but the levels were much higher if the animal had to perform movements (i.e., work, for that reward). Finally, dopamine levels were lower in the ventral striatum if a cue predicted a foot shock, but this reduction was smaller if the animal was actively pressing a lever, even though expending this effort had no consequence on the aversive stimulus.

Dopamine levels in the striatum correlate positively with reward rate.

How can it be that dopamine levels in the striatum are elevated during effort production, whereas dopamine spiking activity shows little or no encoding of effort? The lack of an association between dopamine levels in the striatum and dopamine spiking activity in the projecting neurons was illustrated by Ali Mohebi, Jeffrey Pettibone, Joshua Berke, and colleagues (2019).

The authors trained rats to produce a nose poke in response to a light. A sound acted as go-cue to poke one of two adjacent ports, although the port was not identified. In rewarded trials, this side-in poke was followed by food release at a nearby food port. Left and right choices were rewarded with independent probabilities, which changed without warning. So the rats chose the left or right port on the basis of their history of reward.

When rats were more likely to receive rewards (because of past high reward rate associated with that port), they moved more vigorously, as reflected in shorter latency between light onset and nose poke. So past rewards affected stimulus value, and this, in turn, correlated with vigor.

The authors measured dopamine levels in the NAc at a course time resolution and found that this variable positively correlated with the reward rate. They next recorded from 27 individual dopamine neurons in the ventral tegmental area (VTA). As expected, dopamine spiking activity exhibited strong modulation with reward prediction error; many spikes were produced when food was delivered at random time. However, during the task, tonic dopamine cell firing was indifferent to reward rate, whereas the phasic response encoded RPE. Thus, spiking activity of these VTA dopamine neurons was not entirely responsible for dopamine levels in the NAc.

Next they used dLight, a genetically encoded optical dopamine indicator of D1 receptor activity, to measure dopamine levels in NAc at a finer time resolution. Binding of dopamine caused an increase in the fluorescence of this receptor. The dLight signal showed subsecond fluctuations within each trial, and within these fluctuations there was a clear positive relationship between dopamine levels and reward rate: as reward rate increased, so did dopamine levels. Unlike spiking activity, dLight signals early in each trial were greater when recent trials had been rewarded, consistent with encoding of value.

Thus, in this study, VTA dopamine neurons did not encode stimulus value, and their firing rates did not reflect reward rate. Nevertheless, dopamine concentrations in the NAc varied with reward rate, and when the rats moved more vigorously, dopamine levels tended to be higher.

One possibility is that something other than dopamine cell spiking activity controls dopamine availability in the NAc. For example, serotonin, which we will consider in the next chapter, has the capacity to modulate synaptic dopamine levels. Another possibility is that there is diversity among dopamine cells, and this particular sample did not include dopamine neurons that showed spiking activity that related to average reward rate. With the work of Ben Engelhard, Nathaniel Daw, David Tank, and Ilana Witten (Engelhard et al., 2019), we are a step closer to finding the answer to this question.

Engelhard and colleagues trained mice to walk on a ball and view a screen that displayed discrete visual stimuli to the right and left. As they walked, the mice would need to keep track of the number of stimuli that appeared on the right and left, and at the end of the maze, turn toward the side that had the greater number. They recorded from 303 dopamine neurons in VTA in 20 mice using two-photon imaging of a calcium indicator, thus providing unprecedented access to a very large data set of dopamine neurons during a decision-making task that involved movement.

They found that as the mice walked, variables such as speed, acceleration, and spatial position affected neuronal activity, as did variables such as stimulus appearance. The authors considered a model in which the each kinematic quantity was represented linearly, squared, and cubed during the walk. They then fitted the response of the neurons to the sum of these and other variables such as stimulus appearance and previous reward. Indeed, during the walk period, the highest relative contribution to neuronal activity was from the vigor-related variables (speed and acceleration), accounting for 32% of the variance, followed by position with respect to reward (22%). The history of reward (whether the previous trial was rewarded) also accounted for some of the response variability (18%). In contrast, during the outcome period when the mouse received information regarding the choice that it had made (success or failure), reward outcome (RPE) accounted for about 75% of the response of the dopamine neurons. The kinematic-related dopamine neurons appeared to be clustered anatomically, located more laterally and posteriorly within VTA.

In summary, dopamine levels in the NAc correlate positively with reward rate, and vigor tends to be higher when the animal has experienced a higher reward rate. Dopamine neuronal activity in VTA is not strongly modulated by reward rate. In contrast, dopamine neuronal activity during walking is strongly affected by kinematic variables, such as distance to reward, and vigor variables, such as speed and acceleration.

Dopamine levels in the dorsal striatum show preference for contralateral movements. There are further clues that differentiate dopamine levels from dopamine spiking activity. Whereas dopamine spiking does not appear to encode the direction of the movement needed to acquire reward, dopamine levels in the striatum are sensitive to this variable.

Nathan Parker, Rachel Lee, Nathaniel Daw, Ilana Witten, and colleagues (Parker et al., 2016; Lee et al., 2019) reported that dopamine levels in the dorsal striatum showed a greater increase when the movement that the animal performed was toward the contralateral side than when the movement was toward the ipsilateral side.

In their experiment, mice decided whether to press a lever on the left or right to acquire reward. The trial started with a light at the nose-poke region (figure 6.21). Once the mouse poked its nose, two levers appeared on either side and the mouse had 10 s to press one of the levers. Each lever led to reward with either a high probability (70%) or a low probability (10%), with the probability swapping after a block of variable trial length. After the lever press, there was an additional 0–1 s before an auditory cue indicated whether reward was delivered or not. Because the block signals were not available to the mouse, it learned from the patterns of reward to make its selection, selecting the high-value lever on ~75%

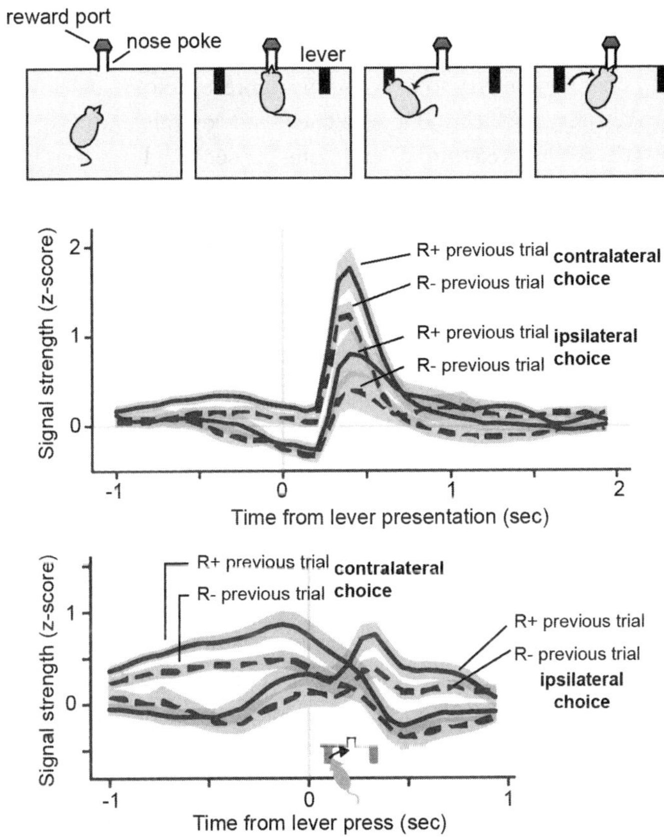

Figure 6.21
Dopamine response in the dorsal striatum encodes both value-related and movement-related information. Mice performed an instrumental reward-reversal task in each trial; they chose one bar to press in order to acquire a reward. The reward probability was high for one lever during one block of trials, and then switched to the other lever without any signals to the animals. Data show calcium-generated fluorescence as recorded from the dorsal striatum. In the trials shown, the animal chose the same side (ipsilateral or contralateral) twice in a row, one following a rewarded outcome (R+), and one following an unrewarded outcome (R-). In response to lever presentation, activity was higher after R+ trials, regardless of the side the choice was on. However, regardless of reward at stake, the response to lever presentation within the trial was higher if the animal chose to press the key on the contralateral side. Dopamine response was also higher as the animal walked to press the contralateral lever, as well as during period of time that the animal turned toward the contralateral side to collect its reward. (From Lee et al., 2019.)

of the trials, and then switching to the other lever at the block transition with a time constant of around five trials. The authors recorded dopamine concentrations in the dorsal striatum by using fiber photometry to measure the fluorescence of a calcium indicator.

They considered trials in which the mouse chose the contralateral lever (i.e., the lever contralateral to the recorded striatum) after having chosen that lever also in the previous trial. In some cases, the previous choice was rewarded (R+ trials), whereas in other cases, the previous choice was not rewarded (R- trials). We can presume that following an R+ trial, the value of the contralateral lever had increased. Indeed, when the levers were presented, in the contralateral choice trials, dopamine response was greater in R+ trials than R- trials (figure 6.21, middle row). A similar pattern was seen in ipsilateral choice trials (i.e., trials in which the mouse chose the ipsilateral lever after also having chosen it in the previous trial). However, in general, the dopamine response was greater in the contralateral trials than in the ipsilateral ones. Therefore, at stimulus presentation, the dopamine response carried a directional signal that indicated a greater response to reward for stimuli that required movements toward the contralateral side. Furthermore, in the ipsilateral choice trials, activity was greater in R+ trials than in the R- trials. This indicated that in addition to encoding a movement-related signal, dopamine response also encoded the value of the trial. Notably, this value signal appeared to encode the reward expected at the chosen target, regardless of whether the target was to the left or right.

As the animals approached each lever, dopamine level was greater when the movement was toward the contralateral side (figure 6.21, bottom row) and greater if the value of that lever press was larger (R+ trials as compared to R- trials). This again illustrates the movement-related response of dopamine, as well as the fact that reward value modulates that response. Although the authors did not record the motion of the animals, we would predict that the movements that coincided with greater dopamine release were performed with greater vigor. That is, movement velocity was likely greater after R+ trials than after R- trials.

After the animal pressed the lever and turned back toward the reward port, dopamine levels again exhibited a movement-related pattern: it increased as the animal turned toward the contralateral side, and decreased as it turned toward the ipsilateral side.

These results suggest that at stimulus onset, dopamine level in the dorsal striatum is a reflection of both the perceived reward at stake and the direction of movement that needs to be performed to acquire that reward. As the movement takes place, dopamine levels are greater when the movement is directed toward the contralateral side and greater if the expected reward is larger. These movement-related responses of dopamine are present in the dorsal striatum (figure 6.21) but not in the ventral striatum (figure 6.20).

Dopamine release in the motor cortex makes it easier to produce muscle contractions. Finally, let us briefly consider the effect of dopamine release on other parts of the motor system, in particularly the motor cortex. Nobuo Kunori, Riichi Kajiwara, and Ichiro Takashima (2014) examined the functional consequences of dopamine release in the motor cortex. They anesthetized rats and provided a single pulse of stimulation (less than 1 ms in duration) to the VTA, which houses dopamine neurons that project to many areas, including the motor cortex. They then used a voltage-sensitive dye to record activity of the motor cortex and found that after a single pulse of VTA stimulation, there was a brief increase in the activity of M1 neurons (bilaterally) that lasted around 50 ms, followed by a sustained period of decreased activity that lasted around 400 ms (figure 6.22A).

To examine whether this brief excitation had a functional significance, the researchers paired VTA stimulation with motor cortex stimulation, and then they recorded from contralateral arm muscles (figure 6.22B). They reduced motor cortex stimulation intensity until by itself it produced little or no electromyographic (EMG) response. They then provided a single pulse of VTA stimulation, followed by motor cortex stimulation at 10 ms latency. Whereas VTA stimulation or motor cortex stimulation alone produced no EMG response, the pairing of the two produced a robust EMG response. Therefore, it appeared that dopamine release just before motor cortex stimulation helped increase the excitability of the motor cortex neurons, allowing them to reach threshold and thus resulting in EMG activity that would be absent without dopamine release. These results suggest that a consequence of dopamine release in the moments before the onset of a movement is to increase the response of the motor structures to their excitatory input, resulting in greater muscle activation.

In summary, there is a strong causal relationship between dopamine release and control of vigor: artificial activation of dopamine neurons invigorates the ensuing movement, and a subject who does not respond with high vigor to large rewards also lacks reward-dependent sensitivity in their release of dopamine. In addition, the effect of reward on vigor is partly mediated by the effect that dopamine has on the D1 receptor neurons in the

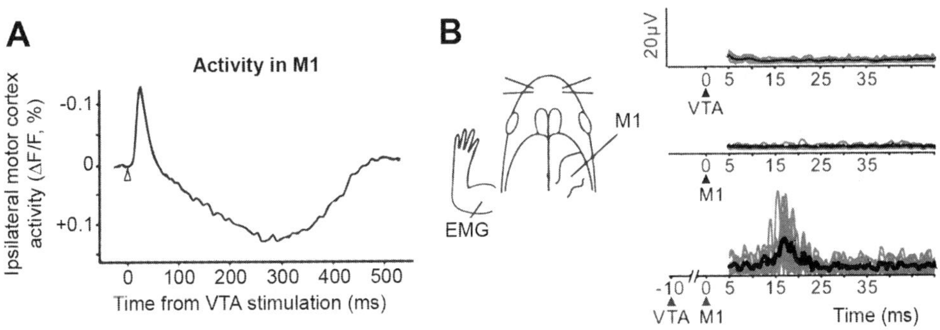

Figure 6.22
Effect of stimulation of ventral tegmental area (VTA) on excitability of the motor cortex. **A**. Voltage-sensitive dye imaging of the contralateral motor cortex following a brief pulse (<1 ms) of excitation to the ventral tegmental area. **B**. Effect of pairing of stimulation of VTA with stimulation of primary motor cortex (M1) on recruitment of arm muscles. VTA stimulation or M1 stimulation by itself did not produce an electromyographic (EMG) response. However, VTA stimulation at 10 ms before M1 stimulation resulted in robust EMG activity. (From Kunori et al., 2014.)

caudate (i.e., cells that are part of the direct pathway to the SNr). However, there is an important puzzle that remains poorly understood: there is a discrepancy between dopamine neuron spiking activity and dopamine levels in the striatum. Dopamine spiking activity encodes reward prediction error, but not the past reward rate or the current stimulus value. In contrast, dopamine levels in the striatum vary with past reward rate, are higher during periods of movement, and reflect the required effort. Dopamine spiking appears indifferent to direction of movement needed to acquire reward, whereas dopamine levels in the dorsal (but not ventral) striatum are greater when movements are toward the contralateral side.

The difference between spiking activity of dopamine neurons and concentration levels of dopamine in the downstream pathways remains an important puzzle that, at this writing, is poorly understood.

6.5 Dopamine and the Willingness to Exert Effort

Does presence of dopamine in the striatum affect only the vigor of the ensuing movement, or does it also affect patterns of decision-making? Some answers have been provided by John Salamone and his students, who have researched this topic during the past three decades. Salamone and his students have found that when decisions involve a cost-benefit analysis of the effort and reward, decision-making depends on dopamine's actions on the neurons of the ventral striatum (Salamone et al., 1991).

Dopamine antagonists in the ventral striatum reduce the willingness to expend effort for a reward.

Salamone's experiments have relied on the fact that rats, like all animals, prefer certain foods and are willing to work in order to acquire their preferred foods. When dopamine antagonists are injected into the nucleus accumbens, rats still prefer the better food, but are not willing to work as hard for it.

In a typical experiment (Aberman and Salamone, 1999), a rat was placed in an experiment box and trained to press a lever one time to receive a pellet of high quality food. After a few days of training, the rat was trained for a few weeks to press this lever four times to receive the same pellet. This was called the "fixed ratio 4" (FR4) schedule. In each session, the rat was provided with 30 minutes in the experiment box and needed to press the lever to acquire food. After a couple of weeks of FR4 training, the animals were trained in FR16, and then FR64. At this point, the animals were divided into four groups (FR1, FR4, FR16, and FR64) and then tested in a decision-making task. In this task, the FR1 group was provided with an experiment box that included the usual lever and the opportunity to acquire high quality food pellets, and also a bowl of low quality lab chow, with no requirement to work for the chow (figure 6.23). In the FR4 group, the option was between the lab chow and four presses of the lever to receive one pellet. As a result, in each session, the animal chose

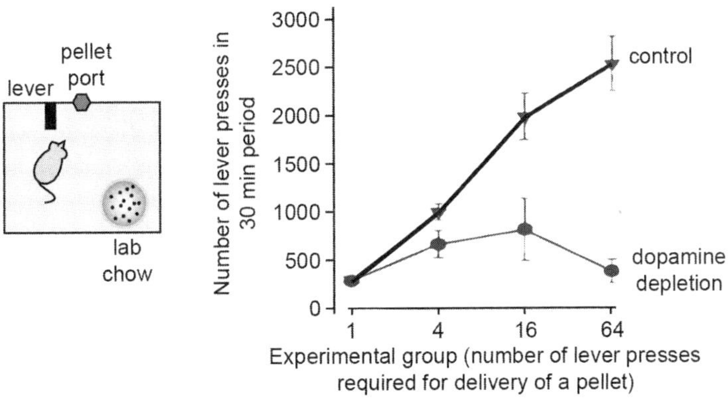

Figure 6.23
Effect of dopamine depletion in the nucleus accumbens on effort-based decision-making. Rats were trained to press a lever to receive a high-quality food (pellets) or consume low-quality lab chow without any work requirement. Four experimental groups were trained to press the lever 1, 4, 16, or 64 times to earn a single pellet. Each experimental session lasted 30 minutes. The triangles show performance of animals in each group in the baseline trials. Circles show performance of the same animals after injection of a drug into the nucleus accumbens that depleted the available dopamine. (From Aberman and Salamone, 1999.)

between working for the preferred food and freely consuming the less preferred food. The measured variable was the number of times the lever was pressed.

The authors found that in the FR1 group, the rats pressed the lever about 300 times, whereas in the FR64 group, they pressed the lever around 2500 times. The rats were willing to work hard to acquire the preferred food.

At this point a drug was injected into the nucleus accumbens to deplete the available dopamine, and the animals were again tested in the experiment box. Dopamine depletion did not affect the FR1 group; this group continued to press the lever roughly the same number of times as before. There was a small decrease in number of lever presses in the FR4 group, but profound decreases in the FR16 and FR64 groups. In fact, whereas before dopamine depletion the FR64 group had pressed the lever 2500 times, after dopamine depletion, this number fell to around 300.

Although the authors did not measure movements of the animals, they did note changes in behavior, including slower response initiation and a greater tendency to pause and rest.

These results illustrated that before dopamine depletion in the striatum, the animals were willing to deposit a large effort payment in order to acquire their preferred food. After dopamine depletion, the groups that had to pay a small cost (FR1 and FR4) were not affected. They continued to choose the more effortful option at similar rates as before. This suggested that dopamine depletion did not devalue the preferred food. However, the groups that had to pay a large cost (FR16 and FR64) were severely affected by dopamine depletion. These groups were no longer willing to exert a large effort to acquire the more valuable food. That is, dopamine depletion appeared to alter the reward-effort balance.

Later experiments confirmed that the effect of dopamine in the nucleus accumbens was a shift in preference of the rats toward low effort options (Farrar et al., 2010;Yohn et al., 2015). For example, Michael Koch and colleagues (Koch et al., 2000) trained rats on the FR5 schedule without lab chow and then presented them with an experiment box that had both lab chow and the lever. After the injection of a D1 dopamine antagonist in nucleus accumbens, the rats reduced their rate of lever pressing and increased their intake of lab chow. A similar result was observed after the injection of a D2 antagonist in nucleus accumbens. In a control experiment in which the animals had free access to both pellets and chow, the animals continued to prefer the pellets over the chow, both before and after D1 or D2 antagonist infusion.

Therefore, blocking dopamine reception in the ventral striatum affected the animals' decision-making process by reducing the animals' willingness to expend effort for a reward.

Dopamine concentration in the ventral striatum decreases at cue onset as the effort required to acquire the reward increases.

Results of Salamone and colleagues imply that if one could manipulate availability of dopamine in the striatum, one could alter behavior, making a subject decide to work hard in order to acquire reward. Results of Scott Schelp, Erik Oleson, and colleagues (Schelp et al., 2017) illustrate both the promise and the pitfalls of this conjecture.

The authors trained food-restricted rats to respond to a light (cue) and press a lever in order to receive a sucrose solution (figure 6.24A). The effort required to acquire sucrose was varied with time, either increasing or decreasing in price (the term price refers to the number of lever presses that were necessary to earn 1 mg of sucrose).

In the task, the animals were rewarded with a bolus of 45 mg of sucrose, but the price changed as the experiment progressed. In the increasing price schedule, the price was initially set to one lever press for 45mg of sucrose (price of 0.022 lever presses/mg of sucrose). Once the cue light came on, the lever was presented and the animal had 30 seconds to press it. If it failed to do so, the lever retracted and there was a 30 s timeout period. The price increased every 10 min, gradually rising from 0.022 to 6.9 lever presses/mg (that is, from 1 lever press to 310 lever presses to acquire the bolus of reward). In contrast, in the decreasing price schedule, the sucrose price was initially quite high (6.9 lever presses/mg, requiring 310 presses to get a single bolus), but then gradually decreased to 0.022 lever presses/mg (1 lever press to acquire one bolus). As the rats performed this task, the authors measured dopamine levels in the nucleus accumbens using fast-scan cyclic voltammetry (roughly a 10 Hz resolution).

The authors reported dopamine levels at two important time periods: when the cue was presented and when the reward was acquired. The cue-period dopamine was a proxy for the predicted subjective value of the stimulus, whereas the reward-period concentration, (i.e., the dopamine that was released as the animal consumed the reward) was a proxy for the difference between the predicted value and the received value (i.e., RPE).

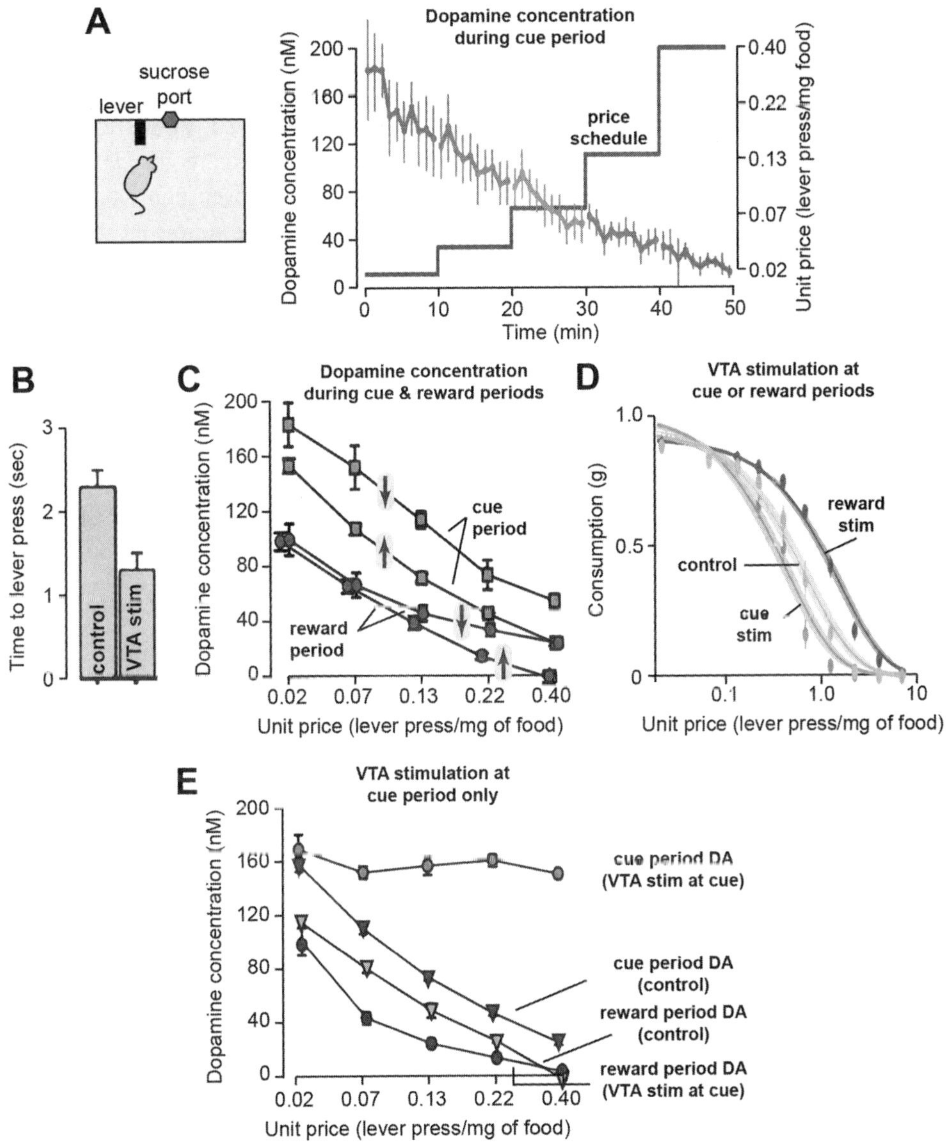

Figure 6.24
Dopamine concentrations in the ventral striatum during effort-based decision-making. **A**. Rats responded to a light (cue) and pressed a lever in order to receive a sucrose solution. Cost was measured as the number of times the rat had to press the lever to get the reward. In this case, cost of sucrose was increased with passage of time; the rat had to work increasingly harder for a reward. Ten costs were examined (five lowest costs are shown). As the cost increased, the dopamine response at cue onset decreased (as measured via voltammetry). **B**. Time from cue onset to lever press decreased when dopamine neurons in VTA were stimulated at cue onset. (VTA, ventral tegmental area) **C**. As cost of reward increased, dopamine concentrations decreased during both the cue and reward periods. The arrows indicate the price schedule of the experiment: decreasing with time (down arrow) or increasing with time. Note that when the price schedule decreased with time, indicating a past history of high cost, cue-response dopamine was higher than after a history of low cost. **D**. Amount of reward that was consumed by the animal as a function of whether the dopamine cells in VTA were stimulated at cue onset or reward onset. Animals were more willing to exert high effort when the stimulation was at reward onset. **E**. Dopamine concentration in the striatum at various time points after VTA stimulation. VTA stimulation was performed only during the cue period. VTA stimulation produced constant and high dopamine concentrations during the cue period. However, during the reward period that followed, dopamine levels fell below those measured during the control trials. (From Schelp et al., 2017.)

Let us consider the schedule in which the price steadily increased (figure 6.24A, and also the up-arrow lines in figure 6.24C). When the price was low, there were substantial amounts of dopamine at cue onset (figure 6.24A) and at reward onset (figure 6.24C). At this low effort level, the subjective value of the reward was high, and so was the dopamine level at cue onset. On average, it took about 2 s for the rats to respond to the cue and press the lever (control, figure 6.24B). Once the reward was delivered, the dopamine level again increased, but was still at substantially lower level than those present at the cue period (figure 6.24C, up-arrow lines). As the effort requirements increased (i.e., price increased), both cue-period and reward-period dopamine levels decreased. Indeed, about halfway into the session, as the price exceeded 0.4 presses/mg, the animals were no longer pressing the bar enough to acquire a significant amount of reward (control curve, figure 6.24D). The effort-dependent (i.e., price sensitive) dopamine levels during cue and reward periods remained consistent when the experiment was repeated during sessions in which the price started high and then steadily decreased (down arrow lines, figure 6.24C).

Regrettably, the authors did not report dopamine levels during the lever-pressing period. In the earlier work that we summarized, researchers found that during periods of greater effort, there were greater levels of dopamine in the striatum (figure 6.19C and figure 6.20B). Those results would predict that during lever presses for high-price rewards, dopamine concentrations were likely higher than lever presses for low-price rewards.

Together, the results suggested that at cue onset, dopamine levels in the accumbens roughly encoded the effort-related price of reward. At cue onset, for the low-price reward, dopamine levels were high, but then they declined as the effort price increased. In contrast, during reward consumption, dopamine levels were always lower, a potential reflection of the difference between the predicted value of the reward and the value of the reward that was actually received. Indeed, if the reward was acquired after a large expenditure of effort, the reward-period dopamine level was equal to or lower than baseline, potentially reflecting a negative RPE. At these high prices, perhaps the reward was not worth the effort, as reflected in the fact that the animal stopped responding.

Dopamine stimulation alters effort-based decision-making.

Schelp and colleagues (Schelp et al., 2017) investigated whether dopamine levels in the accumbens and elsewhere were causally influencing the rats' decision to exert effort for a given reward. The researchers used optogenetics to stimulate the dopamine neurons of the VTA for 0.5 s either during the cue-presentation period or during the reward-consumption period. In this experiment, the price started low and then gradually increased.

Without stimulation, cue-period dopamine levels in the accumbens decreased as the price increased. As the cue was presented, the VTA was stimulated, and as a result, dopamine levels remained high in the ventral striatum regardless of the price of the reward (figure 6.24E, cue-period VTM stimulation). What followed was partly predictable and partly surprising. Cue-period stimulation of the VTA led to reduced time to lever press (figure 6.24B), consistent with the idea that higher levels of dopamine lead to greater movement vigor. However, cue-period stimulation did not increase the rats' willingness to press the lever. In fact, the result was the opposite of what was expected. When great effort was required, the rats reduced the number of times they pressed the lever. However, when the effort requirements were low, there was no change in response (figure 6.24D, cue stimulation curve).

An important clue was that although stimulation at cue onset increased dopamine levels during the cue period, the levels then dipped below normal during the reward period (figure 6.24E, reward period VTA stimulation versus reward period control). This observation suggested that the stimulation during cue period installed higher-than-normal expectations of reward. After the expenditure of effort and the actual acquisition of the reward, the RPE was less positive than before. This post-reward reduction in dopamine levels may have been the critical factor in discouraging the rats from expending effort in future trials. To test for this, the authors stimulated VTA dopamine cells during the reward period; they found that it made the animals more willing to expend effort to acquire high price rewards (figure 6.24D, reward stimulation curve).

In summary, stimulation of VTA dopamine cells during the cue period increased vigor, but in this specific experiment, in which effort requirements increased gradually, cue-period dopamine stimulation in the VTA made the animals less willing to exert effort. In contrast, stimulation during the reward period increased their willingness to accept the high price and exert effort. Therefore, willingness to exert effort may depend on the dopamine response after the expenditure of effort and acquisition of reward, not after the presentation of the stimulus.

After a period of high-effort expenditure, more dopamine is released for low-effort rewards, resulting in higher vigor.

It is noteworthy that in Schelp's (Schelp et al., 2017) experiment, when the price had initially been high and then lowered, dopamine levels remained high for the low prices (figure 6.24C, cue period, down arrow). However, when the price had initially been low and then increased, dopamine levels were low for the same low prices (figure 6.24C, cue period, up arrow). That is, dopamine levels at cue onset to a given effort requirement was history dependent: dopamine levels were higher if the effort price had dropped.

This observation regarding the history-dependent concentration of dopamine in the ventral striatum prompts an interesting prediction: vigor of movements toward a given stimulus should depend on the history of effort that has been expended. Actions should be more vigorous after a period of high-effort expenditure than after one of low-effort expenditure.

Tehrim Yoon (Yoon et al., 2018) in our laboratory performed an experiment that provided experimental evidence in support of this prediction. The experiment was a variant of that shown in figure 2.11A: human subjects viewed an image on a computer monitor while a dot provided them with information regarding position of the next image (figure 6.25A). They were free to look at the image as long as they wished and then make a saccade to the dot to view the next image. The experiment was designed to test a prediction based on the marginal value theorem: the amount of time that will be spent harvesting (i.e., looking at the image) will depend on the subject's history; after a period of high effort, resulting in reduced capture rate, subjects will spend a greater amount of time staying at a particular harvest location. Notably, the theory-based prediction was that after a period of high effort, the reduced capture rate will lower vigor of movements to the next harvest site.

However, in this experiment, marginal value theorem and empirical evidence regarding dopamine led to two different predictions regarding vigor. On the basis of the theory, vigor should decrease after a history of high-effort harvesting. However, the dopamine response to the cue was higher after a history of high costs (figure 6.24C); on the basis of this finding, movements should be more vigorous after a history of high-effort harvesting.

To modulate effort history, in our experiment (Yoon et al., 2018) (figure 6.25B), during a block of 100 trials, all images (except probe images) appeared at high eccentricity in the

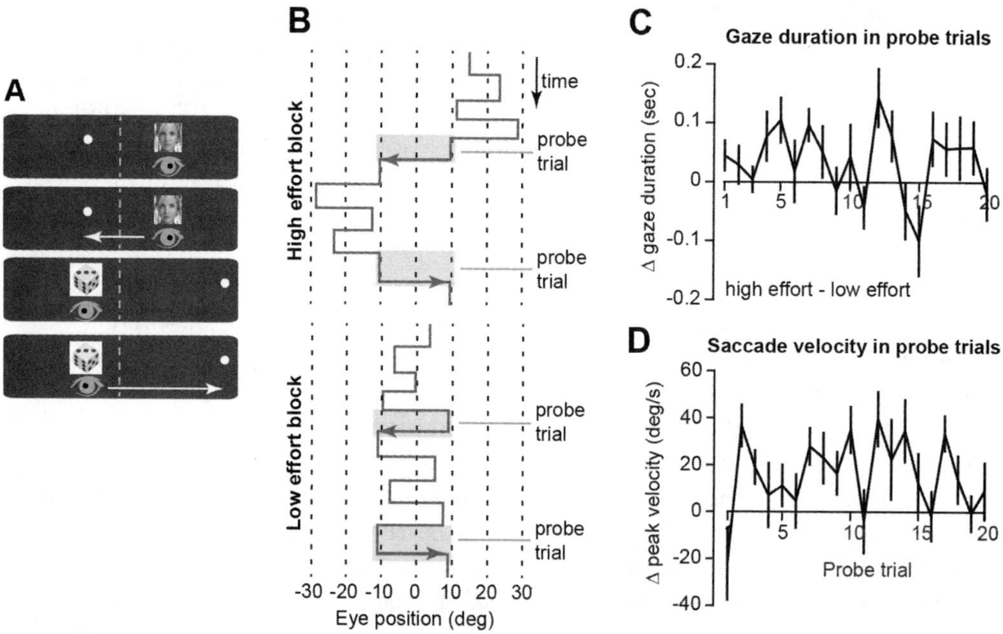

Figure 6.25
After a history of high-effort expenditure, vigor is elevated. **A**. Subjects viewed an image (randomly selected from faces and objects) and maintained gaze for a self-selected period, then made a saccade to another location (dot) to view the next image. Variables of interest were the harvest period (duration of gaze) and saccade vigor. **B**. Effort history was manipulated via image eccentricity, and its effects were quantified in occasional probe trials in which image and dot locations were kept constant. The vertical traces show eye position during gazing, and the horizontal traces show saccades. In the high-effort block, the non-probe images were located in the 10°–30° range. In the low-effort block, the non-probe images were located within the ±10° range. **C**. Within subject change in gaze duration during probe trials in the low- and high-effort blocks. After expenditure of high effort, gaze duration in probe trials increased. **D**. Within subject change saccade peak velocity in the low- and high-effort blocks of probe trials. After expenditure of high effort, vigor in probe trials increased. (From Yoon et al., 2018.)

range of 10°–30° with respect to midline (uniform distribution, mean eccentricity of 20°). In another block, the images appeared at low eccentricity within 10° of the midline (mean eccentricity of 0°). These blocks represented high-effort and low-effort environments, respectively. In each block, 25% of the trials were probe; the image was placed exactly at 10° with respect to the midline, and the dot (indicating position of the next image) also placed exactly at 10° with respect to midline. Images were selected randomly from two categories: faces and inanimate objects. As a result, the probe trials were identical in all blocks and thus allowed us to determine whether past effort expenditure modulated vigor and harvest duration.

We observed unequivocal effects (figure 6.25C): subjects spent greater time gazing at the probe images when those images appeared in the context of the high-effort environment. This confirmed the prediction based on the marginal value theorem regarding effect of past effort on duration of harvest. However, contrary to the theory's prediction, and consistent with predictions of dopamine response, vigor was higher after period of high effort: saccade velocities in the probe trials from the image to the dot had higher velocity in the context of the high-effort environment. Therefore, if past harvests required a high expenditure of effort, subjects behaved in probe trials as if their subjective value of reward had increased: they lengthened their duration of gaze and then moved with greater vigor.

Effort that was expended to acquire reward increases subjective value of that reward. Dopamine release is critical for learning the value of a stimulus: if the stimulus predicts a reward, but upon experience of that reward dopamine response is negative (-RPE, less reward than expected), then the value of that stimulus is decreased. However, we have seen that dopamine levels during expenditure of effort vary positively with the magnitude of effort. For example, in figure 6.20B, dopamine levels are higher when the mouse is pressing levers as compared to when it is standing quietly. Does this increased dopamine levels alter the value of the stimulus that was associated with the movement?

While the answer to this question is not known, it highlights one of the fascinating paradoxes of effort: working hard to acquire an outcome appears to make the subjective value of that outcome higher (Inzlicht et al., 2018). Psychologists call this the *IKEA effect*, referring to the observation that working to put together a piece of furniture increases the subjective value of that furniture.

An example of this effect is from the work of Alex Kacelnik and Barnaby Marsh (Kacelnik and Marsh, 2002) who trained starlings to fly between perches placed at far ends of a cage either 4 or 16 times (labeled as easy and hard trials). To indicate the number of times that travel was required, there was a light on each perch that was lit when the animal had to fly to it. At the conclusion of the easy trial, an orange light was lit on a food dispenser, and at the conclusion of the hard trial, a white light was lit on the same dispenser. In both cases, the animal needed to peck the light to acquire identical food.

The easy and hard trials were intermixed in blocks of 5 trials, after which the animal was given a choice without any requirement for flights: both the orange and white lights were lit and upon pecking either key the animal received food. The authors observed that in 10 out of 12 animals, they preferred the light that was rewarded after expenditure of high effort.

From the results presented in figure 6.24C we can conjecture that in the birds, as they flew in the hard trials and expended greater effort, they experienced a greater total dopamine release than in the easy trials. Perhaps this played a role in their preference toward the light that was associated with high effort trials. The results highlight the fact that whereas in decision-making we view expenditure of effort as a cost, production of effort coincides with increased dopamine levels in the striatum, which in turn affects valuation of the effortful outcome. Thus, we value more the object that we have worked for, as compared to a similar object that we have been given.

Limitations

The discrepancy between discharge of dopamine neurons that project to the striatum and the levels of dopamine in the striatum remains an intriguing puzzle. We saw that whereas some studies that relied on dopamine single cell neurophysiology had not found an encoding of effort (figure 6.18 and figure 6.19), other studies that had used imaging of the striatum had observed this effort coding (figure 6.24). Transient spiking activity of dopamine neurons appeared to reflect reward-prediction error, but was poorly related to effort requirements of the task, and poorly encoded movement direction. In addition, the tonic spiking activity of dopamine neurons appeared to poorly encode the average reward rate of the animal. In contrast, striatal dopamine levels rose during the exertion of effort, reflected the history of effort expenditure, were related to the rate of reward experienced by the animal, and were higher when the movement was toward the contralateral side. Thus, it would appear that in addition to the spiking activity of dopamine neurons, another factor may be critical in regulating levels of dopamine in the striatum. That factor remains elusive.

In order to produce an effortful movement, the brain must not only engage muscles via the central nervous system, but also recruit the cardiovascular system via the autonomic nervous system. We briefly mentioned locus coruleus and release of noradrenaline, noting that in response to the stimulus, noradrenaline release did not vary with reward and only weakly varied with effort (Varazzani et al., 2015). However, just before the act of force production, there was another burst of noradrenaline, and this release was greater for the more effortful act. It seems likely that dopamine and noradrenaline share some common mechanism of activation, as both levels are greater when one is exerting effort. The coordination between noradrenaline release from the locus coruleus and dopamine levels in the striatum remains to be understood.

Vigor is expressed not only when we are working to acquire reward, but also when we are working to avoid punishment. For example, a prey will run fast to avoid being eaten, while at the same time the predator will run fast to acquire the prey. In each case, there is an expression of high vigor, but for different reasons: one to avoid loss, the other to acquire gain. If dopamine levels in the striatum are associated with vigor, then we need dopamine in both cases. This might imply that dopamine response should not be positive for only the stimuli that are better than expected, but in some cases also positive for things that are worse than expected. Perhaps this is why in figure 6.15, one finds that a fraction of dopamine neurons produced a high response to stimuli that predicted a punishment. (In that case, rapid blinking of the eye avoided the punishment.) This speculation regarding importance of these negative-value preferring dopamine neurons and the vigor expressed in avoiding punishment remains to be explored.

We began this chapter with the observation that Parkinsonian patients exhibit bradykinesia, but their slowed movements spontaneously improve as the urgency of the task increases. Perhaps urgency increases dopamine levels in the striatum, possibly because of dopamine neurons that prefer negative-value stimuli (figure 6.15). However, this conjecture remains a speculation.

Summary

SNr cells inhibit the superior colliculus. While this input cannot specify which movement to make, it can bias the selection and vigor of movements. During periods of stillness, SNr cells are highly active at levels that depend on the urgency of making a movement. These cells pause their spiking activity to encourage movement when the stimulus is valuable, but they burst to discourage movement if the action will be too costly. The depth of this pause biases where the saccade is directed and the vigor with which it is made. The reward value of the movement is conveyed to the SNr from the striatum via the direct pathway, while the cost (or risk) is conveyed via the indirect pathway. Striatal cells receive their information about stimulus utility via excitatory inputs from the cortex, but their spiking response to that input is modulated by dopamine. In response to a stimulus that promises reward in exchange for effort expenditure, the spiking activity of dopamine encodes reward-prediction error but is largely insensitive to effort. However, dopamine levels in the striatum rise during effort exertion and reflect the subject's history of reward amounts and effort rates. As dopamine levels rise, the subjects' decisions and their ensuing actions are affected: subjects become more willing to expend more effort for a reward and then perform their movements with greater vigor.

Dopamine is but one of the neurotransmitters that affect how we learn the subjective value of stimuli and control our vigor. Whereas dopamine signals the rewarding value of stimuli and its production promotes vigor, in some ways serotonin functions as its opposite: it preferentially signals the punishment value of stimuli and its production promotes sloth. Critically, serotonin is sensitive to the history of reward that the animal has experienced. In our final chapter, we will turn our attention to serotonin.

7

Serotonin and the Promotion of Sloth

> You are the sum total of everything you've ever seen,
> Heard, eaten, smelled, been told, forgot—it's all there.
> Everything influences each of us, and because of that
> I try to make sure my experiences are positive.
> —Maya Angelou

Recent events can affect us more than we know or like to admit. For example, perhaps you had a good run in the morning and are now feeling pretty happy. Or perhaps you went on that run, but you dropped your phone, the screen shattered, and now you are feeling decidedly less happy. If someone were to ask you for a favor, what happened earlier in the day will affect the generosity of your response. This is the effect of your past reward rate on your current actions. In a healthy individual, this effect may be fleeting and gone within a few hours, whereas in a person suffering from depression, this effect can be long lasting and persist for days.

Depression can be a debilitating disorder in which a person is in an unpleasant mood for an extended period of time, loses interest in the world, and in some cases exhibits reduced vigor. For decades, neuroscientists have sought to understand the neurological basis of depression. While the underlying mechanisms remain poorly understood, the neuromodulator serotonin (5-HT, or 5-hydroxytryptamine) is strongly implicated, a view supported largely by the effectiveness of drugs acting on serotonergic pathways in treating depression. Recent theory suggests that one of serotonin's many roles is to oppose dopamine by signaling the expectation of aversive events and encouraging sloth (Daw et al., 2002; Boureau and Dayan, 2010). Notably, serotonin signals long-term rewards and punishments and keeps track of the average reward rate. This makes it a critical neuromodulator in the framework of the marginal value theorem (MVT); as we saw in chapter 2, the average reward rate plays a central role in determining patterns of decision-making and vigor.

Here, we will review the neural circuitry of the serotonergic system and note that unlike dopamine, firing rates of serotonergic neurons demonstrate the capacity to encode the reward history of the animal. Because of this sensitivity to the history of reward, serotonin in principle has the capability to affect both the decision-making process regarding how long to harvest, and the motor-control process regarding how vigorously to move.

Indeed, modulation of serotonin affects many aspects of decision-making, including the willingness to stay and wait for a reward. For example, when animals are trained to harvest at one of two discrete reward sites, theoretically they should stay at one site, where they continue to harvest until their local capture rate falls below their global capture rate. When this happens, they should move to the other site. As we will see, stimulation of serotonergic neurons increases the length of time that animals are willing to stay and persist in harvesting before abandoning that site. This same stimulation also affects movement vigor, usually resulting in reduced velocity.

A clue that serotonin may interact with dopamine and contribute to Parkinson's disease (PD) is that postmortem studies of PD reveal depletion of serotonin in the caudate and putamen, although not to the same extent to which dopamine is depleted (Kish et al., 2007). Reduced serotonin activity correlates with tremor in PD patients, but not with levodopa-responsive symptoms such as rigidity and bradykinesia; this finding suggests that serotonin may have independent contributions to PD (Loane et al., 2013). However, selective serotonin reuptake inhibitors (SSRIs), which work to increase synaptic serotonin concentration and are the drug type of choice for depression, are not successful in PD (Fox et al., 2009). Thus, serotonin may independently or in conjunction with dopamine influence aspects of PD symptoms and progression.

Further clues regarding the function of this neurotransmitter can be gleaned from a condition called serotonin syndrome, which can result from an adverse reaction to SSRIs (Boyer and Shannon, 2005). In this syndrome, the gain of the stretch reflex is abnormally high, the limbs and the eyes exhibit clonus (rapid, involuntary oscillation of small amplitude that lasts from seconds to minutes), and the individual is easily startled. The syndrome suggests that the threshold for generating a movement has been lowered.

To consider the effects of serotonin on the regulation of decision-making and vigor, in this chapter we will begin with the observation that unlike the activity of dopaminergic neurons, the activity of serotonergic neurons is modulated by reward history. This predicts that changes in serotonin levels should affect harvest duration as well as movement vigor. Indeed, serotonin affects both the ability to wait for a reward, and the vigor with which a movement is expressed.

7.1 Anatomy of the Serotonergic System

The cell bodies of serotonergic neurons are found in the brainstem around the midline raphe nuclei and have projections that are widely distributed throughout the brain and the spinal cord. The raphe neurons in the pons and midbrain project to the forebrain, whereas the raphe neurons in the lower brainstem project to the spinal cord. Of particular interest are the serotonergic projections to the forebrain, with projections to higher-order brain centers such as the amygdala, prefrontal cortex, and nearly every area in the basal ganglia (striatum, globus pallidus external, and substantia nigra pars reticulata). The dorsal raphe nucleus (DRN), one of the more rostrally located nuclei, contains the largest number of serotonergic neurons, accounting for almost 30% of the serotonergic neurons in the brain. This nucleus is particularly relevant for our discussion, because the DRN is the primary source of serotonin projections to the basal ganglia and ventral tegmental area.

Serotonergic system projects widely throughout the brain and spinal cord.
An integral aspect of serotonin's multiple roles is the existence of multiple receptor types expressed both within the brain and on the periphery. In the mammalian brain, 14 receptor subtypes have been identified. A brain region can contain multiple receptor types. Even single cells can contain multiple receptor types. Autoreceptors are found on the neuron that releases the neurotransmitter, and heteroreceptors are found on other neurons. As such, serotonin autoreceptors are located on the serotonergic neurons themselves and modulate serotonin release, whereas heteroreceptors are found on other types of neurons and can thus mediate the release of other types of neurotransmitters from those neurons as well. Adding to this complexity is that some receptors can play a different role in the brain from the one they play in the periphery (Donaldson et al., 2013).

For example, let us take a closer look at the serotonin receptor 5-HT1a, one of the first receptors to be studied extensively. Within the brain, this receptor is expressed both as an autoreceptor and a heteroreceptor (on neurons in the hippocampus, septum, cortex, and hypothalamus). The receptor acts to hyperpolarize the cell. Therefore, as an autoreceptor, it inhibits serotonergic neurons in the dorsal and medial raphe nuclei. Its actions as a heteroceptor on neurons throughout the brain and periphery are less clear, with results indicating that its effects are region-specific. There is a similar diversity of projections found among the remaining 13 receptors, a reflection of the complex nature of serotonin's actions.

In comparison to dopaminergic actions, serotonergic actions are more complex and not as well understood. Dopaminergic neurons have a more focused anatomic range of projections, with three main pathways linking the dopaminergic (DA) nuclei in the substantia nigra pars compacta and ventral tegmental area to the striatum and prefrontal cortex. There are two main types of DA receptors that are largely anatomically distinct, each with a specific function. In contrast, serotonin has much more diffuse projections, a higher number of receptor types, and a greater diversity of functions across the central nervous system, with some receptors having opposing effects on dopamine release. As we will see throughout this chapter, the multifaceted interaction between serotonin and dopamine has made it difficult to identify one theory to encompass all its many functions, and instead should make us cautious against treating serotonin function as one monolithic block.

7.2 Serotonin and the Average Reward Rate

The history of past rewards and efforts, encoded in the global capture rate, is one of the key signals in optimal foraging theory (chapter 2): it affects both the decision regarding harvest duration and the decision regarding movement vigor. How is this history reflected in the activity of neurons?

Nathaniel Daw, Sham Kakade, and Peter Dayan (Daw et al., 2002) proposed that one of serotonin's many roles was to act as an opponent to dopamine. They began with the observation that decisions should result in maximizing an average measure of gain, termed the average-case reinforcement learning model of optimal control. As in the marginal value theorem, this model maximizes net gains in the long-term rather than maximizing net gains over a fixed duration in the immediate future. The advantage of a reinforcement learning framework (as compared to the marginal value theorem) is that it allows for a more dynamic model of learned reward and punishment values in changing environments. For example, in a reinforcement learning framework, one can consider more realistic scenarios in which the value of states in the future depends on the current action.

To implement reinforcement learning, Daw and colleagues (Daw et al. 2002) initially proposed that tonic dopamine firing rates signaled average punishment rate, whereas tonic serotonin firing rates signaled average reward rate. However, in their later work (Boureau and Dayan, 2010), they changed this view and suggested that tonic dopamine activity reflected average reward rate, whereas tonic serotonin activity signaled average level of punishment. As we will see here, the data paint a more complicated picture, providing little or no evidence for an encoding of reward history in the firing rates of dopamine neurons. In contrast, this history is present in the tonic discharge of neurons that express serotonin.

Discharge of serotonin, but not dopamine, is modulated by average reward rate. In a comprehensive study, Jeremiah Cohen, Mackenzie Amoroso, and Naoshige Uchida designed an experiment to measure the effects of short- and long-term reward and punishment on the activity of both serotonin and dopamine neurons (Cohen et al., 2015).

They trained mice in a classical conditioning paradigm in which the animals learned to associate different stimuli, such as odor cues, with the delivery of various unconditioned stimuli: either a reward (water), punishment (air puff), or nothing (figure 7.1A). The conditioned-stimulus period lasted 1 s, and after a 1 s delay, the unconditioned stimulus was delivered. If the cue indicated water, the mice started licking immediately (figure 7.1A). In contrast, if the cue indicated nothing or an air puff, then the mice did not lick.

Trials were blocked in groups of ten of nearly all reward or all punishment, with a tone indicating the beginning of each block (figure 7.1B). The same tone was used regardless of the type of block (reward or punishment). The third cue, the one that predicted a neutral trial, was randomly intermixed in the various blocks.

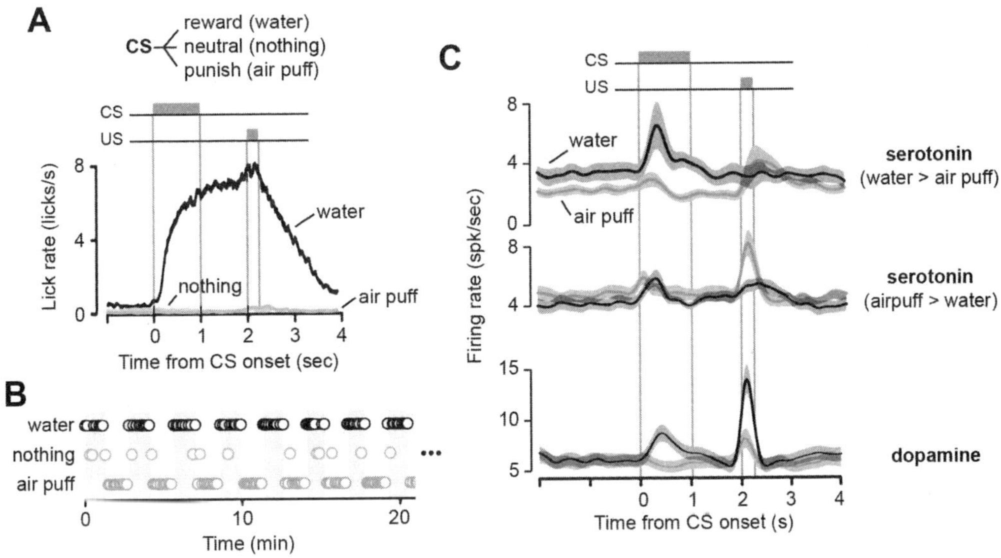

Figure 7.1
Serotonergic activity encodes average reward rate. **A**. Mice received a conditioned stimulus (CS), an odor cue, indicating the nature of the unconditioned stimulus (US): reward, neutral or punish. Lick rates increased when the mouse received the reward odor cue, in anticipation of the impending water. No change in lick rates were observed after the other two CSs. **B**. Schematic of trial blocks. Trials of a single CS type were grouped in blocks of 10, and blocks types alternated between reward and punishment. Some of the trials were selected at random as neutral trials, in which the animal received neither reward nor punishment. **C**. Tonic serotonergic activity was influenced by the type of trial block. Some neurons had greater activity during the reward blocks (top panel), whereas others had greater activity during the punishment blocks (middle panel). In contrast, no such modulation was observed in dopaminergic neurons (bottom panel). (From Cohen et al., 2015.)

As the mice performed the task, the authors recorded activity of serotonergic neurons in the DRN (5 mice, 29 neurons). They were curious about the short-term as well as the long-term encoding of reward and punishment. Specifically, the researchers sought to determine whether serotonergic neurons responded with their phasic discharge to the reward or punishment, and whether they changed their tonic discharge as the animals repeatedly experienced those rewards or punishments over the course of the ten-trial block, and thus experienced a change in the average reward rate.

First, let us look at the responses during the intertrial interval (ITI), the time between the delivery of the unconditioned stimulus (reward/punishment/neutral) and the delivery of the conditioned stimulus that initiated the subsequent trial. The experimenters controlled the ITIs so that they varied unpredictably with an average duration of 6 s, significantly longer than the duration of an individual trial. This design provided them with a sufficient block of data representing tonic neuronal activity within a block, thus allowing the comparison of tonic rates between reward and punishment blocks.

If serotonin was influenced by the average reward rate in the environment, then we would expect to see a difference between the tonic rates in the reward block and those in the punishment block. Among the 29 serotonergic neurons that were recorded, 12 modulated their tonic activity with respect to the block type. Intriguingly, seven of these neurons were more active in the reward blocks, while the remaining five were more active in the punishment blocks (figure 7.1C). This suggested that the average reward rate affected the tonic activity of serotonergic neurons.

The authors also recorded from 28 dopaminergic neurons in the ventral tegmental area (VTA) during the same task. Here, they observed the characteristic increase in the firing rate of dopamine neurons upon the presentation of the odor cue. This increase was greater when the cue indicated a reward than when it indicated a punishment (figure 7.1C, lower plot). However, in contrast to the serotonin neurons, there was no effect of block type on the tonic levels of dopamine discharge (figure 7.1C, lower plot), thus demonstrating that tonic activity of this sample of VTA dopaminergic neurons was not modulated by the history of rewards.

Next the authors turned to the phasic (i.e., transient) response of the serotonergic neurons. As a population, their response when the cue appeared was greater in reward trials than in punishment trials (figure 7.1C, response to water versus air puff). When the unconditioned stimulus was given, the population response was much greater in punishment trials than in reward trials. However, individual neurons differed in their response to the cue; some responded by increasing their discharge only when the cue indicated reward, some increased their discharge only when the cue indicated punishment, and some increased their responses for both cues.

The diversity in the responses of the serotonergic neurons to reward and punishment is puzzling, but it likely indicates that the neurons project to different regions of the brain and thus their signals carry different meanings. In a more recent study, Jing Ren and colleagues (Ren et al., 2018) anatomically defined distinct DRN serotonin neurons that projected to the frontal lobe and amygdala. Serotonin neurons that projected to the frontal lobe were activated by reward and inhibited by punishment. In contrast, serotonin neurons that projected to the amygdala were activated by both reward and punishment.

The researchers also gave the mice rewards at random times without any cues. Dopaminergic cells responded very strongly and positively to the unexpected rewards. Serotonergic cells had weaker responses, but they also showed a transient increase in receiving the unexpected rewards.

In summary, tonic serotonergic activity, but not tonic dopamine activity, is influenced by average reward rate. Among serotonergic neurons, there is diversity in the sensitivity of their tonic response, with some neurons increasing their tonic rates as the average reward rate increases, and other neurons decreasing their tonic rate. In addition, serotonin neurons are phasically activated in response to cues indicating reward as well as those associated with punishment. However, whereas phasic dopamine activity is generally positive for reward-indicating cues, serotonin response is more muted and diverse, with the population response generally positive when the reward-indicating cue appears and also generally positive when the punishment is delivered. These results demonstrate that serotonin does relay information regarding long-term rewards, via average firing rates, and short-term rewards, via phasic firing rates; thus, the neurotransmitter does provide critical signals for cost-benefit analysis that might support decision-making as well as movement vigor.

7.3 Serotonin and the Willingness to Wait

Serotonin is often implicated in impulsivity. Impulsivity is defined as the inability to wait and the tendency to respond prematurely. It can take many forms, including cognitive manifestations related to drug addiction and motor forms related to the inability to withhold a physical response. The mechanisms underlying an impulsive act are diverse. For example, a premature response may be due to an altered valuation of the costs and rewards, or perhaps due to an altered threshold for making a decision. It is intriguing that despite the many potential mechanisms underlying impulsive behavior, as well as the various forms it can take, serotonin plays a role in its modulation.

Ultimately, our goal is to understand the effects of serotonin on control of vigor. A good place to start is the role of serotonin in the motor aspect of impulsivity, termed behavioral inhibition, which refers to the ability to withhold the motor response. To study the role of serotonin in behavioral inhibition, a typical experiment quantifies an animal's willingness to wait. These tasks often involve training the animal to withhold a motor response for a duration of time in order to receive a reward. A premature response indicates that the animal is unable to inhibit its motor response. In the framework of movement vigor, a premature response can be viewed as an inability to inhibit movement, a malfunction that can arise with disorders of the basal ganglia. For example, we saw in chapter 5 that patients with Parkinson's disease or damage to the frontal eye field (FEF) had a hard time inhibiting their habitual response to the visual stimulus when they were instructed to saccade in the opposite direction. Of course, the withholding of a response may be due to increased patience, a reduced ability to move, or both. While the reasons underlying this behavioral phenotype are myriad and complex, one potential explanation is that serotonin release is modulated by the average reward rate that the animal has experienced. Thus, its release influences the balance between exploration and exploitation. In some tasks, but not all, this is reflected in the balance between moving and holding still.

Serotonin modulates the willingness to hold still.

A study by Dawn Eagle, Trevor Robbins, and colleagues used a depletion approach to examine the effects of serotonin on the ability to inhibit action (Eagle et al., 2008). They injected neurotoxin 5,7-dihydroxytryptamine (5,7-DHT) to selectively kill serotonin neurons in a region that likely included the DRN. A control group received saline injections in the same brain location. Both before and after surgery, rats performed a variant of the stop-signal task (a saccade version of this task in humans is shown in figure 3.20). The rats were placed in a cage with two retractable levers on each side of a food well. A trial began with a nose poke into the well, upon which the left lever would extend. Once the rat pressed the left lever, the right lever would extend for a limited amount of time. Successful pressing of the right lever was rewarded with one pellet. If the right lever was not pressed within the allotted time, no pellet was provided, and there was a forced time-out of 5 s in darkness.

The rat's ability to withhold movement was tested with a few stop trials that were interspersed randomly. In stop trials, a stop tone would sound immediately upon extension of the right lever. The stop tone indicated to the rat that the right lever should not be pressed in that trial. Instead, the lever would remain extended for a fixed amount of time, and the rat was required to wait and withhold the lever press. The duration of time the lever was extended was termed the limited hold (LH) duration. If the rat was successful in withholding the right lever press for the LH duration, it was rewarded with one pellet. Failure was punished by withholding the pellet and a 5 s time-out in darkness. Rats performed two sessions on two different days. In one session, the limited hold duration was normal (LH×1), while in the other, the limited hold duration was doubled (LH×2). In this way, the researchers tested the rat's ability to withhold pressing the right lever for an extended period of time.

For both the control group and the serotonin-depletion group, the number of failed stop trials increased with the greater LH duration. Thus, the rats had trouble withholding their motor response when they had to wait for the longer period of time.

The question is whether serotonin depletion affected the ability to withhold the motor response. In accordance with a role for serotonin in behavioral inhibition, the serotonin depletion group exhibited greater failure rates than did the control. This effect did not reach significance for the normal LH duration (LH×1), but it was significant for the longer duration (LH×2). Moreover, with extended testing, this effect remained stable. Over the course of four additional testing sessions, the serotonin-depleted group was still significantly less able to withhold their movement responses.

Thus, after the rats were trained to associate a movement with reward, suppressing that movement when instructed to do so appeared to depend on the function of serotonin. Destruction of serotonin neurons resulted in an impairment in the willingness to hold still and not press the lever.

To understand the mechanisms underlying this effect, the authors used histology to determine the effects of the lesion not only on serotonin levels, but also the levels of other neurotransmitters implicated in behavioral adaptations. Postmortem histology of cortical, striatal, and limbic areas confirmed that the levels of serotonin were more than 90% lower in the lesion group than in the control group. To further investigate the underlying cause, they also measured the levels of dopamine and noradrenaline, other neurotransmitters implicated in behavioral adaptation. In contrast to the profound reduction observed in serotonin levels throughout the brain, levels of dopamine and noradrenaline were similar in the experimental and control groups.

In summary, the rats with diminished serotonin levels were less able to resist pressing the lever that was normally associated with the delivery of a food pellet, but now cued via a light to be avoided. Thus, a reduction in serotonin levels led to lower levels of behavioral inhibition. These animals were less able to withhold a physical response that was normally associated with reward but was now instructed to be otherwise.

Ability to hold still is associated with decreased dopamine in nucleus accumbens. Serotonin-receptor knockout is another method one can employ to probe the role of serotonin in behavior. Among the receptors implicated in impulsive behavior is the 5-HT1b receptor. As an autoreceptor, it inhibits neurotransmitter release from serotonergic neurons. As a heteroreceptor, it modulates neurotransmitter release from non-serotonergic neurons such as those that release GABA, glutamate, and dopamine. As such, this receptor can influence serotonergic as well as dopaminergic activity in many areas of the brain.

Katherine Nautiyal, René Hen, Susanne Ahmari, and colleagues (Nautiyal et al., 2015) employed a receptor knockdown approach to better understand the role of the 5-HT1b receptor in impulsive behavior. In this knockdown study, expression of the gene encoding the 5-HT1b receptor was reduced but not completely eradicated. Importantly, using gene knockdown rather than gene knockout allowed the authors to rescue gene expression in these animals later in life through the administration of doxycycline. Thus, this approach allowed some degree of temporal specificity and could be used to probe the causal relation between 5-HT1b receptor expression and impulsive behavior.

Mice with whole-brain knockdown of the 5-HT1b receptor gene and their littermate controls were tested in two operant paradigms that probed their ability to withhold a motor response. In the first experiment, knockdown mice and controls were tested in a differential reinforcement of low rate paradigm (DRL) in which they were required to press a lever and then wait to receive a reward. In this case, a reward followed a lever press only if the lever press was preceded by 36 s of no lever pressing (figure 7.2A). To characterize the ability to refrain from responding, the authors quantified the number of responses with latencies of less than 3 s (termed burst responses) and the latencies of the remaining responses (termed non-burst). Knockdown mice had 80% more burst responses, and the latencies of the remaining presses were significantly shorter in the knockdown group than in the control group (figure 7.2B top panel). Overall, the distribution of non-burst responses was shifted to significantly earlier in the knockdown mice, indicating that these mice were less able to withhold a response. This led to profound performance deficits in the task when measured as the ratio of lever presses to reward earned. The knockdown mice were drastically less efficient, earning a reward on average after approximately 30 responses, whereas the control group needed only 10 responses to earn the same reward.

These findings suggest a correlation between 5-HT1b receptor expression and the ability to withhold a response, but they do not provide clear evidence for a causal effect of

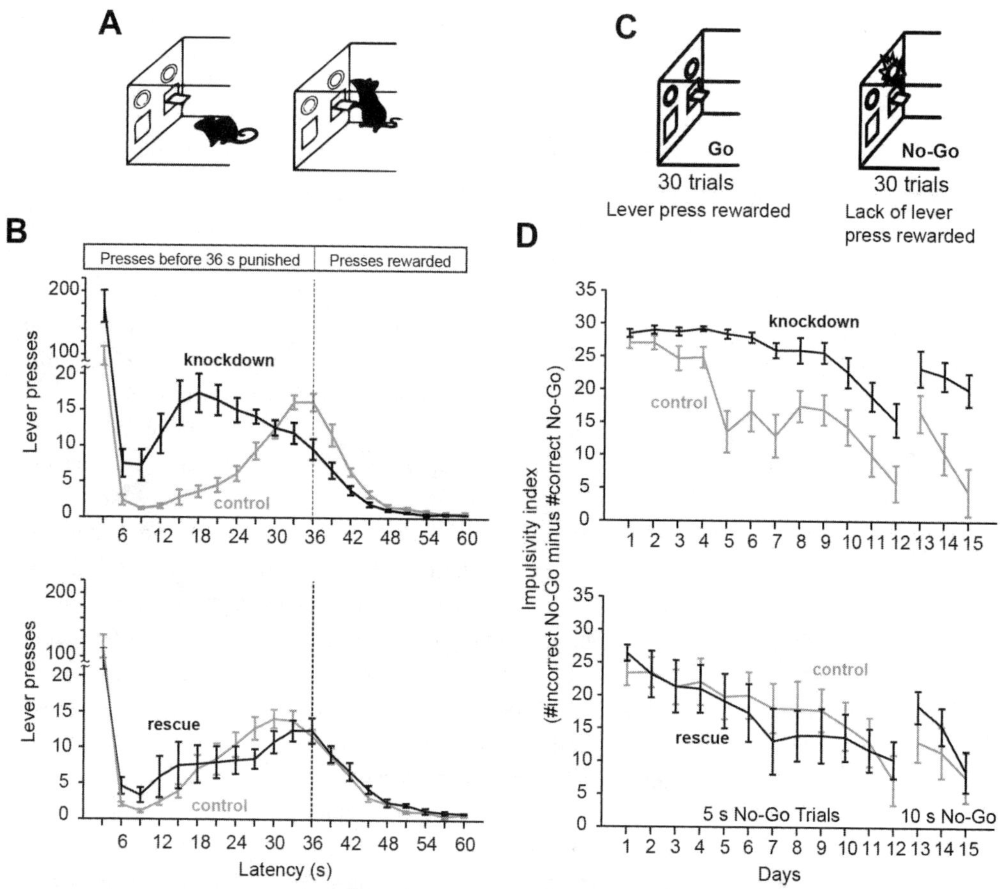

Figure 7.2
Serotonin-receptor knockdown reduces the ability to wait. **A**. Differential reinforcement of low rate (DRL) task. Upon lowering of the lever, mice were required to wait 36 s before pressing the lever to receive a reward. **B**. Lever presses as a function of wait duration (latency). Only lever presses after 36 s (vertical line) were rewarded. Knockdown mice had significantly more incorrect lever presses (waiting less than 36 s) than control mice (top panel). Behavior is normalized with the application of doxycycline (rescue, bottom panel). **C**. Go/no-go task. In half of the trials, the mouse was cued to press the extended lever. In the other half, the mouse was required to withhold from pressing the extended lever to receive a reward. **D**. Impulsivity index, calculated as the difference between incorrect and correct no-go trials, as a function of days of experience with the task. Knockdown mice exhibited significantly greater impulsivity whether the no-go trials lasted 5 s or 10 s (top panel). Behavior was rescued and brought to control levels with the application of doxycycline (bottom panel). (From Nautiyal et al., 2015.)

serotonin on response inhibition. To examine this further, the authors were able to recover 5-HT1b receptor expression in adult knockdown mice with doxycycline treatment. Remarkably, the performance fully recovered: the burst responses and non-burst latencies of the rescued mice were similar to those of the control group (figure 7.2B bottom panel). Accordingly, their performance on the task, assayed with the ratio of number of responses to rewards obtained, matched that of the controls.

A second operant task was used to determine the robustness of these observations. In this second experiment the mice performed a go/no-go task in which go trials required the mice to press a lever to obtain a reward, and no-go trials required the mice to resist pressing a lever to obtain a reward. The go trials were cued by presenting the lever and turning on the house lights. The no-go trials were cued by presenting the lever, turning the house lights off, and turning on a small light above the lever (figure 7.2C). Mice were tested daily over the course of 15 sessions. During the first 12 sessions, no-go trials lasted 5 s. The withholding time was increased to 10 s during the last 3 days of testing. Results in this task supported the observations in the DRL task. Whole-brain knockdown mice exhibited greater motor impulsivity than did the controls: they had more unrewarded no-go trials in which they were unable to withhold the pressing response (figure 7.2D top panel). As in the time-delay study, doxycycline was used to restore 5-HT1b receptor expression when the knockout mice reached adulthood. The go/no-go trials were repeated and, as in the time-delay study, there were no significant differences in performance between the knockout and control groups (figure 7.2D bottom).

Together, the experiments suggest that full-brain knockdown of the serotonin receptor results in impulsive motor behavior (an inability to wait and not press the lever), but rescue of receptor expression in adulthood normalizes impulsivity to control levels.

Knowing that the 5-HT1b receptor is mediating this impulsive phenotype, the next goal is to understand how it does this. Given the role of dopamine in invigorating movement, perhaps this serotonin receptor modulates dopamine levels in the brain. Microdialysis was used to determine the effects of 5-HT1b receptor knockdown on dopamine levels in the nucleus accumbens (NAc) and dorsal striatum. There was an effect of 5-HT1b receptor expression on dopamine levels in the NAc, but not in the dorsal striatum. Whole-brain knockdown led to elevated levels of dopamine in the NAc. Again, adult treatment with doxycycline normalized dopamine levels in the NAc. Together, these findings show that the 5-HT1b receptor plays an important role in modulating dopamine levels in the nucleus accumbens.

Recall that in chapter 3, we noted that there were trait-like differences among healthy people; some tended to consistently move with high vigor, and some with low vigor. We considered an experiment in which they were provided with a target and an instruction to saccade, but the instruction occasionally changed, informing them to withhold their movement (figure 3.20). Individuals who had trouble waiting, and thus made the saccade, tended to also make saccades with high vigor (figure 3.21). Thus, vigor tended to be somewhat higher in people who, in this task, exhibited impatience. Here, experiments in rodents suggested that reduced expression of the 5-HT1b receptor led to a reduced ability to withhold a motor response, and this was linked to elevated levels of dopamine in the nucleus accumbens. The neural basis of trait-like differences in vigor in healthy individuals, and its potential association with the ability to withhold movements, remains unknown.

Optogenetic activation of serotonin neurons increases patience.
Serotonin depletion and receptor knockdown are just two of many ways to probe this neurotransmitter's role in behavior. Although lesion and knockout approaches have their advantages, they also have disadvantages. These techniques may affect additional cell types beyond those being targeted, and they may also lead to adaptations that complicate the interpretation of results. Both these methods also lead to long-term compensatory changes that make it difficult to determine the time course over which serotonin is acting. In contrast, optogenetics provides a powerful approach to transiently and specifically activate targeted cell types. To better determine the causal role serotonin plays in modulating an animal's willingness to wait, Kayoko Miyazaki, Kenji Doya, and colleagues (Miyazaki et al., 2014) optogenetically activated serotonergic DRN neurons during a delayed-reward task. Building on previous observations of serotonin-related behavioral inhibition, the researchers sought to determine the temporal specificity of neuronal activity on the ability to wait and withhold a motor response.

In the delayed-reward task, mice would initiate a trial by traveling to a start site where they were required to continuously poke their noses until a tone was presented. At tone presentation, the mouse was then required to approach the reward site and continuously poke there for a length of time until a reward (food pellet) was presented. Thus, the task required two distinct periods of waiting: the start-site period and the reward-site period. An error occurred when the mouse did not maintain its nose in a fixed posture at the start site or the reward site. In a series of experiments, the authors independently stimulated serotonergic DRN neurons as the mice waited either at the start site or at the reward site.

In the first set of experiments, the DRN neurons were stimulated as the animals waited for the tone at the start site (figure 7.3A). Neurons were stimulated in only half of the trials. In the first experiment, the wait period at the start site was set at 0.8 s. The number of errors was significantly lower in the stimulation trials than in the no-stimulation trials. Thus, stimulating serotonin neurons during the nose-poke period at the start site seemed to improve the mice's ability to keep waiting for the tone.

In another experiment, the authors examined the effect of tone-delay duration. As the tone-wait period at the start site increased from 0.6 s to 1.5 s, the number of errors increased, but to a lesser extent in the stimulation trials than in the non-stimulation trials (figure 7.3B). For the longer duration trials, the effect of the stimulation was more pronounced. Thus, the activation of serotonergic neurons as the mice waited for the tone enabled them to wait longer.

Next, the authors asked whether the timing of the activation was critical. They probed the temporal specificity of this effect by testing different timings of the delay and

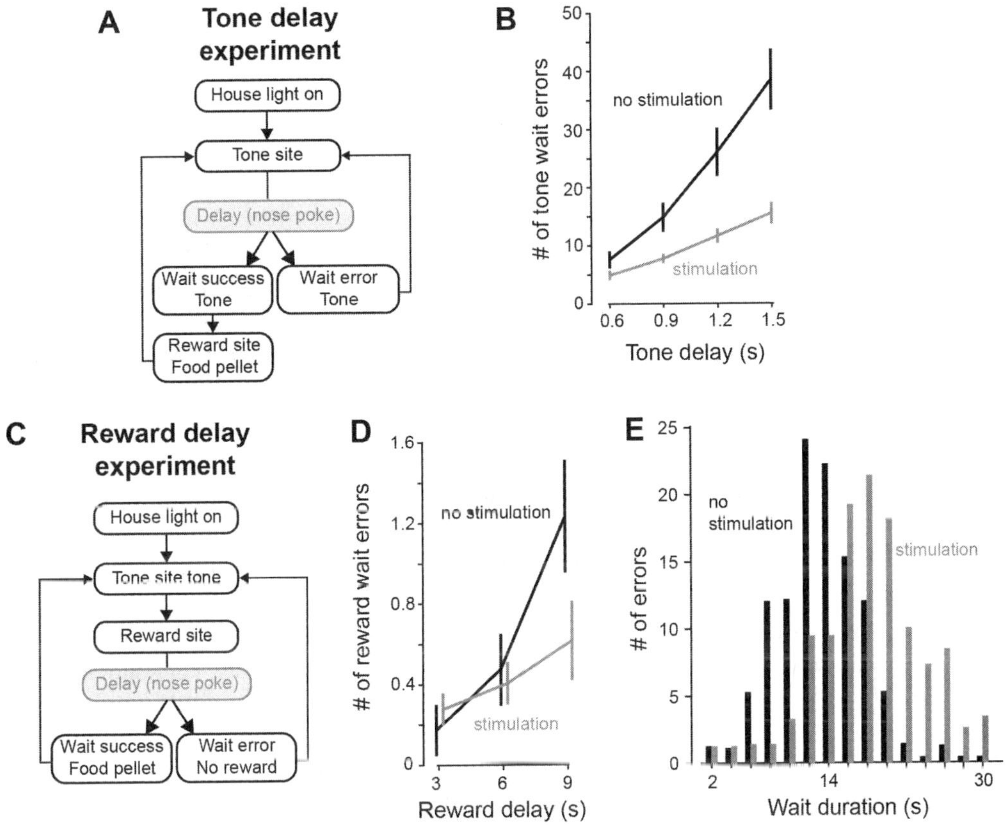

Figure 7.3
Stimulation of serotonin neurons increases the ability to wait. **A**. Flow diagram of tone-delay experiment. After moving to the tone site, subjects were supposed to wait there until they heard a tone, after which they traveled to the reward site for a pellet. **B**. When stimulation was applied during the tone-delay period, subjects made fewer errors (leaving the site prematurely before the tone) than when no stimulation was applied during that period. **C**. Flow diagram of reward-delay experiment. After hearing the tone and traveling to the reward site, subjects were supposed to wait for the reward for durations of varying length. **D**. When subjects received stimulation during the reward-delay period, they exhibited fewer wait errors than when no stimulation was applied during that period. This effect was greater for longer wait durations. **E**. For the infinite wait duration (no reward was provided), errors were made after longer wait periods in the stimulation trials than in the no-stimulation trials. (From Miyazaki et al., 2014.)

stimulation. When the stimulation was applied for 0.8 s before the animals began nose poking, no effect was observed on the number of errors. These results suggested that serotonergic neuron stimulation during the wait period, but not before, increased the animal's ability to keep waiting and withhold a motor response.

This may be because serotonin leads to an overall inhibition of movement that simply manifests as increased patience in these delayed-reward tasks. Alternatively, serotonin may be modulating the willingness to wait in the hope of greater reward in the long term.

Thus, the authors used another experiment to determine whether stimulation of serotonergic neurons led to inhibition of motor behavior, regardless of whether waiting for delayed rewards was involved. In this experiment, rather than apply the stimulation during the tone-wait period, the authors applied it during the tone, and then they quantified the effect of stimulation on both the time it took the mouse to exit the poke hole and the time it took the mouse to reach the reward site. If serotonin release was increasing wait times by inhibiting movement, then there should be a slowing of motor behavior as the mouse exited the start site. The authors found that if the stimulation was applied during the tone, little or no effect was observed on either the time to exit the start site or the time to reach the reward site.

These results suggested that stimulation of the DRN neurons did not cause a significant inhibitory effect on movement. This finding supports the hypothesis that serotonin promoted the willingness to wait, thus allowing for acquisition of higher rewards in the long term. Once the mice heard the tone, they exited the start site, traveled to the reward site, and waited there until a reward was provided after a set delay. Next, the authors examined the effect of serotonin activation during the nose-poke period at the reward site (figure 7.3C). Here, the reward delay was set to 3 s, 6 s, or 9 s. In some trials, the delay was set to infinity (trials on which reward was omitted).

Applying the stimulation during the reward waiting period also reduced the number of wait-period errors. This was true whether the stimulation was applied continuously or transiently for 0.8 s at the start of the reward-delay period (figure 7.3D). When the delay was infinite (leading to reward omission), stimulation led to longer wait durations before failure, an indication that serotonin-neuron activation prolonged the length of time the mice were willing to wait for the reward (figure 7.3E).

A potential reason for the increased patience is that serotonin activation led to increased valuation of reward. Indeed, there is evidence that a subset of serotonergic neurons respond to rewarding stimuli (Cohen et al. 2015, Bromberg-Martin et al., 2010). In the current experiment, an increased valuation of reward would also lead to longer wait times and fewer errors. To check whether the observed effects were due to greater reward valuation, the authors performed a set of control experiments. Mice were trained on the delayed-reward task, but now the reward for one task was one pellet and for the other was two pellets. As expected, mice would wait longer and have fewer errors in the two-pellet task. In a third experiment, the reward was again one pellet, but authors stimulated serotonin neurons upon reward delivery (i.e., not during the tone-wait period). If stimulation was increasing reward value, then we should see longer wait times for the one-pellet task with stimulation than for the one-pellet task without stimulation. However, this was not observed. Stimulation during reward delivery had no effect on wait times for the single pellet, suggesting that serotonin neuron activation did not increase the reward value of the pellet.

Overall, these results suggest that in scenarios in which the animal had to withhold movements, activation of serotonergic DRN neurons increased patience for delayed rewards. This effect was temporally specific (stimulation had to be during the waiting period), and it did not appear to be a result of a reduced ability to move or an increased valuation of reward.

Serotonin promotes extending the harvest period during foraging.
Multiple studies, using a diverse range of approaches, confirm that serotonin influences the ability to withhold a motor response. Higher levels of the neurotransmitter facilitate holding still, whereas lower levels inhibit that ability. The full story, however, is not as straightforward. All of these tasks were designed so that increased patience had the same behavioral manifestation as an increased ability to withhold a response. However, patience and withholding of a motor response are not synonymous. There is a difference between serotonin making someone more patient and making someone less willing to move. Both would lead to similar behavior (i.e. the patient withholding of a response in these specific tasks), but the two explanations are functionally very different.

Miyazaki and colleagues (Miyazaki et al., 2014) considered this distinction. They specifically asked whether activation of serotonergic neurons also led to a general inhibition of movement that could be observed in the mouse reaction times and movement times between locations. Critically, they observed no effect of serotonin activation on these aspects of motor behavior not related to the instrumental lever-pressing response. Thus, they concluded that serotonin did not inhibit movement, but rather it increased the willingness to persevere and wait for the reward.

Eran Lottem, Zachary Mainen, and colleagues (Lottem et al., 2018) realized that a stronger test of this conjecture was to consider a task in which the waiting period also involved movement. Essentially, they wanted to find out whether the withholding of the response was really due to increased patience or actually due to an increased tolerance for inactivity.

To test this, they used a foraging task in which mice harvested reward at a site until they stopped getting rewards, thus forcing them to give up and move to a new site. Critically, patience was represented via the harvest period, which entailed an active process. If activation of serotonergic neurons promoted persistence, then this activation should have led to a longer period of harvesting before moving on.

They designed a task in which mice poked their noses at one of two ports and were rewarded for their pokes according to a random schedule with probabilities that decayed over time. Thirsty mice were placed in a rectangular box with a water port at both ends, representing harvest locations (figure 7.4A). A nose poke into the water port was rewarded with a drop of water in some trials. As the mouse poked repeatedly at the same port, the probability of receiving a drop of water decayed exponentially. At some point the mouse left the port and traveled to the other port and resumed poking. The reward probability was reset to its initial value whenever the mouse left the port, and a trial was only considered successful if the mouse alternated between ports (figure 7.4B). A single foraging trial began when the mouse approached the port (entered its region of interest [ROI]), and ended when it exited the ROI. The measurement of interest here is the number of nose pokes within a foraging trial, which provided a measure of how long the mouse persisted in its pursuit of rewards at that site.

To encourage goal-directed behavior, the authors introduced three types of trials whose reward probabilities started at different levels (low, medium, and high) but decayed at the same rate. The different reward trials were randomly interleaved, and no cue was provided to indicate to the mouse any expectation of reward.

We saw in chapter 2 that in a foraging task, marginal value theorem predicts that one should stop harvesting and leave the patch when the current rate of return matches the global average that one has experienced in the past. Thus, by manipulating the reward contingency, the authors could determine if the mouse was indeed foraging in a goal-directed (i.e., optimal) manner.

As expected, animals received more rewards in the high-reward environment. This simply falls out of the reward probabilities. The more critical question is how long the animal harvested and whether that duration was affected by serotonin levels. Because the trials were interleaved, the authors could make the assumption that the average reward rate was effectively constant across the experiment. This means that the reward rate at the time of leaving should be equal to the average rate across the experiment, and indeed this was the case. An additional prediction was that the mice should leave earlier in the low-reward trials and later in the high-reward trials. The authors quantified the cumulative probability of pokes per trial as a function of reward condition and observed that the curve was shifted to the right in the high-reward trials, an indication that a greater number of pokes was more likely in the high-reward trials. The number of pokes per trial also increased with increasing initial probability of reward. All of these behaviors were consistent with the predictions of the theory.

Lottem and colleagues then proceeded to investigate the effects of serotonin on behavior. In 10 of the 16 mice tested, they optogenetically stimulated serotonergic neurons in the DRN as the animals performed the task. The remaining six mice tested were their wildtype littermates. Light stimulation occurred randomly in 50% of the trials, triggered by the first nose poke in each trial and lasting for 10 s or until the end of the foraging trial, whichever came first (figure 7.4C). The researchers found that the number of nose pokes in the light-on trials was greater than in the control (no stimulation) trials, and that the number of pokes by the light-sensitive mice was greater than the number by the wildtype mice (figure 7.4D). The effect did not influence subsequent trials. Thus, stimulation of serotonergic neurons increased harvest duration, even though during the harvest, the mice were actively moving.

Therefore, the results illustrated that activation of serotonergic neurons promoted persistence and patience, and not necessarily sloth. Indeed, it raised the possibility that serotonin may signal average rate of reward, or the global capture rate.

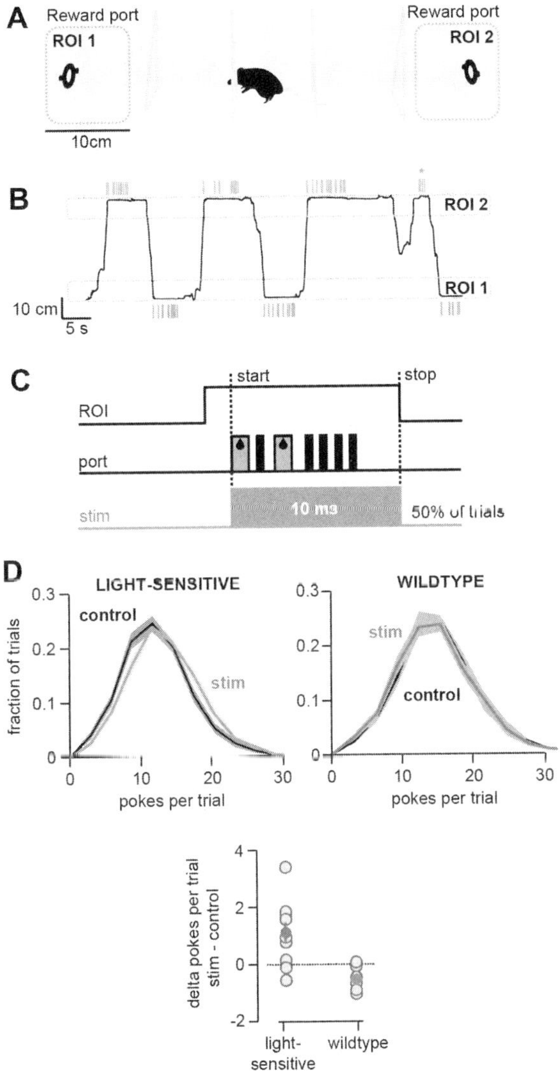

Figure 7.4
Activation of serotonin neurons increases harvest duration. **A**. Foraging experiment. Mouse travels between reward ports on opposite ends of a rectangular walkway. At each port, it pokes its nose to receive a water reward. Rewards are delivered randomly in accordance with an exponentially decaying schedule. **B**. Example data showing the trajectory of a mouse between regions of interest (ROIs, dark line) and nose pokes (short gray vertical lines) in six successful trials and one unsuccessful trial (highlighted with an asterisk). **C**. Schematic of a single trial with stimulation. Stimulation was applied in 50% of all trials. Stimulation started with the first nose poke and continued for 10 s or when the mouse left the ROI, whichever happened first. Rewarded nose pokes are shown in the gray, wider bars. **D**. Fraction of trials as a function of pokes per trial during stimulation (stim) and no-stimulation (control) trials for both the light-sensitive and wildtype mice. Stimulation led to more trials having a greater number of pokes in light sensitive mice, but not in the wildtype mice (bottom panel). (From Lottem et al., 2018.)

These results go against a popular theory of serotonin as a neurotransmitter that inhibits movements. So, as a next step, the researchers focused specifically on the animal's movements. Did activation of serotonergic neurons affect the vigor of movements? First, they looked at the time between the final nose poke in the water port and the time at which the mouse left the ROI. Intriguingly, they found that serotonin activation did indeed delay the time between the final nose poke and the time of ROI exit, an indication of a slower motor response. However, no effect of stimulation was observed on the travel time between foraging sites (the time between leaving one ROI and entering another). The lack of effect may be because stimulation was not applied during the travel time. In a follow-up experiment, serotonergic neurons were stimulated during the travel period. However, again, no effect on travel duration was observed. Taken together, serotonin activation increased harvest duration (both the time spent nose poking and time spent exiting) but not travel duration.

The results thus far indicate that serotonin is involved in managing the tradeoff between exploration and exploitation, rather than having a purely inhibitory effect on movement vigor. Serotonin seems to promote exploitation (staying longer at the harvest site and visiting fewer sites) over exploration (staying for a shorter duration at the harvest site and visiting more sites), independent of the details of the responses themselves, be they waiting or moving. In the framework of optimal foraging theory, this could be described as serotonin encoding the global capture rate in the environment. Reducing the global capture rate promotes greater exploitation over exploration and increased harvest duration, which are consistent with the findings in this study. However, as we saw in chapter 2, a reduction in the global capture rate should also affect vigor. That aspect of the prediction was not seen in this experiment and thus remains to be further examined.

7.4 Serotonin and Movement Vigor

At this point, we have a growing body of evidence that serotonin plays a role in the ability to withhold movement when required, and some support for the theory that this is not only due to a reduced willingness to move. However, most of the tasks examined have explicitly required that animals suppress movement and only initiate it when allowed to do so. Another way of stating this is that in many of these studies, animals needed to be patient in their motor responses. Lottem and colleagues showed in their foraging task that patience need not involve sloth. However, the task was a foraging task and not designed to explicitly measure movement vigor. Thus, the question still remains as to whether serotonin levels relate to the willingness to move. Even more specifically, how do these findings relate to movement vigor?

Serotonin release promotes a reduction in motor activity.
Earlier studies hint at an answer. For example, let us look back at the study by Dawn Eagle and colleagues (Eagle et al., 2008). Recall that they observed that serotonin-depleted rats

were significantly worse at withholding their responses in a stop-signal task. If serotonin was linked with behavioral inhibition, then perhaps it influenced sloth in general. The authors investigated this by testing the spontaneous locomotor activity of the animals in an open arena fitted with light sensors that detected breaks in beams of light. They found that serotonin-depleted rats demonstrated significantly more beam breaks than the control rats, especially during the initial and final 30 min of testing. A beam break indicated that the rat had moved. Hence, it seemed that the serotonin-depleted rats were moving more than the control rats. This finding implies that serotonin in healthy animals promotes a general reduction in motor activity.

While these results are intriguing, recording beam breaks only provides a coarse measure of locomotor activity. A reduction in the number of beam breaks may occur because the rat is resting or grooming, or it could also reflect a slower movement speed. To better understand the effect of serotonin on locomotion, a more careful analysis of kinematics was needed.

Yu Ohmura and colleagues (Ohmura et al., 2014) utilized an optogenetic approach to target serotonin release in neurons in the DRN. They used blue light to activate serotonergic DRN neurons as mice performed an elevated plus maze test and measured locomotor activity as the total distance traveled during the test. The elevated plus maze test is typically used to measure anxiety levels. It consists of an apparatus with two open and two closed arms, all of which is elevated 40 cm above the ground. The closed arms are surrounded by high walls, so the mice generally prefer them to the open arms. Less time spent in the open arm is taken as a measure of increased anxiety, and total distance traveled is a measure of locomotor activity. Each trial began by placing the subject in the center of the maze and recording its behavior for the next 4 min. Blue light, which optogenetically activated the DRN serotonergic neurons, was applied alternately in 1 min intervals for a duration of 500 ms each. As a control condition, mice also received exposure to yellow light, which did not lead to activation of serotonergic neurons. Compared to the yellow light treatment (control), blue light activation led to lower levels of locomotor activity as well as higher levels of serotonin in the dorsal striatum. Dopamine levels were unaffected. This result indicates that increased levels of serotonin led to a reduction in movement vigor. However, the results may have been confounded by increased anxiety levels and the nature of the elevated maze test.

Serotonin reduces movement vigor.

Patricia Correia, Eran Lottem, Zachary Mainen, and colleagues (Correia et al., 2017) approached this question by using a combination of optogenetics and precise motion capture. They used optogenetics to activate serotonin-producing neurons in the DRN in mice. The light-sensitive mice and their wildtype littermates were tested as they freely moved about in an open arena devoid of any cues. Both groups of mice received alternating 5 min blocks of stimulation and no stimulation. During the stimulation blocks, the researchers applied blue light in pulses of 3 s on followed by 7 s off. As a first pass at quantifying the effects of serotonin, the authors categorized and quantified the types of motor behaviors exhibited by the mice. Behavior was classified as walking, rearing, grooming, digging, scratching, or resting. Resting did not mean the mouse was sleeping; only that the mouse was awake but mostly stationary. They then calculated the probability that the mouse would be in a given state for the final 2 s of the stimulation interval as well as the final 2 s of the preceding interval when the stimulation was off. There was a remarkable and rapid decrease in locomotor activity upon onset of the optogenetic stimulation, which reversed upon its offset. During stimulation, the genetically treated mice were less likely to walk and more likely to be in a resting state (figure 7.5B). In contrast, no effect of light delivery was observed in the wildtype mice.

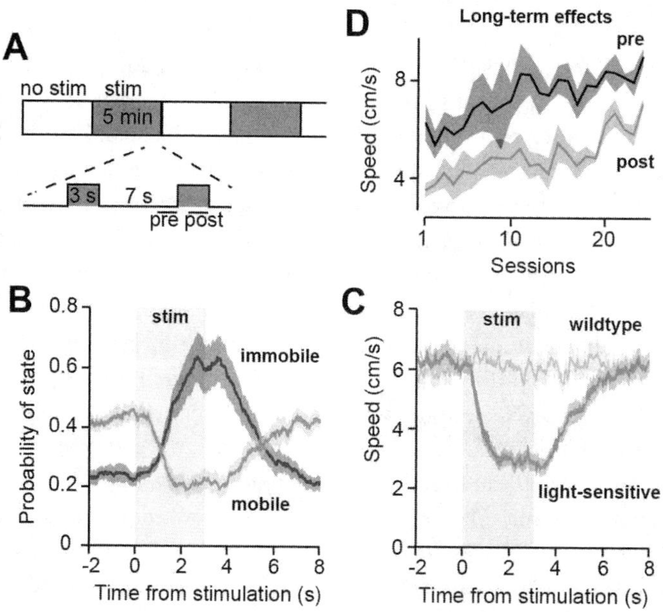

Figure 7.5
Activation of serotonergic neurons leads to slower movements. **A**. Schematic of a testing session. Sessions consisted of alternating 5 min blocks of stimulation and no stimulation. In the stimulation blocks, stimulation was applied in intervals of 3 s on followed by 7 s off for the duration of the 5 min. Movement measures for intervals before (pre) and after (post) stimulation were quantified, and the resultant values were compared. The pre period consisted of the 2 s preceding the start of stimulation; the post period comprised the final 2 s of the stimulation period. **B**. The probability of a mouse being in either a mobile or immobile state averaged over all stimulation intervals and aligned to the onset of stimulation. With stimulation, the probability of being in an immobile state sharply rose, while the probability of a mobile state decreased. Effects were eliminated once the stimulation ceased. **C**. Movement speeds of both light-sensitive and wildtype mice while they were freely walking in the arena. **D**. Movement speed with increasing testing sessions. The black line is movement speed during the pre period, and the gray line is the movement speed during the post period. In B, C, D, lines represent the average across all subjects and shaded areas depict 1 standard error of the mean (SEM). (From Correia et al., 2017.)

To more precisely characterize the effects on the speed of movement, the researchers used automated tracking methods to capture the mouse's spatial position over time from movies recorded during the experiment. Here again, the blue light optogenetic activation led to a near-immediate reduction in speed of locomotion in the light-sensitive mice, but not in the wildtype mice (figure 7.5C).

The inhibitory effect of the stimulation was context specific. The mice were also tested on an accelerating rotarod task for which the goal was to remain on the rod for as long as possible. This is a common test of motor coordination wherein the mouse is placed on a rotating rod that is high enough to motivate them to try to avoid falling. Mice were tested over 3 days and on the 3rd day of testing, both groups received light stimulation. The time spent on the rotarod, or the latency to fall, is the major performance criterion of interest. Both before and during light exposure, the two groups exhibited similar performance. Interestingly, it seemed that when there was increased urgency, serotonin stimulation had no observable effects on vigor.

The mice were also tested in a second task in which they were required to traverse a brightly lit walkway to obtain a food reward. The walkway was instrumented to allow for the measurement of the details of the mouse's locomotor kinematics. In addition to the mouse's average speed, the researchers quantified the speed and position of the individual limbs during the swing phase to determine any effects on the quality of gait and the position of each paw relative to the body center of mass. General locomotor kinematics were unaffected by the stimulation. Surprisingly, in this walking task, there was no effect of stimulation on walking speed. Thus, the stimulation-induced reduction in vigor observed in the open field arena was not observed in a slightly different walking task. Perhaps this could be attributed to the differing motivational aspects of the tasks. In the walkway task, the mice were given strong motivation to traverse the walkway. When they completed the task, they received a food reward and also avoided exposure to the brightly light walkway. Even in the rotarod task, there was strong motivation to remain on the rod and avoid a fall. In contrast, in the open field arena, there was no explicit motivational component to really do anything. One hypothesis to explain the disparate results is that serotonin plays a role in motivating action when there is a balance between the competing forces of whether to act or to wait. When the motivational aspects of the task lead to one force overwhelming the other, such as in the rotarod and walkway tasks, serotonin has a weaker effect.

The authors then examined the effect of repeated stimulation on locomotor behavior. Two sets of light-sensitive mice were tested in the open field arena over a period of a few weeks. One group of mice received repeated stimulation for 24 consecutive days (figure 7.5A). The other group received stimulation only on the 24th day. As the authors had found earlier, the stimulation reduced movement speed in the open field arena. However, the stimulated group progressively increased their average speed over the course of the 24 days (figure 7.5D). Sensitivity to the stimulation nonetheless remained the same over the entire period, with the stimulation leading to constant reductions in velocity when calculated as the difference between the speeds before and after stimulation. However, the overall average velocity continued to increase. In contrast, the group that received stimulation only on the 24th day showed no increase in velocity over the same time period. Thus, repeated stimulation led to long-term locomotor enhancement, a result that resembles the timeline of the therapeutic effects of SSRI antidepressant treatment in humans.

Serotonin reduces vigor under low-threat conditions, but increases vigor when the animal is under severe threat.

Under normal conditions, stimulation of serotonergic neurons appears to produce behavioral inhibition manifested in an increased willingness to wait and a reduction in movement vigor. However, serotonin is strongly implicated in modulating anxiety-related responses. This has been seen in the therapeutic effects of SSRIs on anxiety in humans and in the effects of 5-HT lesions and serotonin agonists on anxiety responses in animals. If serotonin influences both anxiety levels and the willingness to move and exert effort, do anxiety levels alter how serotonin modulates vigor?

To clarify the context-specific effects that serotonin has on vigor, Changwon Seo, Melissa Warden, and colleagues (Seo et al., 2019) used fiber photometry to image calcium and record population activity of serotonin neurons in the DRN while mice moved freely in an open field. As the animals transitioned from stillness to locomotion, activity of serotonin neurons dropped and remained low during the periods of movement (figure 7.6A, middle plot), a pattern roughly the opposite of what is seen with dopamine neurons.

The researchers next trained the mice to associate a tone with the availability of water: they placed the mice on one side of the experiment box and then played a tone. The tone remained on for 8 s, during which time the mice needed to walk across the box and lick the tube to acquire the water. As the mice began walking, activity in serotonin neurons dropped (figure 7.6B, middle plot).

To test whether there was a causal relationship between serotonin release and reduction in vigor, the researchers optogenetically stimulated the neurons for 3 s during open field locomotion (figure 7.6A, right subplot). This produced an immediate reduction in the speed with which the mice walked, and the speed did not recover until at least 2 s after stimulation ended. Similarly, when the mice were placed in the chamber where they had learned to associate a tone with availability of water, stimulation of the serotonin neurons reduced the mice's approach speed toward the water spout (figure 7.6B, right subplot). Note that here, upon termination of stimulation, speed rapidly recovered, a result suggesting that when movements were goal driven (i.e., must reach the water spout before the 8 s deadline), the effect of serotonin on vigor had a time constant of 0.5 s or less.

In contrast to these low-threat scenarios in which increases in serotonin reduced vigor, the effect of serotonin appeared to reverse during high-threat scenarios. To produce a high-threat situation that encouraged an attempt to escape, the authors hung the animals by

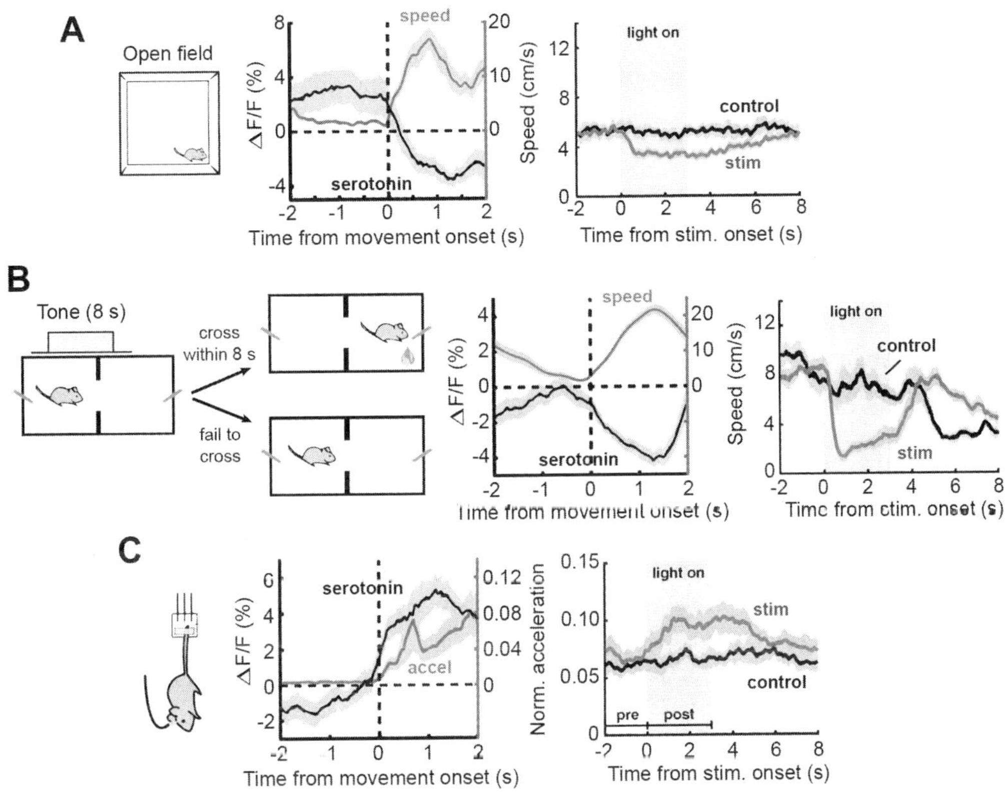

Figure 7.6
Serotonin release tends to lessen vigor, except in high-threat scenarios. **A**. In the open field, activity of serotonin neurons was measured via a genetically encoded calcium indicator. The middle subplot shows speed and serotonin levels aligned to movement onset. The right subplot shows the effect of optogenetic stimulation of serotonin neurons on spontaneous movements. **B**. The mice were trained to move from one side of the chamber to another to receive liquid rewards. **C**. Mice were hung with their tail to a horizontal bar and motion was measured via an accelerometer taped to their tail. In this high-threat scenario, movements accompany increase in serotonin activity, and stimulation of serotonin neurons increases vigor. (From Seo et al., 2019.)

their tail to a horizontal bar and then measured the resulting struggle using an accelerometer. Being hung upside down naturally induced occasional bursts of rapid movements. Surprisingly, these high-vigor movements accompanied high serotonin activity (figure 7.6C, middle subplot). In this context, the release of serotonin increased the vigor of the escape movements (figure 7.6C, right subplot).

In summary, whereas suppression of serotonin enhanced vigor under normal conditions, this effect was reversed under threat conditions. Under low-threat conditions, activation of 5-HT neurons led to immediate and dramatic reductions in locomotor activity, although more habitual movements, such as grooming, were spared. The effects were context specific and not observed in tasks with highly motivational or highly threatening components. This may explain the inconsistent effects of serotonin modulation in the more goal-directed behavioral inhibition tasks described earlier. Intriguingly, repeated stimulation led to a long-term increase in the vigor of walking, revealing a complex interaction between phasic serotonin release and synaptic plasticity in these serotonergic circuits.

7.5 Serotonin, Dopamine, and the Willingness to Work

If increasing serotonin leads to a reduction in vigor, then lower serotonin levels should increase vigor and perhaps also influence the willingness to work. One way to do this is via serotonin receptor antagonists. In mice, the serotonin 5-HT2c receptor is expressed in several brain areas involved in many aspects of motivated behaviors, including the prefrontal cortex, the limbic area, the nucleus accumbens, the dorsal striatum, and the dopaminergic nuclei. The molecule SB242084 is a highly selective 5-HT2c receptor ligand, commonly described as a receptor antagonist, that blocks serotonin-mediated activity. The activity of the 5-HT2c receptor can modulate the activity of dopamine neurons and the release of dopamine in nigrostriatal pathways (Alex and Pehek, 2007). The 5-HT2c receptor exerts an inhibitory influence on dopamine release, while serotonin and 5-HT2c receptor agonists decrease the firing rates of dopamine neurons in the VTA and substantia nigra pars compacta; as a result, dopamine levels are lowered in the nucleus accumbens and the striatum. Conversely, antagonists increase dopamine firing rates and dopamine levels in these areas. Thus, the 5-HT2c receptor seems like an ideal target to manipulate in order to understand the effects of serotonin on vigor and its interactions with dopamine.

Serotonin antagonists increase the willingness to work.
Matthew Bailey, Eleanor Simpson, and colleagues used the 5-HT2c receptor antagonist SB242084 to investigate the effects of serotonin on the willingness to work (Bailey et al., 2016). The team tested mice in a progressive ratio task (PRT) in which they were required to press a lever for a drop of milk. After each reinforcement, the number of lever presses required for the next reinforcement would increase. A session ended after 2 hours or when a period greater than 3 min elapsed after reinforcement. The willingness to work was measured as the duration of each session. Mice were tested for 5 consecutive days; they received either an injection of SB242084 or saline (control group). The mice that had received the receptor antagonist exhibited a greater number of lever presses and longer session durations than the control group, and thus earned a greater amount of reward overall. The lever press rate was also slightly greater in the SB242084 group.

Similar to the outcomes of the willingness-to-wait studies, the results of this PRT study could be ascribed to competing reasons. The findings could be explained by either an increased willingness to work or simply an increase in hyperactivity. It was not clear whether SB242084 led to a specific increase in goal-directed action or simply a general increase in movement. To test this, the authors designed a novel progressive hold down ratio test; the requirement was to hold a lever down for a required amount of time, rather than press a lever for a required number of presses. After each trial, the hold time requirement was increased. Thus, earning more rewards required holding the lever for a longer length of time rather than pressing the lever a greater number of times. Mice were tested for 3 days with SB242084 injections, then for 3 days with saline injections, and then another 3 days with SB242084. Injection of the receptor antagonist led to improved performance on this novel task: more rewards were earned, session durations were longer, as were the hold-down durations. Importantly, if SB242084 were driving a general increase in activity levels, the mice would have been less capable of holding the lever down for extended periods of time. Thus, a serotonin receptor antagonist appeared to increase the animal's willingness to work, by either moving or waiting, for rewarding outcomes.

Serotonin antagonist reduces effort costs through elevation of dopamine.
The increased willingness to work may be due to an altered representation of either the reward or effort costs. To better understand the mechanisms underlying increased motivation, Matthew Bailey, Eleanor Simpson, and colleagues (Bailey et al., 2018) followed up with a series of experiments in which they specifically examined and compared the effects of the receptor antagonist on effort-related and reward-related behavior. Since dopamine has been implicated in driving goal-directed behavior, they also measured the effect of SB242084 on extracellular concentrations of dopamine in the nucleus accumbens and dorsal striatum.

An important first step was to determine whether the receptor antagonist altered the value of reward. To determine whether the treatment led to a general increase in sensitivity to reward, the authors measured dopamine release by using chronically implanted microelectrodes in subjects as they performed a Pavlovian conditioning task. Mice were trained in the conditioning paradigm in which a tone was associated with either a reward (milk) or no reward. After 14 days of training, mice received either injections of SB242084 or injections of the vehicle, an inert compound that served as a control. On the following day, the mice received the opposite treatment. Results revealed that dopamine levels in response to either stimulus (reward or no-reward) were unaffected by treatment. Thus, the 5-HT2c receptor antagonist SB242084 did not seem to alter the dopamine response to reward.

To examine the effect of SB242084 on the willingness to work in more detail, the researchers turned again to the progressive ratio task. Fifteen mice performed a PRT in which the first reward required three lever presses, and the next press requirement would be multiplied by two (i.e., 6 presses to acquire reward). The sessions would last 2 hours or until 3 min had elapsed without a single press. In the PRT, mice were tested during the first 5 days with SB24084 injections and during the next 5 days with vehicle injections. The order of treatment was counterbalanced across subjects. They replicated their previous findings (Bailey et al., 2016) and again observed that SB242084 treatment led to a greater willingness to work in that subjects would work longer and with an increased rate of response (figure 7.7B). The researchers then looked at the finer details of the response and found that bout length, defined as consecutive presses made with < 2 s elapsing between responses, increased with treatment (figure 7.7A). They also found that the duration of pauses in responding between bouts decreased. Both bout length and pause duration generally varied with the press requirement (figure 7.7C, D). Treatment effect seemed consistent, except for bout length at the lower press requirements. The average inter-response time within a bout (IRT) also declined with treatment and was modulated by press requirement (figure 7.7E). Post-reinforcement pauses (PRPs) were unaffected by the treatment, a finding consistent with the lack of an effect of treatment on reward valuation.

In summary, the authors observed that with serotonin receptor antagonist treatment, there was an increased willingness to work, resulting in longer bouts of responding, shorter pauses between bouts, and faster responses within a bout.

If SB242084 treatment increases response vigor, does the underlying mechanism involve dopamine? To answer this question, they used *in vivo* microdialysis to measure dopamine levels in the dorsomedial striatum and nucleus accumbens in subjects as they performed an effort-based choice task. Eight mice performed three sessions on separate days, approximately 1 week apart. Each session was 60 min during which a single milk reward was earned after an average of 15 presses. Alternatively, they could eat home cage chow at will. In each session, the mice underwent one of three procedures: (1) SB242084 injections, (2) vehicle injections, or (3) SB242084 injections without behavioral testing. Confirming the earlier behavioral findings, the treatment group exhibited greater vigor with a greater number of presses and shorter IRTs. Treatment also led to a significant increase in extracellular dopamine concentrations in the dorsomedial striatum, and this occurred only when the subjects were actively engaged in the task. Dopamine levels did not increase in the group that received treatment without behavioral testing. In contrast to the marked effects observed on dopamine levels in the dorsomedial striatum, no effects were observed in the nucleus accumbens.

To determine whether dopamine levels were responsible for the increased motivation, the researchers infused a dopamine antagonist (flupenthixol) into the dorsomedial striatum as the subjects performed another lever-pressing task in which rewards were earned after 15 presses. Injection of the dopamine receptor antagonist resulted in slower responses and fewer rewards earned. To test the hypothesis that SB242084 invigorates movement by increasing dopamine levels, they next tested if the dopamine receptor antagonist would cancel the positive effects of SB242084. Receiving the SB242084 injection alongside the dopamine receptor antagonist nullified the effect of SB242084, with effort-based responses the same as those of the control group. Taken together, blocking the dopamine activity in the dorsal striatum prevented the invigorating effects of the serotonin receptor antagonist.

SB242084 does not alter reward preference and has no effect on dopamine levels in the nucleus accumbens; these results suggest that the drug is not affecting reward valuation.

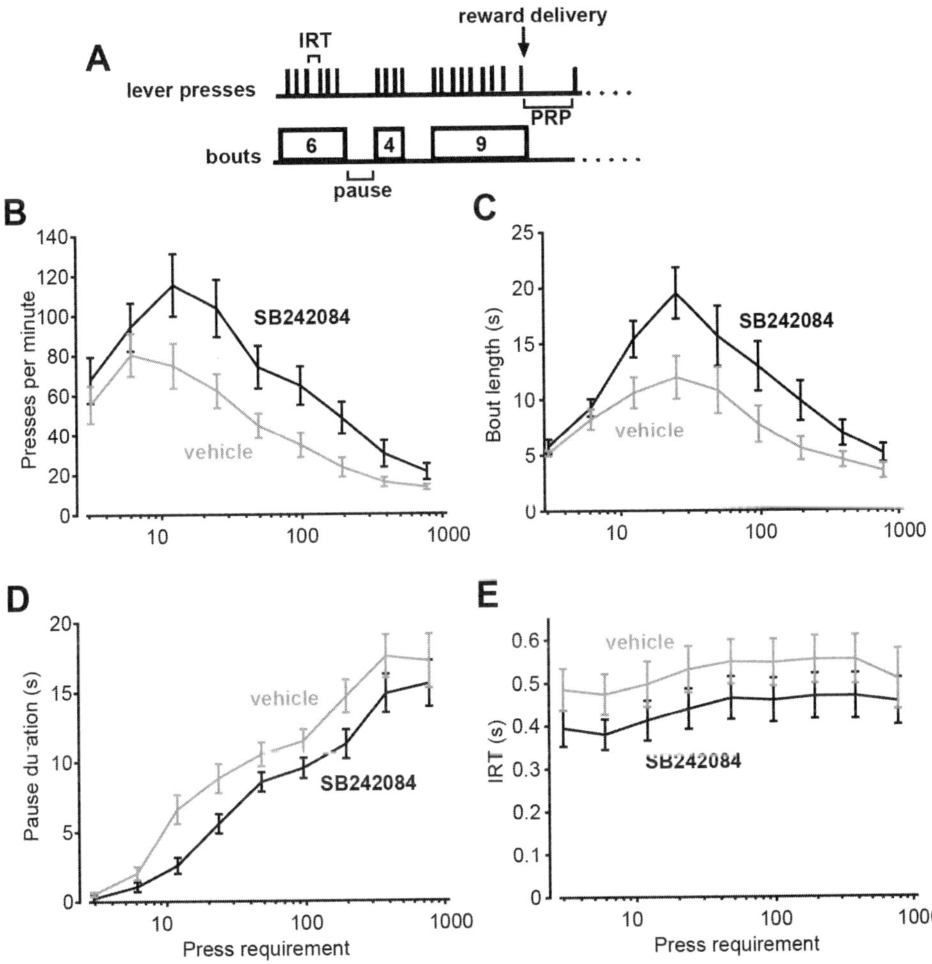

Figure 7.7
5-HT antagonist increases the willingness to work for reward. **A**. Movement vigor analysis. (PRP, postreinforcement pause) **B**. Average press rate as a function of press requirement with systemic 5-HT antagonist SB242084 or vehicle (control solution). **C**. Average bout length with increasing press requirement. **D**. Average duration of pauses between bouts with increasing press requirement. **E**. Average interresponse time (IRT) as a function of press requirement. Data are averages across mice. (From Bailey et al., 2018.)

In contrast, SB242084 allowed subjects to work harder, longer, and faster. This was associated with greater levels of dopamine in the dorsal striatum, and the effect was only observed when subjects needed to work for rewards. An infusion of dopamine antagonists blocked this effect, suggesting that the 5-HT2c receptor mediates response vigor by facilitating dopamine release into the dorsomedial striatum.

In summary, a serotonin receptor antagonist increases the vigor of movements through potentiation of dopamine release in the dorsal striatum. Perhaps this means that under normal circumstances, serotonin leads to an inhibition of dopamine release, but this remains to be better studied.

7.6 Serotonin and Motoneuron Activity

One of the defining characteristics of serotonin neurons is their set of diffuse projections throughout both the brain and the spinal cord. Thus far we have focused on the DRN, which along with the medial raphe nuclei, is located in the rostral cluster and consists of serotonin neurons that project primarily to the forebrain. However, serotonin neurons are also located caudally in the brainstem and medulla. This caudal group of serotonin neurons, which includes the nuclei raphe magnus (NRM), nuclei raphe obscurus (NRO), and nuclei raphe pallidus (NRP), innervates the spinal cord, including motoneurons.

Serotonergic firing rates track motor activity.

Barry Jacobs and colleagues (Veasey et al., 1995) studied the role of serotonergic neurons that projected to the spinal cord by recording chronically from single neurons in freely moving cats. Their results were intriguing: in contrast to the movement slowing observed in the studies that had stimulated the DRN, Jacobs observed that faster movements led to increased activity among the serotonin neurons in the NRO and NRP projecting to the spinal cord.

To record neuronal activity over long periods of time as well as adjust for large movements of the animal, they developed an innovative technique in which chronically implanted microwires were attached to a mechanical microdrive. After surgery, the cats were trained to walk or run across a range of speeds on a treadmill. Data were collected for 1 min at each speed, with a 1 min rest between each trial. To ensure consistency in the recorded neuronal activity, the researchers repeated the tests for several speeds.

The authors were able to obtain successful locomotor trials for 24 serotonergic neurons. Overall, during locomotion, these neurons significantly increased their activity with respect to the baseline levels. Moreover, there was a significant positive correlation between treadmill speed and serotonergic activity (figure 7.8). Thus, whereas serotonergic neurons in the DRN reduced their activity during locomotion (Seo et al., 2019), here we see increased activity in 5-HT neurons projecting to the spinal cord.

Serotonin increases spinal reflex excitability.

What might be the functional consequences of increased serotonin release to the motoneurons of the spinal cord during voluntary movements? Kunlin Wei, Konrad Kording, CJ Heckman, Zev Rymer, and colleagues (Wei et al., 2014) examined this question by measuring the effects of serotonin on reflexes that reside in the spinal cord.

One of the simplest reflexes is the stretch reflex, mediated via Ia spinal afferents that monosynaptically project to motoneuron cell bodies in the spinal cord. Wei and colleagues

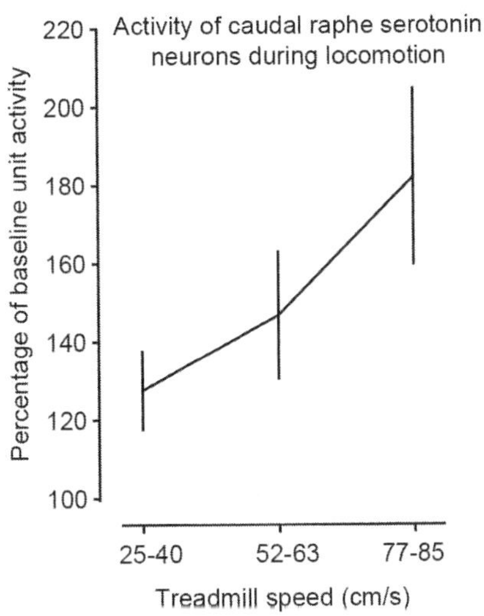

Figure 7.8
Caudal serotonergic activity increases with movement speed. Activity of twelve 5-HT neurons in the nuclei raphe obscurus (NRO) and the nuclei raphe pallidus (NRP), recorded from cats walking on a treadmill at each of the three speeds. The 5-HT activity correlates with movement speed. Line is the mean across the 12 neurons and vertical bars are 1 SEM. Baseline is quiet standing. (From Veasey et al. 1995, Copyright [1995] Society for Neuroscience.)

performed a series of experiments in which they tested the reflex responses after human subjects ingested either a serotonin agonist or a serotonin antagonist. They then measured the effect of the drug on a motor task during which the subject's tendon was mechanically tapped to elicit an extension reflex.

In the first experiment, subjects either took a placebo or a tablet of escitalopram 5 hours before testing. Escitalopram is an SSRI that elevates serotonin levels in the brain. The test was a probe of reflex gains in the wrist extensor muscles. Subjects placed the middle finger in a metal ring, with the wrist supinated. The finger and hand were splinted and immobilized to eliminate confounding movements. A round metal head attached to a linear motor was aligned with the wrist tendon. As the motor drove movement of the metal head into the tendon, a reflex response was elicited, the strength of which was measured via a force transducer attached to a metal ring supporting the middle finger. So, the reflex response was stimulated via a tendon tap and measured via the reflexive force generated at the finger as the wrist extended.

After calibrating for initial force level at the metal head, the reflex was elicited by applying a 100 Hz vibration for 8 s, after which the vibration came to a stop and any reflexive force was allowed to decay. Twenty-four trials were conducted in each session, in four blocks of eight trials, with approximately 3 min between blocks. Importantly, the test probed reflex gains, so beyond performing a maximum voluntary contraction (MVC)

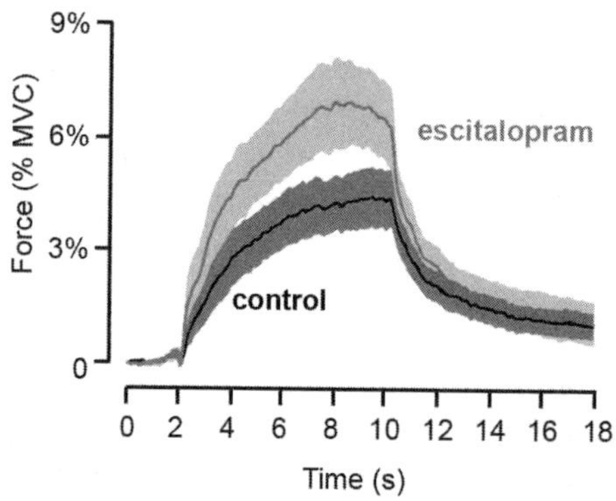

Figure 7.9
Selective serotonin reuptake inhibitors (SSRIs) increase reflex gain in healthy adults. Reflex force elicited in healthy human adults measured after ingesting an SSRI (escitalopram) or a placebo tablet. The SSRI led to a greater reflexive force. Lines represent means across subjects. Shaded areas are 1 SEM. (From Wei et al., 2014.)

at the start of each session, subjects were not asked to voluntarily develop force. Among the subjects that took an SSRI, the force at the finger increased more rapidly and reached a higher magnitude, as assessed in the final 2 s of vibration (figure 7.9). In other words, the SSRI led to stronger reflex gains in healthy subjects.

The increased force response could be attributed to a stronger reflex gain through increased excitability of the motoneuron or a stronger descending drive. Therefore, in their next experiment, they tested subjects with chronic spinal cord injury who exhibit low levels of descending drive. The authors tested the strength of the knee-extension reflex as subjects sat upright with the knee flexed to 90 degrees. Each patient performed two sessions, each on a separate day, with at least 5 days between sessions. In one session, patients took an SSRI (escitalopram), and in the other they took a 5-HT2 receptor antagonist, cyproheptadine. The patellar tendon was tapped with a rubber head attached to a linear servomotor that would apply a single position-controlled tap to the tendon (in contrast to the continuous vibration in the previous experiment). In each session, patients were tested both before and after taking the assigned treatment. Electromyographic (EMG) responses in the three monoarticular knee extensor muscles (vastus lateralis, vastus intermedius, and vastus medialis) were measured. As with the healthy subjects, when the patients took an SSRI, significant increases were seen in reflex amplitudes, as assessed via EMG magnitudes before and after treatment (figure 7.10A, C, E). The reflex gain, calculated as the slope of the relation between EMG responses and force applied, also increased. In contrast, the 5-HT2 receptor antagonist cyproheptadine led to a weaker reflex response and lower gain (figure 7.10B, D, F).

In summary, use of an SSRI, which elevates levels of serotonin, led to stronger spinal reflex responses in both healthy subjects and patients with chronic spinal cord injuries.

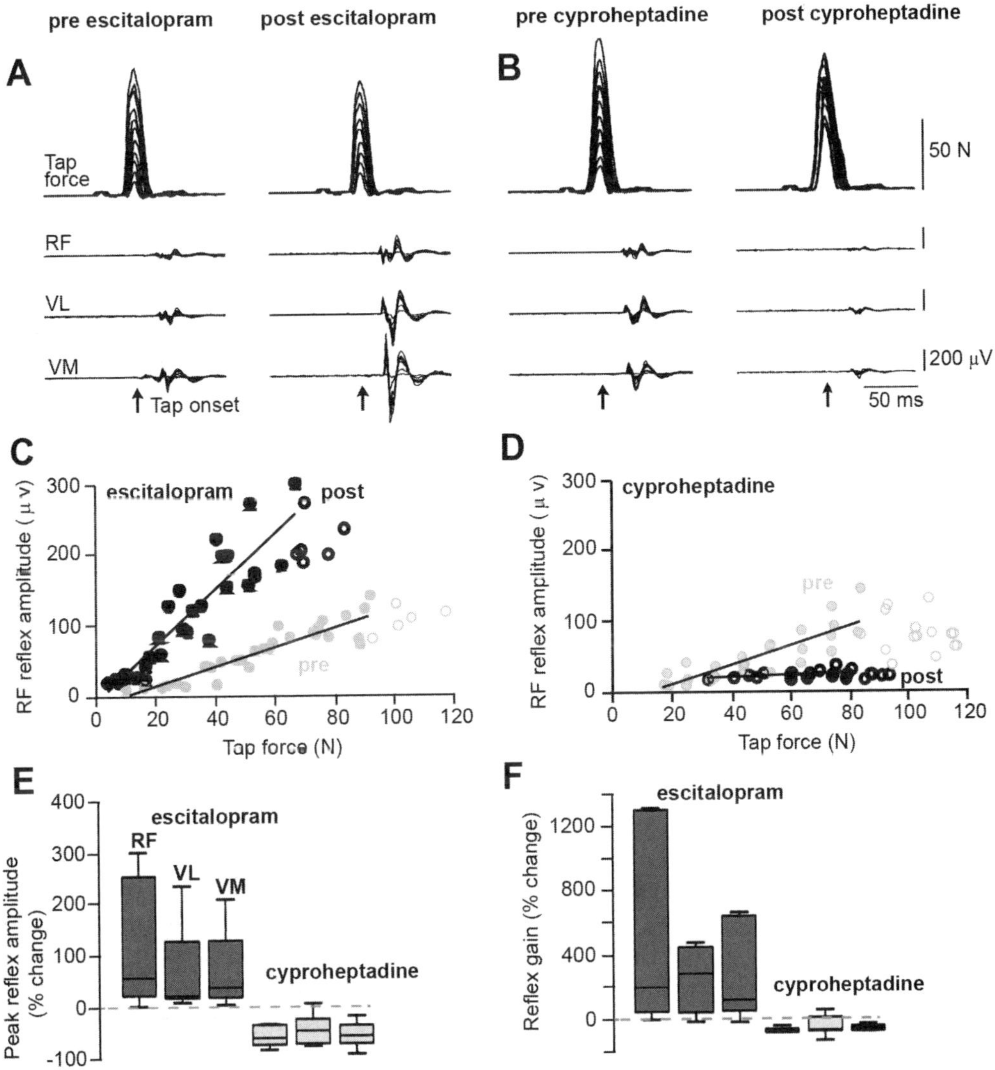

Figure 7.10

Serotonin antagonists decrease spinal reflex excitability. **A**. Muscle activity recorded in three knee extensor muscles: rectus femoris (RF), vastus lateralis (VL), and vastus medialis (VM). The activity was in response to a patellar tendon tap force of increasing magnitude, before and after having taken the selective serotonin reuptake inhibitor (SSRI) escitalopram. Muscle activity increased after taking the SSRI. Data shown for a single subject. **B**: Same as in A, but before and after taking the serotonin receptor antagonist cyproheptadine. Muscle activity reduced after taking cyproheptadine. **C,D**. RF reflex amplitude measured as a function of tap force both before and after escitalopram (C) or cyproheptadine (D). Reflex gain is calculated as the slope for the relationship between amplitude and tap force (black line). Reflex amplitude and gain are increased with escitalopram and decreased with cyproheptadine. Data shown for one subject. **E,F**. Percentage change in peak reflex amplitude and reflex gain for each muscle in response to each treatment. Data averaged across all subjects. Across subjects, escitalopram increased reflex amplitude and gain in all three muscles, while cyproheptadine led to a reduction in reflex amplitude and reflex gain. (From Wei et al., 2014.)

Use of a drug that reduces the serotonin-mediated activity of the 5-HT2 receptor led to the opposite effect: a reduction in reflex response. Overall, these results demonstrate the powerful manner in which serotonin can influence movements through spinal pathways.

Limitations

Diversity is a defining feature of the serotonergic system. Most of the limitations discussed below arise from the breadth of 5-HT projections throughout the central nervous system and multitude of receptor types that complicate our understanding of the direct effects of 5-HT on movement vigor. In many instances, it is nearly impossible to identify the direct effect of a given type of serotonergic activity on dopamine levels in the brain, and more generally, on movement vigor.

Many of the studies we have described either optogenetically stimulated 5-HT neurons in the DRN, or measured their activity (Miyazaki et al., 2014; Cohen et al., 2015; Correia et al., 2017; Lottem et al., 2018; Seo et al., 2019). It is difficult to interpret their findings because not all DRN neurons project to the same brain area, nor do all of them have the same function (Ren et al., 2018). Each study is only recording from a subset of neurons in the DRN; therefore, their results are specific to those neurons and caution should be used when generalizing those results to the function of DRN 5-HT neurons as a whole.

The range of responses observed by Cohen et al. (Cohen et al., 2015) may be in part due to the fact that the neurons were projecting to different areas and had different functions. This may also help explain some of the discordant results. For example, Miyazaki et al. and Lottem et al. observed no effect of stimulation on speed of movement (Miyazaki et al., 2014; Lottem et al., 2018), whereas others did (Correia et al., 2017; Seo et al., 2019). The difference may be due in part to the different effects of various subgroups of DRN neurons. Future work will need to increase the specificity of stimulation and identify the neuronal targets to better understand the relationship between 5-HT activity and vigor of movement.

In addition to the anatomical diversity, there is a multitude of 5-HT receptor types found in cells across the central nervous system (CNS), a reflection of the neurotransmitter's broad range of functions. Approaches that target 5-HT receptors systemically (Nautiyal et al., 2015; Bailey et al., 2016, 2018) are complicated by the fact that a given receptor type can be found on multiple cell types. For example, the 5-HT2c receptor is found on both inhibitory GABA cells and dopaminergic neurons. Therefore, it is difficult to identify the neural circuit leading to the increased levels observed by Bailey et al. (Bailey et al., 2018). They administered a systemic 5-HT2c antagonist, SB242084, that may have acted by reducing the GABAergic inhibitory control over dopaminergic neurons, or by reducing GABAergic control over dopamine-mediated striatal activity. The density and breadth of 5-HT2c receptor expression throughout the CNS makes it difficult to identify the circuit and ultimately the mechanism underlying the effect of 5-HT on the willingness to work.

The nearly opposite roles of serotonin in the brain and in the spinal cord is a fascinating observation that highlights the diversity of functions of the 5-HT system. Not only are there multiple effects of serotonin on the control of vigor, but serotonin is implicated in a range of neurological disorders that involve mood, anxiety, and affect. Indeed, to the lay public, serotonin is known as the happiness hormone. We do not touch on these roles, but it is intriguing that serotonin, like dopamine, is involved in both feelings of happiness and control of vigor.

Summary

If you have a "spring in your step," odds are you have had a good day, which means that you have experienced a high average reward rate. From a theoretical perspective, a high reward rate implies a high capture rate, which translates into shorter harvest duration and increased vigor. Serotonergic neurons are sensitive to average reward rate: they change their activity to reflect the history of reward and punishment. These neurons relay this information to the basal ganglia, which then affects control of vigor. Marginal value theorem predicts that a lower average reward rate should lead to slower movements and longer harvests. The relationship between serotonergic activity and motor behavior across a range of experiments supports such a role in signaling average reward rate. Increased serotonergic activity reduces impulsivity and increases the willingness to wait, thereby prolonging the harvest period. In contrast, serotonin receptor knockout leads to greater impulsivity, the promotion of impatience, and a desire to seek a better harvest elsewhere. During the expenditure of effort, serotonergic activity is generally lowered. Artificial activation of serotonergic neurons increases harvest duration and reduces movement vigor. Thus, serotonin portrays some of the key signatures needed in a marker of average reward rate. However, serotonergic projections are broad and multifunctional, with occasionally opposing effects on movement vigor. For example, in contrast to its motor-slowing effects in the brain, serotonin's spinal projections invigorate movement, lowering the threshold for action initiation. Taken together, these results suggest that serotonin, along with dopamine, are modulating the balance between exploration and exploitation, thereby affecting both the process of decision-making and the vigor of the ensuing action. Together, these neuromodulators influence the cost-benefit analysis required for the optimization of the global capture rate.

Conclusions

> Forget your perfect offering.
> There is a crack in everything.
> That's how the light gets in.
> —Leonard Cohen

Why do we run toward the people we love, but only walk toward others? If we view the experience of being with our loved one as a form of reward, then by moving faster we are expending energy to shorten the time to reward. The magnitude of this reward minus the effort expended to acquire it, divided by time, describes the utility of our act, which is a currency termed capture rate.

Capture rate has been important to the process of evolution: animals that behave in ways that increase their capture rate tend to live longer and have more offspring. By running, our brain has decided that the reward, if acquired sooner, is worth the expenditure of effort.

Thus, in the context of an expected reward, running has a greater impact on the denominator of the capture rate (by saving time) than on the numerator (by increasing effort expenditure). By running and then subsequently harvesting reward, we increase our capture rate.

But, occasionally, we get to our destination and find it disappointing; we experience a negative reward-prediction error. Curiously, we walk away with low vigor. Why does experience of a bad outcome reduce the vigor of our subsequent movements?

If we think of our global capture rate as the grand sum of all of our efforts and rewards divided by time, and we view patterns of decision-making as an attempt to increase our supply of this currency, then our behavior cannot focus only on what is in front of us. Rather, we must also pay attention to our history. If our past actions have produced rewards that have been low, reducing our global capture rate, then the best current action is to use the past capture rate as a rudder to maximize the global rate: we should move in such a way that when we arrive at our destination, the short-term losses that we are about to endure are no greater than the average gains that we have experienced in the near past.

Thus, if we have had a tough day, we should not (and typically will not) run toward our loved ones. This does not mean that we value them any less. Rather, the low vigor is a reflection of our recent history of reduced capture rates. By walking, we save some energy, achieving a local capture rate that is lower than if we had run. Over the long term, this is the policy that will lead to the maximum global capture rate.

We inferred these ideas by constructing our theory of vigor in the framework of optimal foraging. The theory allowed us to consider decisions and movements in a unified framework. With it, we had a way to consider how long an animal should stay to harvest the reward and how fast it should travel to its next reward opportunity. We had a way to consider how reward and effort at the harvest site affected harvest duration and how the effort of travel to the next reward site affected vigor. Finally, we had a way to link the history of rewards acquired and efforts expended with the vigor of the movement that was about to be produced.

However, in order to test this theory, we had to abandon the example of walking patterns as people moved toward their loved ones and instead focus on traits more easily measurable: eyes as the sensor that harvests visual information for the brain, and saccades as the mode of travel that brings that sensor to a location where reward (information) is found.

In examining saccades, we found that some, but not all, of the theoretical predictions were confirmed. People and other animals reacted sooner and moved their eyes with greater velocity toward images that they valued. Once there, they spent a greater amount of time gazing at the valuable image and then moved with high vigor to the next image (something akin to leaving a good meeting with a spring in their step). Thus, both their history of past rewards (what they left behind) and their expectation of the current reward (what they moved toward) affected saccade vigor in accordance with the theory.

However, contrary to the theory, after an expenditure of high effort, people did not reduce their vigor. Rather, if their past movements had required high effort, they moved with greater vigor toward their next reward opportunity. This was puzzling because past effort should have reduced their long-term capture rate and thus compelled them, in accordance with the theory, to slow down.

Instead, people acted as if their past effort expenditure had increased their valuation of the current reward opportunity. That is, after having expended high effort to acquire a reward, they behaved as if that reward had become more valuable.

This was yet another example of the IKEA effect: if we have spent effort to build it, we value it more. To explore this and other puzzles like it, we turned our attention to the neural basis of effort expenditure and vigor control.

We began in the superior colliculus, where there are neurons that reflect the utility of staying still (fovea-related neurons in the rostral pole) and neurons that reflect the utility of moving to a specific location (saccade and buildup neurons in the caudal region). Among the fovea-related neurons, activity was higher when fixation was associated with greater reward (e.g., while looking at a more valuable image). These neurons reduced their discharge when there was a rewarding stimulus elsewhere and then paused near saccade onset. In the caudal regions of the colliculus, there were neurons that built up their activity during the reaction time period, then produced a burst near saccade onset. The reward increased the rate of rise in the activity of these neurons, enabling them to burst sooner and produce a more vigorous saccade.

Together, the rate of decline in the activity of collicular neurons that encoded the utility of staying still (fixate and harvest) and the rate of rise in the neurons that encoded the utility of moving (abandon harvest and travel) corresponded to the reaction time of the saccade.

However, although the colliculus receives direct projections from the retina and has privileged access to the machinery that made saccades, it is not trusted with the task of directing gaze. Rather, its activity is largely under the control of the cerebral cortex and the basal ganglia. Through excitation, the cortical projections specified where the saccade should be directed and through inhibition, the basal ganglia biased that selection and influenced vigor. Together, the net difference between the sum of excitation and the sum of inhibition became an instruction that defined the timing and vigor of the movement that was ultimately directed by the colliculus.

The decision of where to make a saccade appeared to be largely controlled by cortical regions such as the frontal eye field (FEF), lateral intraparietal cortex (LIP), and orbitofrontal cortex. In these regions, some neurons encoded the utility of staying, whereas other neurons encoded the utility of moving. The neurons that encoded the utility of moving often did so in terms of the magnitude of the expected reward and the probability of receiving that reward (i.e., variables that affected expected value of the stimulus).

As a result, cortical neurons associated with a particular movement displayed a faster rise in activity when the expected value of the stimulus was larger. Thus, the decision of where to direct the saccade appeared to be due to the utility-dependent activity of neurons that increased their discharge, thereby increasing the excitation that they imposed on neurons in the colliculus that directed that movement.

This cortical activity, however, did not need to reach a fixed level before the movement was generated. Sometimes the movement started when the cortical activity was relatively low, whereas other times deliberation was longer, resulting in a later, slower saccade. For example, when the urgency of the task was increased, encouraging a shorter period of deliberation, cortical activity started at a higher level even before the onset of the stimulus, seemingly priming the decision-making system to rapidly arrive at a choice. Indeed, in trials in which the movement had high vigor, the level of activity reached by the cortical neurons that preferred that movement tended to be higher than when the same movement was selected but the vigor was lower.

From this, we inferred that the decision to move was not made because of the activity of neurons in any particular region of the cortex. Rather, it seemed possible that the threshold of activity needed to start the movement was set by a brainstem structure much closer to the motoneurons (i.e., the omnipause neurons). Here, inputs converged from the various cortical and brainstem regions, including neurons that preferred holding still and neurons that preferred movement; a saccade was made only if the sum of all activities passed a threshold. Thus, the decision to move the eyes was a collaborative process involving the cortex, the basal ganglia, and the colliculus.

Whereas the cortical regions provided utility-reflecting excitation to the colliculus, specifying a particular movement, the basal ganglia's output to the colliculus was inhibitory. As a result, the decision to abandon the current harvest and to move, along with the vigor of that movement, was due to a combination of increased excitation from the cortex and reduced inhibition from the basal ganglia.

However, unlike the movement-related cortical neurons that fired to indicate their preference for a specific saccade, the SNr neurons, providing an output from the basal ganglia, had broad response fields and showed little specificity for amplitude. (These neurons in the SNr were more active for contralateral saccades than for ipsilateral ones, but not particularly well tuned for amplitude). The SNr neurons biased the decision-making process by modulating their inhibitory activity during deliberation, pausing their discharge to affect the vigor of the chosen movement.

The magnitude of the inhibition imposed by the SNr depended on the expected risks and reward value of the stimulus, which in turn were transmitted to them from cells in the striatum. The striatal cells received excitatory inputs from the cerebral cortex, but their output to the SNr via the direct and indirect pathways depended on the amount of dopamine that was present in the striatum. That is, dopamine influenced how the striatal cells responded to their cortical inputs; the neurotransmitter either impeded or encouraged the transition of striatal cells to an active state, a state in which the striatal cells were more responsive to their excitatory cortical inputs. Dopamine levels, in turn, were influenced by serotonin. Together, dopamine and serotonin, two ancient neurotransmitters, influenced both the decision of where to move to and the vigor with which the movement would occur.

Although dopamine neurons fired in response to stimuli that promised reward, their firing rate was not modulated by the effort required to acquire that reward. Furthermore, the firing rates appeared insensitive to one of the critical variables that the theory had identified: global capture rate. However, during effortful movements, there was greater concentration of dopamine in the striatum.

Thus, dopamine spiking activity at the time of stimulus presentation appeared to encode reward-prediction error, whereas during production of effort, dopamine levels appeared to support production of vigorous movements. In this way, dopamine played the role of Janus the Roman god: one face looking to the reward, the other looking toward the effort needed to acquire that reward.

This provided an important clue as to the neurobiology that may underlie the IKEA effect. After the expenditure of high effort, humans and other animals value rewards acquired through great effort more than if the same rewards were acquired with less effort. During the period of high effort, dopamine levels increased, thus making it easier for cortical inputs to alter the activity of the striatal neurons; in turn, the inhibition imposed on the colliculus by the SNr was effectively reduced, and a vigorous movement resulted. It seems likely that another effect of dopamine is an increase in the value of the stimuli that were present at the conclusion of the effortful act.

But, if firing rates of dopamine neurons did not provide a signal that reflected the global capture rate, how was it that the brain controlled vigor as a function of reward history? One possible answer was serotonin, which is sensitive to reward history and in many ways appears to act as an antagonist to dopamine, encouraging sloth. The elevated presence of serotonin in the brain under normal conditions coincided with reduced movement vigor and a reluctance to expend effort in exchange for a reward. In addition, serotonin increased the tendency to linger and harvest for a longer period of time, thus encouraging persistence. In some of these experiments, the effect of serotonin on behavior was mediated via the modulation of dopamine.

An exciting implication of the research on vigor is that movements can provide an easily measured proxy for hidden variables such as utility. Economists have strived for decades to estimate utility on the basis of preferences that people have expressed through their decisions. Vigor may provide an implicit measure of this elusive variable, thus providing a new way to estimate how much people value each of their various options.

Because the neurobiology of vigor is based on many of the same neurotransmitters that malfunction in diseases such as Parkinson's and depression, tracking vigor may provide a real-time proxy for the state of these chemicals. Thus, vigor assessment could aid in the administration of interventions and provide an objective measure of treatment efficacy.

However, there is also a danger in examining the mechanics of vigor. Vigor may unveil the secret of how much we value our destination; it inadvertently unmasks our subjective views of people and things around us. A better understanding of vigor could reveal information that we do not intend to share.

> If we assume we've arrived, we stop searching. We stop developing.
> —Jocelyn Bell Burnell

Appendix A: Effective Mass of the Human Arm

The human arm has a mass distribution that resembles a heavy object when it moves in some directions (major axis of the inertia ellipse), and a light object when it moves in other directions (minor axis of this ellipse). For example, when the arm is in the configuration shown in figure 1.11A, it has a large effective mass during reaches toward 135°, but it has a much smaller effective mass during reaches toward 45°. Let us compute this effective mass.

We begin with the inertia of the arm, which, for the planar configuration of the arm, is a 2×2 position-dependent matrix $I(\theta)$, where $\theta=[\theta_s\ \theta_e]^T$ and represents the angular positions of the shoulder and elbow joints. At rest, inertia represents the relationship between a vector of joint accelerations and the resulting torque:

$$\tau = I(\theta)\ddot{\theta} \tag{A.1}$$

We are interested in computing the mass matrix $M(\theta)$, which represents the relationship between the acceleration vector $\ddot{x}$ and the force vector f as measured at the hand at rest:

$$f = M(\theta)\ddot{x} \tag{A.2}$$

We use the Jacobian matrix:

$$\Lambda = \frac{dx}{d\theta} \tag{A.3}$$

We then use the principle of virtual work to relate force to torque and acceleration in joint coordinates to acceleration in hand coordinates:

$$\begin{aligned} \tau &= \Lambda^T f \\ \dot{x} &= \Lambda\dot{\theta} \\ \ddot{x} &= \frac{d\Lambda}{d\theta}\dot{\theta}\dot{\theta} + \Lambda\ddot{\theta} \end{aligned} \tag{A.4}$$

Using the previous equalities, we can write the relationship between hand acceleration and force:

$$f \approx (\Lambda^{-1})^T I(\theta)\Lambda^{-1}\ddot{x} \tag{A.5}$$

Equation A.5 is an approximation because we are neglecting the first term in the equality, which relates angular accelerations (at the joints) to hand accelerations. As a result, we can define the mass matrix at the hand as follows:

$$M(\theta) = (\Lambda^{-1})^T I(\theta)\Lambda^{-1} \tag{A.6}$$

In the case of the planar arm that we are considering, the mass matrix $M(\theta)$ is a 2×2 and describes the relationship between accelerations and forces at the hand. To compute the effective mass, we apply an acceleration of length unity in a given direction and compute the length of the resulting force vector. This length provides us with an estimate of the effective mass of the arm for movements in that direction.

Appendix B: Alternate Forms of Utility

In optimal foraging, the objective is to maximize a utility called the global capture rate. This utility is defined as the sum of all rewards acquired minus all efforts expended, divided by total time. Here, we briefly review properties of the global capture rate and then consider some alternative formulations.

Suppose we travel to patch n and harvest reward there. During the travel period, we spend the amount of time $t_s^{(n)}$ and the amount of energy $g^{(n)}$. Once we arrive at the patch, we spend the period $t_h^{(n)}$ harvesting reward, thus accumulating the amount $f^{(n)}$. This is the local capture rate:

$$J^{(n)} = \frac{f^{(n)} - g^{(n)}}{t_h^{(n)} + t_s^{(n)}} \tag{B.1}$$

The global capture rate, as defined in optimal foraging, is the ratio of rewards acquired, divided by time, as summed over all experiences:

$$\bar{J} = \left(\sum_{n=1}^{N} f^{(n)} - g^{(n)} \right) \left(\sum_{n=1}^{N} t_h^{(n)} + t_s^{(n)} \right)^{-1} \tag{B.2}$$

To find the optimum duration of harvest, we take the derivative of the capture rate with respect to harvest duration:

$$\frac{d\bar{J}}{dt_h^{(n)}} = \frac{df^{(n)}}{dt_h^{(n)}} \frac{1}{t_h^{(n)} + B} - \frac{f^{(n)}(t_h^{(n)}) + A}{(t_h^{(n)} + B)^2} \tag{B.3}$$

At the optimum harvest period $t_h^{(n)*}$, the above expression is equal to 0, which gives us the following expression:

$$\left. \frac{df^{(n)}}{dt_h^{(n)}} \right|_{t_h^{(n)*}} = \bar{J} \Big|_{t_h^{(n)*}} \tag{B.4}$$

Therefore, the optimum time to leave the patch, $t_h^{(n)*}$, is when the marginal capture rate (left side of equation B.4) is equal to the global capture rate.

Note that in the formulation of the global capture rate, the global utility $\bar{J}$ is not the average value of the local capture rate. That is, $\bar{J} \neq N^{-1}\sum_{i=1}^{N} J^{(i)}$. What if the global capture rate was defined as the average value of the local capture rate?

Suppose we defined our global utility as follows:

$$\bar{J} = \frac{1}{N}\sum_{i=1}^{N} J^{(i)} \tag{B.5}$$

To find the optimum harvest duration at patch n, we note the following:

$$\frac{d\bar{J}}{dt_h^{(n)}} = \frac{1}{N}\frac{dJ^{(n)}}{dt_h^{(n)}} = \frac{1}{N(t_h^{(n)} + t_s^{(n)})}\left(\frac{df^{(n)}}{dt_h^{(n)}} - J^{(n)}\right) \tag{B.6}$$

At the optimum harvest period $t_h^{(n)*}$, the above equation is equal to 0. We then have the following:

$$\left.\frac{df^{(n)}}{dt_h^{(n)}}\right|_{t_h^{(n)*}} = \left.J^{(n)}\right|_{t_h^{(n)*}} \tag{B.7}$$

Therefore, if we define the global utility as the average value of the local utility, then the harvest duration and period of travel are dependent only on the local utility. In contrast, when we define the global utility as the sum of rewards and efforts over sum of time, the harvest duration and period of travel depend on both the local conditions and the history of the subject.

A second alternative that we might consider is to weigh the utility of each patch on the basis of the length of time we spent there. Suppose that utility of event n is defined as $J^{(n)} = e^{(n)}/t^{(n)}$. In a sequence of N events that take a total time of $\sum_{j=1}^{N} t^{(j)}$, assume that the probability of a single event is proportional to its duration $p^{(n)} = t^{(n)} \big/ \sum_{j=1}^{N} t^{(j)}$. That is, in this formulation, the assumption is that longer duration events are more likely. In this sequence of events, the expected value of an event is $E[J] = \sum_{n=1}^{N} J^{(n)} p^{(n)} = \sum_{n=1}^{N} J^{(n)} \big/ \sum_{j=1}^{N} t^{(j)}$. This expression simplifies to $E[J] = \sum_{n=1}^{N} e^{(n)} \big/ \sum_{j=1}^{N} t^{(j)}$ which is the definition of $\bar{J}$ in optimal foraging.

In summary, optimal foraging describes a global utility that depends on the sum of all rewards acquired minus all efforts expended, divided by total time. An important consequence of this definition is that behavior depends on the local reward and effort contingencies as well as the past history and future expectations of the subject. However, if we define the global utility as the average of the utilities at each harvest location, then behavior will no longer depend on the history of the subject.

Appendix C: Algebra of Random Variables

Our aim here is to derive the probability density of one variable that is algebraically related to a second variable whose probability density we already know.

Linearly Related Random Variables

Suppose that we have a random variable X, described by the probability density function p_X. The random variable Y is linearly related to X as follows:

$$Y = aX \tag{C.1}$$

Given the above relationship, we can compute the relationship between probabilities (see figure C.1):

$$\Pr[Y \leq y] = \Pr\left[X \leq \frac{y}{a}\right] \tag{C.2}$$

The probability $\Pr[Y \leq y]$ is a function of y, which we can label this way:

$$P_Y(y) = \Pr[Y \leq y] \tag{C.3}$$

Similarly, we can name the probability $\Pr[X \leq x]$ this way:

$$P_x(x) = \Pr[X \leq x] \tag{C.4}$$

The probability density of Y is the derivative of its probability function:

$$p_Y(y) = \frac{dP_Y(y)}{dy} \tag{C.5}$$

Similarly, we write the probability density of X:

$$p_X(x) = \frac{dP_X(x)}{dx} \tag{C.6}$$

Using definitions in equation C.3 and equation C.4, we can rewrite equation C.2 as follows:

$$P_Y(y) = P_X(y/a) \tag{C.7}$$

We insert the previous equality into equation (A3.5):

$$p_Y(y) = \frac{dP_X(y/a)}{dy} = \frac{dP_X(y/a)}{dx}\frac{dx}{dy} = p_X(y/a)\frac{dx}{dy} \tag{C.8}$$

From equation C.1 we have this:

$$\frac{dx}{dy} = \frac{1}{a} \tag{C.9}$$

Inserting equation C.9 into equation C.8, we have the probability density of Y as a function of the probability density of X:

$$p_Y(y) = \frac{1}{a} p_X(y/a) \tag{C.10}$$

Inversely Related Random Variables

We have the following relationship between two random variables:

$$Y = \frac{1}{X} \quad X > 0 \tag{C.11}$$

Using figure C.1, we can relate the probability Pr[$Y \leq y$] to the probability of X:

$$\Pr[Y \leq y] = \Pr\left[X \geq \frac{1}{y}\right] = 1 - \Pr\left[X < \frac{1}{y}\right] \tag{C.12}$$

Using the labels $P_Y(y) = \Pr[Y \leq y]$ and $P_x(x) = \Pr[X \leq x]$, we rewrite equation C.12 as follows:

$$P_Y(y) = 1 - P_X(1/y) \tag{C.13}$$

Probability density is the derivative of the probability function:

$$p_Y(y) = \frac{dP_Y(y)}{dy} = \frac{d\left(1 - P_X(1/y)\right)}{dy} = -\frac{dP_X(1/y)}{dy} = -\frac{dP_X(1/y)}{dx}\frac{dx}{dy} \tag{C.14}$$

From equation C.11, we have this:

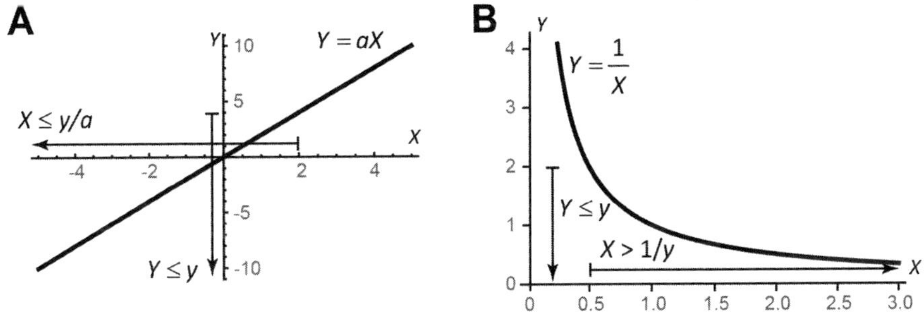

Figure C.1
Relationship between probabilities of two random variables. **A**. Random variables X and Y are linearly related. $\Pr[Y \leq y] = \Pr[X \leq y / a]$. **B**. Random variables X and Y are inversely related. $\Pr[Y \leq y] = \Pr[X \geq 1 / y]$.

$$\frac{dx}{dy} = -\frac{1}{y^2} \tag{C.15}$$

We have the probability density of Y as a function of the probability density of X:

$$p_Y(y) = \frac{1}{y^2} p_X(1 / y) \tag{C.16}$$

References

Aberman JE, Salamone JD. 1999. Nucleus accumbens dopamine depletions make rats more sensitive to high ratio requirements but do not impair primary food reinforcement. *Neuroscience* 92:545–552.

Abrahams MV, Dill LM. 1989. A determination of the energetic equivalence of the risk of predation. *Ecology* 70:999–1007.

Aharon I, Etcoff N, Ariely D, Chabris CF, O'Connor E, Breiter HC. 2001. Beautiful faces have variable reward value: fMRI and behavioral evidence. *Neuron* 32:537–551.

Alex KD, Pehek EA. 2007. Pharmacologic mechanisms of serotonergic regulation of dopamine neurotransmission. *Pharmacol Ther* 113:296–320.

Ariely D, Gneezy U, Loewenstein G, Mazar N. 2009 Large stakes and big mistakes. *Review of Economic Studies* 76:451–469.

Aronson E, Mills J. 1959. The effect of severity of initiation on liking for a group. *Journal of Abnormal and Social Psychology* 59:177–181.

Attwell D, Laughlin SB. 2001. An energy budget for signaling in the grey matter of the brain. *J Cereb Blood Flow Metab* 21:1133–1145.

Atzler E, Herbst R. 1927. Arbeitsphysiologische Studien. *Pflügers Arch* 215:291–328.

Bailey MR, Goldman O, Bello EP, Chohan MO, Jeong N, Winiger V, Chun E, Schipani E, Kalmbach A, Cheer JF, et al. 2018. An interaction between serotonin receptor signaling and dopamine enhances goal-directed vigor and persistence in mice. *J Neurosci* 38:2149–2162.

Bailey MR, Williamson C, Mezias C, Winiger V, Silver R, Balsam PD, Simpson EH. 2016. The effects of pharmacological modulation of the serotonin 2C receptor on goal-directed behavior in mice. *Psychopharmacology (Berl)* 233:615–624.

Bargary G, Bosten JM, Goodbourn PT, Lawrence-Owen AJ, Hogg RE, Mollon JD. 2017. Individual differences in human eye movements: An oculomotor signature? *Vision Res* 141:157–169.

Basso MA, Wurtz RH. 1997. Modulation of neuronal activity by target uncertainty. *Nature* 389:66–69.

Bateson M, Kacelnik A. 1995. Preferences for fixed and variable food sources: variability in amount and delay. *J Exp Analysis Behav* 63:313–329.

Bateson M, Kacelnik A. 1996. Rate currencies and the foraging starling: the fallacy of the averages revisited. *Behavioral Ecology* 7:341–352.

Bautista LM, Tinbergen J, Kacelnik A. 2001. To walk or to fly? How birds choose among foraging modes. *Proc Natl Acad Sci USA* 98:108–91094.

Bayer HM, Glimcher PW. 2005. Midbrain dopamine neurons encode a quantitative reward prediction error signal. *Neuron* 47:129–141.

Berardelli A, Rothwell JC, Thompson PD, Hallett M. 2001. Pathophysiology of bradykinesia in Parkinson's disease. *Brain* 124:2131–2146.

Berret B, Castanier C, Bastide S, Deroche T. 2018. Vigour of self-paced reaching movement: cost of time and individual traits. *Sci Rep* 8:10655.

Berrios GE. 1985. Positive and negative symptoms and Jackson. A conceptual history. *Arch Gen Psychiatry* 42:95–97.

Bizzi E. 1968. Discharge of frontal eye field neurons during saccadic and following eye movements in unanesthetized monkeys. *Exp Brain Res* 6:69–80.

Bogacz R, Brown E, Moehlis J, Holmes P, Cohen JD. 2006. The physics of optimal decision making: a formal analysis of models of performance in two-alternative forced-choice tasks. *Psychol Rev* 113:700–765.

Bornstein MH, Bornstein HG. 1976. The pace of life. *Nature* 259:557–559.

Boureau YL, Dayan P. 2010. Opponency revisited: competition and cooperation between dopamine and serotonin. *Neuropsychopharmacology* 36:74–97.

Boyer EW, Shannon M. 2005. The serotonin syndrome. *N Engl J Med* 352:1112–1120.

Britten KH, Shadlen MN, Newsome WT, Movshon JA. 1992. The analysis of visual motion: a comparison of neuronal and psychophysical performance. *J Neurosci* 12:4745–4765.

Bromberg-Martin ES, Hikosaka O, Nakamura K. 2010. Coding of Task Reward Value in the Dorsal Raphe Nucleus. *J Neurosci* 30:6262–6272.

Bruce CJ, Goldberg ME. 1985. Primate frontal eye fields. I. Single neurons discharging before saccades. *J Neurophysiol* 53:603–635.

Bruce CJ, Goldberg ME, Bushnell MC, Stanton GB. 1985. Primate frontal eye fields. II. Physiological and anatomical correlates of electrically evoked eye movements. *J Neurophysiol* 54:714–734.

Busemeyer JR, Townsend JT. 1993. Decision field theory: a dynamic-cognitive approach to decision making in an uncertain environment. *Psychol Rev* 100:432–459.

Buttner-Ennever JA, Horn AK, Henn V, Cohen B. 1999. Projections from the superior colliculus motor map to omnipause neurons in monkey. *J Comp Neurol* 413:55–67.

Carello CD, Krauzlis RJ. 2004. Manipulating intent: evidence for a causal role of the superior colliculus in target selection. *Neuron* 43:575–583.

Carpenter RH. 1999. A neural mechanism that randomises behaviour. *J Conscious Stud* 6:13–22.

Case DA, Nichols P, Fantino E. 1995. Pigeon's preference for variable-interval water reinforcement under widely varied water budgets. *J Exp Analysis Behav* 64:299–311.

Castrellon JJ, Seaman KL, Crawford JL, Young JS, Smith CT, Dang LC, Hsu M, Cowan RL, Zald DH, Samanez-Larkin GR. 2019. Individual differences in dopamine are associated with reward discounting in clinical groups but not in healthy adults. *J Neurosci* 39:321–332.

Cattell JMK. 1886. The influence of the intensity of the stimulus on the length of the reaction time. *Brain* 8:512–515.

Cavanagh PR, Williams KR. 1982. The effect of stride length variation on oxygen uptake during distance running. *Med Sci Sports Exerc* 14:30–35.

Chan F, Armstrong IT, Pari G, Riopelle RJ, Munoz DP. 2005. Deficits in saccadic eye-movement control in Parkinson's disease. *Neuropsychologia* 43:784–796.

Charnov EL. 1976. Optimal foraging, the marginal value theorem. *Theor Popul Biol* 9:129–136.

Choi JE, Vaswani PA, Shadmehr R. 2014. Vigor of movements and the cost of time in decision making. *J Neurosci* 34:1212–1223.

Cisek P, Puskas GA, El-Murr S. 2009. Decisions in changing conditions: the urgency-gating model. *J Neurosci* 29:11560–11571.

Clement TS, Feltus JR, Kaiser DH, Zentall TR. 2000. Work ethic in pigeons: reward value is directly related to the effort or time required to obtain the reward. *Psychon Bull Rev* 7:100–106.

Cohen JY, Amoroso MW, Uchida N. 2015. Serotonergic neurons signal reward and punishment on multiple timescales. *eLife* 4:463.

Constantino SM, Daw ND. 2015. Learning the opportunity cost of time in a patch-foraging task. *Cogn Affect Behav Neurosci* 15:837–853.

Correia PA, Lottem E, Banerjee D, Machado AS, Carey MR, Mainen ZF. 2017. Transient inhibition and long-term facilitation of locomotion by phasic optogenetic activation of serotonin neurons. *eLife* 6:365.

Cos I, Belanger N, Cisek P. 2011. The influence of predicted arm biomechanics on decision making. *J Neurophysiol* 105:3022–3033.

Cowie RJ. 1977. Optimal foraging in great tits (Parus major). *Nature* 268:137–139.

Crammond DJ, Kalaska JF. 1996. Differential relation of discharge in primary motor cortex and premotor cortex to movements versus actively maintained postures during a reaching task. *Exp Brain Res* 108:45–61.

da Silva JA, Tecuapetla F, Paixao V, Costa RM. 2018. Dopamine neuron activity before action initiation gates and invigorates future movements. *Nature* 554:244–248.

de Grosbois J, Heath M, Tremblay L. 2015. Augmented feedback influences upper limb reaching movement times but does not explain violations of Fitts' Law. *Front Psychol* 6:800.

Daw ND, Kakade S, Dayan P. 2002 Opponent interactions between serotonin and dopamine. *Neural Netw* 15:603–616.

Donaldson ZR, Nautiyal KM, Ahmari SE, Hen R. 2013. Genetic approaches for understanding the role of serotonin receptors in mood and behavior. *Curr Opin Neurobiol* 23:399–406.

Donelan JM, Kram R, Kuo AD. 2001. Mechanical and metabolic determinants of the preferred step width in human walking. *Proc Biol Sci* 268:1985–1992.

Dorris MC, Munoz DP. 1995. A neural correlate for the gap effect on saccadic reaction times in monkey. *J Neurophysiol* 73:2558–2562.

Dorris MC, Pare M, Munoz DP. 1997. Neuronal activity in monkey superior colliculus related to the initiation of saccadic eye movements. *J Neurosci* 17:8566–8579.

Eagle DM, Lehmann O, Theobald DE, Pena Y, Zakaria R, Ghosh R, Dalley JW, Robbins TW. 2008. Serotonin depletion impairs waiting but not stop-signal reaction time in rats: implications for theories of the role of 5-HT in behavioral inhibition. *Neuropsychopharmacology* 34:1311–1321.

Elftman H. 1966. Biomechanics of muscle with particular application to studies of gait. *J Bone Joint Surg Am* 48:363–377.

Engelhard B, Finkelstein J, Cox J, Fleming W, Jang HJ, Ornelas S, Koay SA, et al. 2019. Specialized coding of sensory, motor and cognitive variables in VTA dopamine neurons. *Nature* 570:509–513.

Everling S, Dorris MC, Klein RM, Munoz DP. 1999. Role of primate superior colliculus in preparation and execution of anti-saccades and pro-saccades. *J Neurosci* 19:2740–2754.

Everling S, Munoz DP. 2000. Neuronal correlates for preparatory set associated with pro-saccades and anti-saccades in the primate frontal eye field. *J Neurosci* 20:387–400.

Everling S, Pare M, Dorris MC, Munoz DP. 1998. Comparison of the discharge characteristics of brain stem omnipause neurons and superior colliculus fixation neurons in monkey: implications for control of fixation and saccade behavior. *J Neurophysiol* 79:511–528.

Farrar AM, Segovia KN, Randall PA, Nunes EJ, Collins LE, Stopper CM, Port RG, et al. 2010. Nucleus accumbens and effort-related functions: behavioral and neural markers of the interactions between adenosine A2A and dopamine D2 receptors. *Neuroscience* 166:1056–1067.

Fitts PM. 1954. The information capacity of the human motor system in controlling the amplitude of movement. *J Exp Psychol* 47:381–391.

Fleuriet J, Goffart L. 2012. Saccadic interception of a moving visual target after a spatiotemporal perturbation. *J Neurosci* 32:452–461.

Foley NC, Jangraw DC, Peck C, Gottlieb J. 2014. Novelty enhances visual salience independently of reward in the parietal lobe. *J Neurosci* 34:7947–7957.

Fox SH, Chuang R, Brotchie JM. 2009. Serotonin and Parkinson's disease: On movement, mood, and madness. *Mov Disord* 24:1255–1266.

Gandhi NJ, Keller EL. 1997. Spatial distribution and discharge characteristics of superior colliculus neurons antidromically activated from the omnipause region in monkey. *J Neurophysiol* 78:2221–2225.

Gandhi NJ, Sparks DL. 2007. Dissociation of eye and head components of gaze shifts by stimulation of the omnipause neuron region. *J Neurophysiol* 98:360–373.

Georgopoulos AP, Kalaska JF, Caminiti R, Massey JT. 1982. On the relations between the direction of two-dimensional arm movements and cell discharge in primate motor cortex. *J Neurosci* 2:1527–1537.

Gerfen CR, Surmeier DJ. 2011. Modulation of striatal projection systems by dopamine. *Annu Rev Neurosci* 34:441–466.

Gertler TS, Chan CS, Surmeier DJ. 2008. Dichotomous anatomical properties of adult striatal medium spiny neurons. *J Neurosci* 28:10814–10824.

Glaser JI, Wood DK, Lawlor PN, Ramkumar P, Kording KP, Segraves MA. 2016. Role of expected reward in frontal eye field during natural scene search. *J Neurophysiol* 116:645–657.

Goffart L, Hafed ZM, Krauzlis RJ. 2012. Visual fixation as equilibrium: evidence from superior colliculus inactivation. *J Neurosci* 32:10627–10636.

Goossens HH, Van Opstal AJ. 2006. Dynamic ensemble coding of saccades in the monkey superior colliculus. *J Neurophysiol* 95:2326–2341.

Gordon J, Ghilardi MF, Ghez C. 1994. Accuracy of planar reaching movements. I. Independence of direction and extent variability. *Exp Brain Res* 99:97–111.

Green L, Myerson J, Ostaszewski P. 1999. Discounting of delayed rewards across the life span: age differences in individual discounting functions. *Behav Processes* 46:89–96.

Grice GR. 1968. Stimulus intensity and response evocation. *Psychol Rev* 75:359–373.

Grice GR, Nullmeyer R, Schnizlein JM. 1979. Variable criterion analysis of brightness effects in simple reaction time. *J Exp Psychol Hum Percept Perform* 5:303–314.

Gu C, Wood DK, Gribble PL, Corneil BD. 2016. A trial-by-trial window into sensorimotor transformations in the human motor periphery. *J Neurosci* 36:8273–8282.

Guitart-Masip M, Beierholm UR, Dolan R, Duzel E, Dayan P. 2011. Vigor in the face of fluctuating rates of reward: an experimental examination. *J Cogn Neurosci* 23:3933–3938.

Hagura N, Haggard P, Diedrichsen J. 2017. Perceptual decisions are biased by the cost to act. *eLife* 6.

Handel A, Glimcher PW. 1999. Quantitative analysis of substantia nigra pars reticulata activity during a visually guided saccade task. *J Neurophysiol* 82:3458–3475.

Hanes DP, Patterson WF, Schall JD. 1998. Role of frontal eye fields in countermanding saccades: visual, movement, and fixation activity. *J Neurophysiol* 79:817–834.

Hanes DP, Schall JD. 1996. Neural control of voluntary movement initiation. *Science* 274:427–430.

Hanes DP, Wurtz RH. 2001. Interaction of the frontal eye field and superior colliculus for saccade generation. *J Neurophysiol* 85:804–815.

Harting JK. 1977. Descending pathways from the superior collicullus: an autoradiographic analysis in the rhesus monkey (Macaca mulatta). *J Comp Neurol* 173:583–612.

Hayden BY, Pearson JM, Platt ML. 2011. Neuronal basis of sequential foraging decisions in a patchy environment. *Nat Neurosci* 14:933–939.

Heitz RP, Schall JD. 2012. Neural mechanisms of speed-accuracy tradeoff. *Neuron* 76:616–628.

Hikosaka O, Sakamoto M, Usui S. 1989. Functional properties of monkey caudate neurons. I. Activities related to saccadic eye movements. *J Neurophysiol* 61:780–798.

Hikosaka O, Wurtz RH. 1983. Visual and oculomotor functions of monkey substantia nigra pars reticulata. I. Relation of visual and auditory responses to saccades. *J Neurophysiol* 49:1230–1253.

Hikosaka O, Wurtz RH. 1985. Modification of saccadic eye movements by GABA-related substances. II. Effects of muscimol in monkey substantia nigra pars reticulata. *J Neurophysiol* 53:292–308.

Hodgson T, Chamberlain M, Parris B, James M, Gutowski N, Husain M, Kennard C. 2007. The role of the ventrolateral frontal cortex in inhibitory oculomotor control. *Brain* 130:1525–1537.

Hoefort A, Hofer S. 2006. *Price and Earnings: A comparison of purchasing power around the globe.* Zurich, Switzerland: Union Bank of Switzerland AG.

Hogberg P. 1952. How do stride length and stride frequency influence the energy-output during running? *Arbeitsphysiologie* 14:437–441.

Horwitz GD, Batista AP, Newsome WT. 2004. Representation of an abstract perceptual decision in macaque superior colliculus. *J Neurophysiol* 91:2281–2296.

Hoyt DF, Taylor CR. 1981. Gait and the energetics of locomotion in horses. *Nature* 292:239–240.

Hunter LC, Hendrix EC, Dean JC. 2010. The cost of walking downhill: is the preferred gait energetically optimal? *J Biomech* 43:1910–1915.

Ikeda T, Hikosaka O. 2003. Reward-dependent gain and bias of visual responses in primate superior colliculus. *Neuron* 39:693–700.

Ikeda T, Hikosaka O. 2007. Positive and negative modulation of motor response in primate superior colliculus by reward expectation. *J Neurophysiol* 98:3163–3170.

Inzlicht M, Shenhav A, Olivola CY. 2018. The Effort Paradox: Effort is both costly and valued. *Trends Cogn Sci* 22:337–349.

Irving EL, Steinbach MJ, Lillakas L, Babu RJ, Hutchings N. 2006. Horizontal saccade dynamics across the human life span. *Invest Ophthalmol Vis Sci* 47:2478–2484.

Ivry RB. 1986. Force and timing components of the motor program. *J Mot Behav* 18:449–474.

Jantz JJ, Watanabe M, Everling S, Munoz DP. 2013. Threshold mechanism for saccade initiation in frontal eye field and superior colliculus. *J Neurophysiol* 109:2767–2780.

Jimura K, Myerson J, Hilgard J, Braver TS, Green L. 2009. Are people really more patient than other animals? Evidence from human discounting of real liquid rewards. *Psychon Bull Rev* 16:1071–1075.

Kacelnik A, Bateson M. 1996. Risky theories—the effects of variance on foraging decisions. *Amer Zool* 36:402–434.

Kacelnik A, Marsh B. 2002. Cost can increase preference in starlings. *Anim Behav* 63:245–250.

Kato R, Takaura K, Ikeda T, Yoshida M, Isa T. 2011. Contribution of the retino-tectal pathway to visually guided saccades after lesion of the primary visual cortex in monkeys. *Eur J Neurosci* 33:1952–1960.

Kawagoe R, Takikawa Y, Hikosaka O. 1998. Expectation of reward modulates cognitive signals in the basal ganglia. *Nature Neurosci* 1:411–416.

Kawagoe R, Takikawa Y, Hikosaka O. 2004. Reward-predicting activity of dopamine and caudate neurons—a possible mechanism of motivational control of saccadic eye movement. *J Neurophysiol* 91:1013–1024.

Keller EL, Edelman JA. 1994. Use of interrupted saccade paradigm to study spatial and temporal dynamics of saccadic burst cells in superior colliculus in monkey. *J Neurophysiol* 72:2754–2770.

Keller EL, McPeek RM, Salz T. 2000. Evidence against direct connections to PPRF EBNs from SC in the monkey. *J Neurophysiol* 84:1303–1313.

Kim B, Basso MA. 2010. A probabilistic strategy for understanding action selection. *J Neurosci* 30:2340–2355.

Kim HF, Amita H, Hikosaka O. 2017. Indirect pathway of caudal basal ganglia for rejection of valueless visual objects. *Neuron* 94:920–930.

Kim HF, Hikosaka O. 2013. Distinct basal ganglia circuits controlling behaviors guided by flexible and stable values. *Neuron* 79:1001–1010.

Kish SJ, Tong J, Hornykiewicz O, Rajput A, Chang LJ, Guttman M, Furukawa Y. 2007. Preferential loss of serotonin markers in caudate versus putamen in Parkinson's disease. *Brain* 131:1–12.

Klein ED, Bhatt RS, Zentall TR. 2005. Contrast and the justification of effort. *Psychon Bull Rev* 12:335–339.

Klein-Flugge MC, Kennerley SW, Saraiva AC, Penny WD, Bestmann S. 2015. Behavioral modeling of human choices reveals dissociable effects of physical effort and temporal delay on reward devaluation. *PLoS Comput Biol* 11:e1004116.

Kloppel S, Draganski B, Golding CV, Chu C, Nagy Z, Cook PA, Hicks SL, Kennard C, Alexander DC, Parker GJ, et al. 2008. White matter connections reflect changes in voluntary-guided saccades in pre-symptomatic Huntington's disease. *Brain* 131:196–204.

Kobayashi S, Nomoto K, Watanabe M, Hikosaka O, Schultz W, Sakagami M. 2006. Influences of rewarding and aversive outcomes on activity in macaque lateral prefrontal cortex. *Neuron* 51:861–870.

Kobayashi S, Schultz W. 2008. Influence of reward delays on responses of dopamine neurons. *J Neurosci* 28:7837–7846.

Koch M, Schmid A, Schnitzler HU. 2000. Role of nucleus accumbens dopamine D1 and D2 receptors in instrumental and Pavlovian paradigms of conditioned reward. *Psychopharmacology (Berl)* 152:67–73.

Kording KP, Fukunaga I, Howard IS, Ingram JN, Wolpert DM. 2004. A neuroeconomics approach to inferring utility functions in sensorimotor control. *PLoS Biol* 2:e330.

Krauzlis RJ, Basso MA, Wurtz RH. 2000. Discharge properties of neurons in the rostral superior colliculus of the monkey during smooth-pursuit eye movements. *J Neurophysiol* 84:876–891.

Kravitz AV, Freeze BS, Parker PR, Kay K, Thwin MT, Deisseroth K, Kreitzer AC. 2010. Regulation of parkinsonian motor behaviours by optogenetic control of basal ganglia circuitry. *Nature* 466:622–626.

Kunori N, Kajiwara R, Takashima I. 2014. Voltage-sensitive dye imaging of primary motor cortex activity produced by ventral tegmental area stimulation. *J Neurosci* 34:8894–8903.

Lauwereyns J, Watanabe K, Coe B, Hikosaka O. 2002. A neural correlate of response bias in monkey caudate nucleus. *Nature* 418:413–417.

Leathers ML, Olson CR. 2012. In monkeys making value-based decisions, LIP neurons encode cue salience and not action value. *Science* 338:132–135.

Lee RS, Mattar MG, Parker NF, Witten IB, Daw ND. 2019. Reward prediction error does not explain movement selectivity in DMS-projecting dopamine neurons. *eLife* 8.

Lemon WC. 1991. Fitness consequences of foraging behaviour in the zebra finch. *Nature* 352:153–155.

Leopold DA. 2012. Primary visual cortex: awareness and blindsight. *Annu Rev Neurosci* 35:91–109.

Levine RV, Norenzayan A. 1999. The pace of life in 31 countries. *J Cross-Cultural Psychol* 30:178–205.

Loane C, Wu K, Bain P, Brooks DJ, Piccini P, Politis M. 2013. Serotonergic loss in motor circuitries correlates with severity of action-postural tremor in PD. *Neurology* 80:1850–1855.

Lottem E, Banerjee D, Pietro Vertechi, Sarra D, Lohuis MO, Mainen ZF. 2018. Activation of serotonin neurons promotes active persistence in a probabilistic foraging task. *Nat Commun* 9:1–12.

Louie K, Glimcher PW. 2010. Separating value from choice: delay discounting activity in the lateral intraparietal area. *J Neurosci* 30:5498–5507.

Lovejoy LP, Krauzlis RJ. 2010. Inactivation of primate superior colliculus impairs covert selection of signals for perceptual judgments. *Nat Neurosci* 13:261–266.

Machado L, Rafal RD. 2004. Control of fixation and saccades during an anti-saccade task: an investigation in humans with chronic lesions of oculomotor cortex. *Exp Brain Res* 156:55–63.

Mahlberg R, Steinacher B, Mackert A, Flechtner KM. 2001. Basic parameters of saccadic eye movements—differences between unmedicated schizophrenia and affective disorder patients. *Eur Arch Psychiatry Clin Neurosci* 251:205–210.

Manohar SG, Chong TT, Apps MA, Batla A, Stamelou M, Jarman PR, Bhatia KP, Husain M. 2015. Reward pays the cost of noise reduction in motor and cognitive control. *Curr Biol* 25:1707–1716.

Matsumoto M, Hikosaka O. 2009. Two types of dopamine neuron distinctly convey positive and negative motivational signals. *Nature* 459:837–841.

Mays LE, Sparks DL. 1980. Dissociation of visual and saccade-related responses in superior colliculus neurons. *J Neurophysiol* 43:207–232.

Mazzoni P, Hristova A, Krakauer JW. 2007. Why don't we move faster? Parkinson's disease, movement vigor, and implicit motivation. *J Neurosci* 27:7105–7116.

McGill WJ. 1961. Loudness and reaction time: A guided tour of the Listener's private world. *Acta Psychol* 19:193–199.

McPeek RM, Keller EL. 2002. Saccade target selection in the superior colliculus during a visual search task. *J Neurophysiol* 88:2019–2034.

Millar A, Navarick DJ. 1984. Self-control and choice in humans: effects of video game playing as a positive reinforcer. *Learn Motiv* 15:203–218.

Milosavljevic M, Navalpakkam V, Koch C, Rangel A. 2012. Relative visual saliency differences induce sizable bias in consumer choice. *J Consum Psychol* 22:67–74.

Milstein DM, Dorris MC. 2007. The influence of expected value on saccadic preparation. *J Neurosci* 27:4810–4818.

Miyazaki KW, Miyazaki K, Tanaka KF, Yamanaka A, Takahashi A, Tabuchi S, Doya K. 2014. Optogenetic activation of dorsal raphe serotonin neurons enhances patience for future rewards. *Curr Biol* 24:2033–2040.

Mohebi A, Pettibone JR, Hamid AA, Wong JT, Vinson LT, Patriarchi T, Tian L, Kennedy RT, Berke JD. 2019. Dissociable dopamine dynamics for learning and motivation. *Nature* 570:65–70.

Mohler CW, Wurtz RH. 1976. Organization of monkey superior colliculus: intermediate layer cells discharging before eye movements. *J Neurophysiol* 39:722–744.

Morel P, Ulbrich P, Gail A. 2017. What makes a reach movement effortful? Physical effort discounting supports common minimization principles in decision making and motor control. *PLoS Biol* 15:e2001323.

Moschovakis AK, Scudder CA, Highstein SM. 1996. The microscopic anatomy and physiology of the mammalian saccadic system. *Prog Neurobiol* 50:133–254.

Munoz DP, Schall JD. 2004. "Concurrent, distributed control of saccade initiation in the frontal eye field and superior colliculus." In *The Superior Colliculus: New Approaches for Studying Sensorimotor Integration*, edited by WC Hall, A Moschovakis, 55–82, Boca Raton: CRC Press.

Munoz DP, Wurtz RH. 1993a. Fixation cells in monkey superior colliculus. I. Characteristics of cell discharge. *J Neurophysiol* 70:559–575.

Munoz DP, Wurtz RH. 1993b. Fixation cells in monkey superior colliculus. II. Reversible activation and deactivation. *J Neurophysiol* 70:576–589.

Munoz DP, Wurtz RH. 1995. Saccade-related activity in monkey superior colliculus. I. Characteristics of burst and buildup cells. *J Neurophysiol* 73:2313–2333.

Myerson J, Green L. 1995. Discounting of delayed rewards: Models of individual choice. *J Exp Anal Behav* 64:263–276.

Nagasaki H, Aoki F, Nakamura R. 1983. Premotor and motor reaction time as a function of force output. *Percept Mot Skills* 57:859–867.

Nakamura K, Hikosaka O. 2006. Role of dopamine in the primate caudate nucleus in reward modulation of saccades. *J Neurosci* 26:5360–5369.

Nautiyal KM, Tanaka KF, Barr MM, Tritschler L, Le Dantec Y, David DJ, Gardier AM, Blanco C, Hen R, Ahmari SE. 2015. Distinct circuits underlie the effects of 5-HT1B receptors on aggression and impulsivity. *Neuron* 86:813–826.

Navalpakkam V, Koch C, Rangel A, Perona P. 2010. Optimal reward harvesting in complex perceptual environments. *Proc Natl Acad Sci USA* 107:5232–5237.

Navarick DJ. 2004. Discounting of delayed reinforcers: measurement by questionnaires versus operant choice procedures. *Psychol Record* 54:85–94.

Neuringer AJ. 1970. Many responses per food reward with free food present. *Science* 169:503–504.

Niv Y, Daw ND, Joel D, Dayan P. 2007. Tonic dopamine: opportunity costs and the control of response vigor. *Psychopharmacology (Berl)* 191:507–520.

Nonacs P. 2001. State dependent behavior and the marginal value theorem. *Behavioral Ecology* 12:71–83.

Nummela SU, Krauzlis RJ. 2010. Inactivation of primate superior colliculus biases target choice for smooth pursuit, saccades, and button press responses. *J Neurophysiol* 104:1538–1548.

O'Doherty J, Winston J, Critchley H, Perrett D, Burt DM, Dolan RJ. 2003. Beauty in a smile: the role of medial orbitofrontal cortex in facial attractiveness. *Neuropsychologia* 41:147–155.

Ohmura Y, Tanaka KF, Tsunematsu T, Yamanaka A, Yoshioka M. 2014. Optogenetic activation of serotonergic neurons enhances anxiety-like behaviour in mice. *Int J Neuropsychopharmacol* 17:1777–1783.

Opris I, Lebedev M, Nelson RJ. 2011. Motor planning under unpredictable reward: modulations of movement vigor and primate striatum activity. *Front Neurosci* 5:1–12.

Padoa-Schioppa C, Assad JA. 2006. Neurons in the orbitofrontal cortex encode economic value. *Nature* 441:223–226.

Palmer J, Huk AC, Shadlen MN. 2005. The effect of stimulus strength on the speed and accuracy of a perceptual decision. *J Vis* 5:376–404.

Panigrahi B, Martin KA, Li Y, Graves AR, Vollmer A, Olson L, Mensh BD, Karpova AY, Dudman JT. 2015. Dopamine Is Required for the Neural Representation and Control of Movement Vigor. *Cell* 162:1418–1430.

Pare M, Munoz DP. 1996. Saccadic reaction time in the monkey: advanced preparation of oculomotor programs is primarily responsible for express saccade occurrence. *J Neurophysiol* 76:3666–3681.

Parker NF, Cameron CM, Taliaferro JP, Lee J, Choi JY, Davidson TJ, Daw ND, Witten IB. 2016. Reward and choice encoding in terminals of midbrain dopamine neurons depends on striatal target. *Nat Neurosci* 19:845–854.

Pasquereau B, Tremblay L, Turner RS. 2019. Local field potentials reflect dopaminergic and non-dopaminergic activities within the primate midbrain. *Neuroscience* 399:167–183.

Pasquereau B, Turner RS. 2013. Limited encoding of effort by dopamine neurons in a cost-benefit trade-off task. *J Neurosci* 33:8288–8300.

Pasquereau B, Turner RS. 2015. Dopamine neurons encode errors in predicting movement trigger occurrence. *J Neurophysiol* 113:1110–1123.

Philipp R, Hoffmann KP. 2014. Arm movements induced by electrical microstimulation in the superior colliculus of the macaque monkey. *J Neurosci* 34:3350–3363.

Platt ML, Glimcher PW. 1999. Neural correlates of decision variables in parietal cortex. *Nature* 400:233–238.

Poppel E, Held R, Frost D. 1973. Residual visual function after brain wounds involving the central visual pathways in man. *Nature* 243:295–296.

Prevost C, Pessiglione M, Metereau E, Clery-Melin ML, Dreher JC. 2010. Separate valuation subsystems for delay and effort decision costs. *J Neurosci* 30:14080–14090.

Pruszynski JA, King GL, Boisse L, Scott SH, Flanagan JR, Munoz DP. 2010. Stimulus-locked responses on human arm muscles reveal a rapid neural pathway linking visual input to arm motor output. *Eur J Neurosci* 32:1049–1057.

Ralston HJ. 1958. Energy-speed relation and optimal speed during level walking. *Int Z Angew Physiol* 17:277–283.

Ratcliff R. 1978. A theory of memory retrieval. *Psychol Rev* 83:59–108.

Ratcliff R, Cherian A, Segraves M. 2003. A comparison of macaque behavior and superior colliculus neuronal activity to predictions from models of two-choice decisions. *J Neurophysiol* 90:1392–1407.

Ratcliff R, Hasegawa YT, Hasegawa RP, Smith PL, Segraves MA. 2007. Dual diffusion model for single-cell recording data from the superior colliculus in a brightness-discrimination task. *J Neurophysiol* 97:1756–1774.

Ren J, Friedmann D, Xiong J, Liu CD, Ferguson BR, Weerakkody T, DeLoach KE, Ran C, Pun A, Sun Y, et al. 2018. Anatomically defined and functionally distinct dorsal raphe serotonin sub-systems. *Cell* 175:472–487.e20.

Reppert TR, Lempert KM, Glimcher PW, Shadmehr R. 2015. Modulation of saccade vigor during value-based decision making. *J Neurosci* 35:15369–15378.

Reppert TR, Rigas I, Herzfeld D, Sedaghat-Nejad E, Komogortsev O, Shadmehr R. 2018. Movement vigor as a trait-like attribute of individuality. *J Neurophysiol* 120:741–757.

Richardson H, Verbeek NAM. 1986. Diet selection and optimization by northwestern crows feeding on Japanese littleneck clams. *Ecology* 67:1219–1226.

Rigas I, Komogortsev O, Shadmehr R. 2016. Biometric recognition via eye movements: Saccadic vigor and acceleration cues. *ACM Trans Appl Percept* 13:6.

Robinson DA. 1970. Oculomotor unit behavior in the monkey. *J Neurophysiol* 33:393–403.

Rosenbaum DA. 1980. Human movement initiation: specification of arm, direction, and extent. *J Exp Psychol Gen* 109:444–474.

Russ DW, Elliott MA, Vandenborne K, Walter GA, Binder-Macleod SA. 2002. Metabolic costs of isometric force generation and maintenance of human skeletal muscle. *Am J Physiol Endocrinol Metab* 282:E448–E457.

Sackaloo K, Strouse E, Rice MS. 2015. Degree of preference and its influence on motor control when reaching for most preferred, neutrally preferred, and least preferred candy. *OTJR* 35:81–88.

Salamone JD, Steinpreis RE, McCullough LD, Smith P, Grebel D, Mahan K. 1991. Haloperidol and nucleus accumbens dopamine depletion suppress lever pressing for food but increase free food consumption in a novel food choice procedure. *Psychopharmacology (Berl)* 104:515–521.

Sato M, Hikosaka O. 2002. Role of primate substantia nigra pars reticulata in reward-oriented saccadic eye movement. *J Neurosci* 22:2363–2373.

Schall JD, Hanes DP. 1993. Neural basis of saccade target selection in frontal eye field during visual search. *Nature* 366:467–469.

Schelp SA, Pultorak KJ, Rakowski DR, Gomez DM, Krzystyniak G, Das R, Oleson EB. 2017. A transient dopamine signal encodes subjective value and causally influences demand in an economic context. *Proc Natl Acad Sci USA* 114:E11303–E11312.

Schiller PH, Sandell JH, Maunsell JH. 1987. The effect of frontal eye field and superior colliculus lesions on saccadic latencies in the rhesus monkey. *J Neurophysiol* 57:1033–1049.

Schultz W, Dayan P, Montague PR. 1997. A neural substrate of prediction and reward. *Science* 275:1593–1599.

Schwab RS, Zieper I. 1965. Effects of mood, motivation, stress and alertness on the performance in Parkinson's disease. *Psychiatr Neurol (Basel)* 150:345–357.

Schweighofer N, Xiao Y, Kim S, Yoshioka T, Gordon J, Osu R. 2015. Effort, success, and nonuse determine arm choice. *J Neurophysiol* 114:551–559.

Sebastian-Gonzalez E, Hiraldo F, Blanco G, Hernández-Brito D, Romero-Vidal P, Carrete M, Gómez-Llanos E, Pacífico EC, Díaz-Luque JA, Dénes FV et al. 2019. The extent, frequency and ecological functions of food wasting by parrots. *Sci Rep* 9:15280.

Sedaghat-Nejad E, Herzfeld DJ, Shadmehr R. 2019. Reward prediction error modulates saccade vigor. *J Neurosci* 39:5010–5017.

Segraves MA. 1992. Activity of monkey frontal eye field neurons projecting to oculomotor regions of the pons. *J Neurophysiol* 68:1967–1985.

Seideman JA, Stanford TR, Salinas E. 2018. Saccade metrics reflect decision-making dynamics during urgent choices. *Nat Commun* 9:2907.

Selinger JC, O'Connor SM, Wong JD, Donelan JM. 2015. humans can continuously optimize energetic cost during walking. *Curr Biol* 25:2452–2456.

Seo C, Guru A, Jin M, Ito B, Sleezer BJ, Ho YY, Wang E, Boada C, Krupa NA, Kullakanda DS, et al. 2019. Intense threat switches dorsal raphe serotonin neurons to a paradoxical operational mode. *Science* 363:538–542.

Shadlen MN, Newsome WT. 2001. Neural basis of a perceptual decision in the parietal cortex (area LIP) of the rhesus monkey. *J Neurophysiol* 86:1916–1936.

Shadmehr R. 2017. Distinct neural circuits for control of movement vs. holding still. *J Neurophysiol* 117:1431–1460.

Shadmehr R, Huang HJ, Ahmed AA. 2015. Effort, reward, and vigor in decision-making and motor control. *Translational and Computational Motor Control* 11:1–2.

Shadmehr R, Huang HJ, Ahmed AA. 2016. A representation of effort in decision-making and motor control. *Curr Biol* 26:1929–1934.

Shadmehr R, Mussa-Ivaldi S. 2012. *Biological Learning and Control: How the Brain Builds Representations, Predicts Events, and Makes Decisions.* Cambridge, MA: MIT Press.

Shadmehr R, Orban de Xivry JJ, Xu-Wilson M, Shih TY. 2010. Temporal discounting of reward and the cost of time in motor control. *J Neurosci* 30:10507–10516.

Shenhav A, Straccia MA, Cohen JD, Botvinick MM. 2014. Anterior cingulate engagement in a foraging context reflects choice difficulty, not foraging value. *Nat Neurosci* 17:1249–1254.

Smalianchuk I, Jagadisan UK, Gandhi NJ. 2018. Instantaneous midbrain control of saccade velocity. *J Neurosci* 38:10156–10167.

Soetedjo R, Kaneko CR, Fuchs AF. 2002. Evidence that the superior colliculus participates in the feedback control of saccadic eye movements. *J Neurophysiol* 87:679–695.

Sparks DL. 1978. Functional properties of neurons in the monkey superior colliculus: coupling of neuronal activity and saccade onset. *Brain Res* 156:1–16.

Sparks DL, Hu X. 1999. Saccade initiation and the reliability of motor signals involved in the generation of saccadic eye movements. In *Novartis Foundation Symposium*, 75. Chichester, NY: John Wiley.

Stelmach GE, Worringham CJ. 1988. The preparation and production of isometric force in Parkinson's disease. *Neuropsychologia* 26:93–103.

Stelly CE, Haug GC, Fonzi KM, Garcia MA, Tritley SC, Magnon AP, Ramos MAP, Wanat MJ. 2019. Pattern of dopamine signaling during aversive events predicts active avoidance learning. *Proc Natl Acad Sci USA* 116:13641–13650.

Stephens DW, Krebs JR. 1986. *Foraging Theory.* Princeton, NJ: Princeton University Press.

Steudel-Numbers KL, Wall-Scheffler CM. 2009. Optimal running speed and the evolution of hominin hunting strategies. *J Hum Evol* 56:355–360.

Sugiwaka H, Okouchi H. 2004. Reformative self-control and discounting of reward value by delay or effort. *Jpn Psychol Res* 46:1–9.

Summerside EM, Shadmehr R, Ahmed AA. 2018. Vigor of reaching movements: reward discounts the cost of effort. *J Neurophys* 119:2347–2357.

Syed EC, Grima LL, Magill PJ, Bogacz R, Brown P, Walton ME. 2016. Action initiation shapes mesolimbic dopamine encoding of future rewards. *Nat Neurosci* 19:34–36.

Sylvestre PA, Cullen KE. 1999. Quantitative analysis of abducens neuron discharge dynamics during saccadic and slow eye movements. *J Neurophysiol* 82:2612–2632.

Takikawa Y, Kawagoe R, Hikosaka O. 2002a. Reward-dependent spatial selectivity of anticipatory activity in monkey caudate neurons. *J Neurophysiol* 87:508–515.

Takikawa Y, Kawagoe R, Itoh H, Nakahara H, Hikosaka O. 2002b. Modulation of saccadic eye movements by predicted reward outcome. *Exp Brain Res* 142:284–291.

Tang YY, Posner MI, Rothbart MK, Volkow ND. 2015. Circuitry of self-control and its role in reducing addiction. *Trends Cogn Sci* 19:439–444.

Tsianos GA, Rustin C, Loeb GE. 2012. Mammalian muscle model for predicting force and energetics during physiological behaviors. *IEEE Trans Neural Syst Rehabil Eng* 20:117–133.

Theeuwes J, Kramer AF, Hahn S, Irwin DE, Zelinsky GJ. 1999. Influence of attentional capture on oculomotor control. *J Exp Psychol Hum Percept Perform* 25:1595–1608.

Thura D, Cisek P. 2014. Deliberation and commitment in the premotor and primary motor cortex during dynamic decision making. *Neuron* 81:1401–1416.

Thura D, Cisek P. 2016. Modulation of premotor and primary motor cortical activity during volitional adjustments of speed-accuracy trade-offs. *J Neurosci* 36:938–956.

Thura D, Cisek P. 2017. The basal ganglia do not select reach targets but control the urgency of commitment. *Neuron* 95:1160–1170.

Thura D, Cos I, Trung J, Cisek P. 2014. Context-dependent urgency influences speed-accuracy trade-offs in decision-making and movement execution. *J Neurosci* 34:16442–16454.

Todorov E, Jordan MI. 2002. Optimal feedback control as a theory of motor coordination. *Nat Neurosci* 5:1226–1235.

Treadway MT, Buckholtz JW, Cowan RL, Woodward ND, Li R, Ansari MS, Baldwin RM, Schwartzman AN, Kessler RM, Zald DH. 2012. Dopaminergic mechanisms of individual differences in human effort-based decision-making. *J Neurosci* 32:6170–6176.

Treadway MT, Buckholtz JW, Schwartzman AN, Lambert WE, Zald DH. 2009. Worth the 'EEfRT'? The effort expenditure for rewards task as an objective measure of motivation and anhedonia. *PLoS One* 4:e6598.

Uno Y, Kawato M, Suzuki R. 1989. Formation and control of optimal trajectory in human multijoint arm movement. Minimum torque-change model. *Biol Cybern* 61:89–101.

Usher M, McClelland JL. 2001. The time course of perceptual choice: the leaky, competing accumulator model. *Psychol Rev* 108:550–592.

Van der Stigchel S, van KM, Nijboer TC, List A, Rafal RD. 2012. The role of the frontal eye fields in the oculomotor inhibition of reflexive saccades: evidence from lesion patients. *Neuropsychologia* 50:198–203.

Varazzani C, San-Galli A, Gilardeau S, Bouret S. 2015. Noradrenaline and dopamine neurons in the reward/effort trade-off: a direct electrophysiological comparison in behaving monkeys. *J Neurosci* 35:7866–7877.

Veasey SC, Fornal CA, Metzler CW, Jacobs BL. 1995. Response of serotonergic caudal raphe neurons in relation to specific motor activities in freely moving cats. *J Neurosci* 15:5346–5359.

Vickers D. 1970. Evidence for an accumulator model of psychophysical discrimination. *Ergonomics* 13:37–58.

Waddington KD, Allen T, Heinrich B. 1981. Floral preferences of bumblebees (Bombus edwardsii) in relation to intermittent versus continuous rewards. *Anim Behav* 29:779–784.

Wei K, Glaser JI, Deng L, Thompson CK, Stevenson IH, Wang Q, Hornby TG, Heckman CJ, Körding KP. 2014. Serotonin affects movement gain control in the spinal cord. *J Neurosci* 34:12690–12700.

Westbrook A, Braver TS. 2015. Cognitive effort: a neuroeconomic approach. *Cogn Affect Behav Neurosci* 15:395–415.

Westbrook A, Kester D, Braver TS. 2013. What is the subjective cost of cognitive effort? Load, trait, and aging effects revealed by economic preference. *PLoS One* 8:e68210.

Wickler SJ, Hoyt DF, Cogger EA, Hirschbein MH. 2000. Preferred speed and cost of transport: the effect of incline. *J Exp Biol* 203:2195–2200.

Wong AL, Lindquist MA, Haith AM, Krakauer JW. 2015. Explicit knowledge enhances motor vigor and performance: motivation versus practice in sequence tasks. *J Neurophysiol* 114:219–232.

Xu-Wilson M, Chen-Harris H, Zee DS, Shadmehr R. 2009a. Cerebellar contributions to adaptive control of saccades in humans. *J Neurosci* 29:12930–12939.

Xu-Wilson M, Zee DS, Shadmehr R. 2009b. The intrinsic value of visual information affects saccade velocities. *Exp Brain Res* 196:475–481.

Yasuda M, Hikosaka O. 2017. To wait or not to wait-separate mechanisms in the oculomotor circuit of basal ganglia. *Front Neuroanat* 11:35.

Yasuda M, Yamamoto S, Hikosaka O. 2012. Robust representation of stable object values in the oculomotor basal ganglia. *J Neurosci* 32:16917–16932.

Yohn SE, Santerre JL, Nunes EJ, Kozak R, Podurgiel SJ, Correa M, Salamone JD. 2015. The role of dopamine D1 receptor transmission in effort-related choice behavior: Effects of D1 agonists. *Pharmacol Biochem Behav* 135:217–226.

Yoon T, Geary RB, Ahmed AA, Shadmehr R. 2018. Control of movement vigor and decision making during foraging. *Proc Natl Acad Sci USA* 115:E10476–E10485.

Yoshida K, Iwamoto Y, Chimoto S, Shimazu H. 1999. Saccade-related inhibitory input to pontine omnipause neurons: an intracellular study in alert cats. *J Neurophysiol* 82:1198–1208.

Zalocusky KA, Ramakrishnan C, Lerner TN, Davidson TJ, Knutson B, Deisseroth K. 2016. Nucleus accumbens D2R cells signal prior outcomes and control risky decision-making. *Nature* 531:642–646.

Zarrugh MY, Radcliffe CW. 1978. Predicting metabolic cost of level walking. *Eur J Appl Physiol Occup Physiol* 38:215–223.

Zenon A, Krauzlis RJ. 2012. Attention deficits without cortical neuronal deficits. *Nature* 489:434–437.

Zentall TR. 2013. Animals prefer reinforcement that follows greater effort: justification of effort or within-trial contrast? *Comp Cog Behav Rev* 8:60–77.

Index

5, 7-DHT, 295
5-hydroxytryptamine (5-HT), 290. *See also* Serotonin
  receptor 5-HT1a, 291
  receptor 5-HT1b, 297
  receptor 5-HT2c, 312

Abrahams, Mark, 76
Ahmari, Susanne, 297
Amit, Hidetoshi, 248
Amoroso, Mackenzie, 292
Antisaccade task, 189, 196
Ariely, Dan, 29
Assad, John, 224
Attention
  covert, 180, 183
  interaction with effort, 108, 183
  LIP, 212
Autoreceptor, 291

Bailey, Matthew, 312
Basal ganglia anatomy, 233
Basso, Michele, 157, 160, 162
Bateson, Melissa, 74
Batista, Aaron, 221
Behavioral inhibition, 295
Berke, Joshua, 273
Berret, Bastien, 118, 123
Bizzi, Emilio, 188
Blindsight, 178
Bogacz, Rafal, 104
Bornstein, Marc and Helen, 33
Bouret, Sebastien, 268
Bruce, Charles, 188
Buttner-Ennever, Jean, 205

Capture rate, 35, 41
Carello, Christopher, 169
Carpenter, Roger, 80
Cattell, James, 79
Caudate
  control of vigor, 248–251, 258
  encoding of visual space, 245
  encoding reward, 247, 249, 258
Cavanagh, Peter, 8
Chan, Florence, 184
Charnov, Eric, 44
Choi, Jenny, 115
Cingulate cortex, 229
  harvest duration, 225
Cisek, Paul, 26, 208
Cohen, Jeremiah, 292
Constantino, Sara, 48
Corneil, Brian, 166
Correia, Patricia, 307
Cos, Ignasi, 26
Cost of time, 47, 52, 118, 123, 207
Costa, Rui, 261
Countermanding task, 195
Cowie, Richard, 48
Cyproheptadine, 318

da Silva, Joaquim Alves, 263
Daw, Nathaniel, 50, 273, 292
Dayan, Peter, 51, 292
Dean, Jesse, 11
Decision making
  action-based model, 220
  computational models, 80, 102
  effect of effort, 106
  good-based model, 220
Deisseroth, Karl, 253
Depression, 289
Descending drive, 318
Diedrichsen, Jörn, 105
Differential reinforcement of low rate (DRL), 297
Dill, Lawrence, 75
Donelan, Max, 9
Dopamine
  effort, 227, 269, 272, 283
  history of rewards, 294
  motor cortex, 276

Dopamine (cont.)
movement direction, 276
movement onset, 261
striatal neurons, 234
vigor, 273–274
willingness to work, 278
Dorris, Michael, 91, 144
Dorsal raphe nucleus (DRN), 291
Doxycycline, 297, 299
Doya, Kenji, 300
Drift diffusion model, 107, 209
Dudman, Joshua, 251

Eagle, Dawn, 295, 306
Efficiency, 42
Effort
cognitive, 27
temporal discounting, 21
Elevated plus maze test, 307
Energetic cost
adaptation, 9
driving, 12
reaching, 10
running, 5
walking, 4
Engelhard, Ben, 273
Escitalopram, 317
Everling, Stefan, 147, 190, 202
Express saccades, 137, 144, 163

Flupenthixol, 314
Foley, Nicholas, 217
Food wasting, 75
Frontal eye field
damage, 186
fovea related neurons, 192
neuron types, 189
reward, 197
Frontal lobe damage, 133
Frost, Douglas, 177

Gandhi, Neeraj, 137, 205
Gap saccade task, 142, 147, 165, 202
Geary, Robert, 62
Ghez, Claude, 98
Ghilardi, Felice, 98
Glaser, Joshua, 197
Glimcher, Paul, 108, 212, 214, 217, 236
Goffart, Laurent, 172
Goldberg, Mickey, 188
Gordon, James, 98
Gottlieb, Jacqueline, 217
Green, Leonard, 121
Gribble, Paul, 166
Grice, Robert, 83
Gu, Chao, 166

Haggard, Patrick, 105
Hagura, Nobuhiro, 105
Handel, Ari, 236
Hanes, Doug, 188, 190, 193, 196
Harvest duration
anterior cingulate, 224
depletion rate, 50
effort, 50, 65, 67, 227
history of effort, 65
history of reward, 67
Hayden, Benjamin, 224
Heckman, CJ, 316
Heitz, Richard, 202
Held, Richard, 177
Hen, René, 297
Heteroreceptor, 291
Hikosaka, Okihide, 88, 159, 235–238, 241, 246, 252, 257, 259–260
Hogberg, Paul, 8
Horwitz, Gregory, 221
Hoyt, Donald, 6
Hristova, Anna, 231
Huang, Helen, 10

IKEA effect, 285
Impulsivity, 295
Inter-response time, 314
Intertrial interval, 294
Isoutility curves, 24
Ivry, Richard, 99

Jackson, John Hughlings, 187
Jacobs, Barry, 316

Kacelnik, Alejandro, 39, 74, 285
Kajiwara, Riichi, 276
Kakade, Sham, 292
Kato, Rikako, 178
Kawagoe, Reiko, 91, 252, 257
Kawato, Mitsuo, 25
Keller, Edward, 156, 205
Kim, Hyoung, 246, 248
Kobayashi, Shunsuke, 94
Koch, Michael, 280
Komogortsev, Oleg, 115
Kording, Konrad, 22, 316
Krakauer, John, 231
Kram, Rodger, 9
Krauzlis, Richard, 169, 171, 180, 183
Kunori, Nobuo, 276
Kuo, Art, 9

lateral intraparietal sulcus (LIP)
attention, 215
damage, 186
reward, 212
salience, 219
LATER model, 83
Lauwereyns, Johan, 245
Leathers, Marvin, 215
Lee, Rachel, 274

Lemon, William, 35
Lempert, Karolina, 108
Levine, Robert, 61
Limited hold duration, 296
Locus coeruleus, 269
Lottem, Eran, 303, 307
Louie, Kenway, 214
Lovejoy, Lee, 180

Machado, Liana, 186
Mainen, Zachary, 303, 307
Marginal value theorem, 43, 88, 225, 289, 304
Marsh, Barnaby, 285
Matsumoto, Masayuki, 260
Mazzoni, Pietro, 231
McGill, William, 80
McPeek, Robert, 156
Medial superior temporal cortex (MST), 182
Middle temporal cortex (MT), 184
Milosavljevic, Milica, 73
Milstein, David, 91
Miyazaki, Kayoko, 300, 303
Mohebi, Ali, 273
Motor cortex
  decision to reach, 208
  dopamine, 276
Munoz, Doug, 138, 144, 147, 154, 164, 184, 190, 204

Nagasaki, Hiroshi, 97
Nakamura, Kae, 259
Nautiyal, Katherine, 297
Navalpakkam, Vidhya, 71
Newsome, Bill, 221
Niv, Yael, 51
Noradrenaline, 269
Norenzayan, Ara, 61
Normative, 34
Nuclei raphe magnus (NRM), 316
Nuclei raphe obscurus (NRO), 316
Nuclei raphe pallidus (NRP), 316
Nucleus accumbens (NAc), 233, 254–255
  dopamine and effort, 272, 283
  dopamine and history of effort, 283
  dopamine and reward, 272
  dopamine and vigor, 273
  dopamine and willingness to work, 278
Nummela, Samuel, 171

Ohmura, Yu, 307
Oleson, Erik, 280
Olson, Carl, 215
Omnipause neurons, 202
Opportunity cost, 52
Optimal foraging theory, 35, 225, 292
Orbitofrontal cortex, 222
Overlap saccade task, 148, 184

Padoa-Schioppa, Camillo, 220
Panigrahi, Babita, 251
Pare, Martin, 164
Parker, Nathan, 274
Parkinson's disease, 290
  inhibition of habitual responses, 184
  serotonin, 290
  urgency, 231
Pasquereau, Benjamin, 266
Pearson, John, 224
Perona, Pietro, 71
Pettibone, Jeffrey, 273
Platt, Michael, 212, 224
Poppel, Ernst, 177
Post-reinforcement pause, 314
Premotor cortex, decision to reach, 206, 208
Progressive ratio task (PRT), 312

Race model, 101
Rafal, Robert, 186
Ralston, Henry, 3
Rangel, Antonio, 73
Ratcliff, Roger, 80
Ratcliffe, CW, 6
Reaching
  effect of reward, 92, 267
  natural speed, 19
  subcortical pathway, 168
Reaction time
  capture rate, 88
  effort, 90, 98
  reward, 90–91, 94
Reflex gain, 318
Ren, Jing, 294
Reppert, Thomas, 110, 108, 115, 116
Reward, temporal discounting, 13, 215
Reward prediction error, 261, 264
Richardson, Howard, 37
Risk seeking, D2 receptor agonist, 255
Robbins, Trevor, 295, 306
Rosenbaum, David, 98
Rotarod task, 309
Russ, David, 21
Rymer, Zev, 316

Sackaloo, Kareisha, 94
Salamone, John, 278
Salience, 69, 156, 190, 217
Salinas, Emilio, 113
Sato, Makoto, 238
Schall, Jeffrey, 191, 193, 196, 199
Schelp, Scott, 280
Schiller, Peter, 166
Schweighofer, Nicholas, 25
Scotoma, cortical, 177
Sedaghat-Nejad, Ehsan, 264
Segraves, Mark, 149, 197, 205
Selective serotonin reuptake inhibitor (SSRI), 290, 310
  Escitalopram effect on reflex gains, 317
Seo, Changwoo, 310

Serotonin
  5-hydroxytryptomaine, 289
  autoreceptor, 291
  behavioral inhibition, 295
  depression, 289
  dopamine, 299, 314
  heteroreceptors, 291
  history of rewards, 294
  impulsivity, 295
  interresponse time, 314
  locomotor activity, 307
  receptor 5-HT1a, 291
  receptor 5-HT1b, 297
  receptor 5-HT1c, 312
  reflex gain, 316, 318
  willingness to wait, 295
Shimazu, Hiroshi, 204
Simpson, Eleanor, 312
Smalianchuk, Ivan, 137
Sparks, David, 134, 137, 206
Speed-accuracy tradeoff, 119, 199, 207, 232
Spinal cord, 316
  injury, 318
  serotonin, 316
Stelly, Claire, 272
Stelmach, George, 98
Step length, 8
Steudel-Numbers, Karen, 5
Stop-signal task, 296
Stretch reflex, 316
Striatum
  dopamine and movement direction, 276
  risk seeking, 255
Substantia nigra pars compacta (SNc)
  dopamine and effort, 267, 269
  dopamine and movement onset, 261
  dopamine and reward, 257, 268
  dopamine and reward prediction error, 261
  vigor, 241
Substantia nigra pars reticulata (SNr)
  deactivation, 239
  encoding reward, 239
  encoding risk, 238
  encoding of visual space, 236
Summerside, Erik, 93
Superior colliculus
  attention, 181
  buildup cells, 151
  decision making, 170
  express saccades, 165
  fovea-related neurons, 139
  inputs and outputs, 131
  muscimol, 154
  reward, 157, 160
  saccade velocity, 139
  visual neurons, 135
  visuomotor neurons, 135
Syed, Emilie, 270

Takashima, Ichiro, 276
Takikawa, Yoriko, 252, 257
Tank, David, 273
Taylor, Richard, 6
Theeuwes, Jan, 70
Thura, David, 207
Treadway, Michael, 119
Turner, Robert, 266

Uchida, Naoshige, 292
Urgency model, 103
Urgency signal, 152, 209

Varazzani, Chiara, 268
Vaswani, Pavan, 114
Ventral tegmental area (VTA), 294
  dopamine and reward prediction error, 274
  dopamine and vigor, 274
Verbeek, Nicolaas, 37
Vickers, Douglas, 102
Vigor
  aging, 120, 123
  between-subject differences, 114, 117
  decision-making traits, 123
  effort, 68
  history of effort, 284
  loss avoidance, 97
  rate of reward, 97
  reward, 63, 65, 91, 98
  reward prediction error, 264
  salience, 96, 217
  SNc dopamine activation, 263
  subjective valuation, 112
  uncertainty of reward, 115, 207
  within-subject measure, 109

Walking speed, 4, 60
Wall-Scheffler, Cara, 5
Walton, Mark, 270
Wanat, Matthew, 272
Warden, Melissa, 310
Wei, Kunlin, 316
Williams, Keith, 8
Willingness to wait, 295
Witten, Ilana, 273–274
Wood, Daniel, 166
Worringham, Charles, 98
Wurtz, Robert, 138, 146, 154, 162, 192, 235

Yamamoto, Shinya, 241
Yasuda, Masaharu, 237
Yoon, Tehrim, 62, 102, 284
Yoshida, Kaoru, 204

Zalocusky, Kelly, 254
Zarrugh, Mohamed, 6
Zelinsky, Gregory, 71
Zenon, Alexandre, 183